FAST METHODS IN PHYSICAL BIOCHEMISTRY
AND CELL BIOLOGY

Fast Methods in Physical Biochemistry and Cell Biology

edited by

RAMADAN I. SHA'AFI *and* SALVADOR M. FERNANDEZ

Department of Physiology, University of Connecticut Health Center,
Farmington, CT 060032 (U.S.A.)

1983

ELSEVIER
AMSTERDAM · NEW YORK · OXFORD

7115-1436

© 1983 Elsevier Science Publishers

ISBN for this volume: 0 444 80470 6

Published by:
Elsevier Biomedical Press
1 Molenwerf, P.O. Box 211
1000 AE Amsterdam, The Netherlands

Sole distributors for the U.S.A. and Canada:
Elsevier North-Holland Inc.,
52 Vanderbilt Avenue
New York, NY 10017

Library of Congress Cataloging in Publication Data

Main entry under title:

Fast methods in physical biochemistry and cell biology.

    Includes index.
    Contents: Some general aspects of signal analysis /
by Salvador M. Fernandez and Ramadan I. Sha'afi -- Stopped-flow, temperature-jump, and flash photolysis techniques / by H.B. Dunford -- Recent applications or the
stopped-flow and pressure-jump relaxation techniques in
the biological sciences / by Yves Engelborghs and Karel
Heremans -- [etc.]
    1. Biological chemistry--Methodology--Addresses,
essays, lectures. 2. Cytochemistry--Methodology--
Addresses, essays, lectures. 3. Chemistry, Physical
organic--Methodology--Addresses, essays, lectures.

I. Sha'afi, Ramadan I. (Ramadan Issa), 1938-   .
II. Fernandez, Salvador M. [DNLM: 1. Biochemistry--
Methods. 2. Cytological technics. QU 25 F251]
QH345.F38 1983   574.19'2'028   83-1692
ISBN 0-444-80470-6

PRINTED IN THE NETHERLANDS

With gratitude to the late Charles N. Loeser who initiated our interest in the field of fluorescence relaxation spectroscopy and to Richard D. Berlin for continued support and encouragement.

Preface

As its name implies, this book is concerned with methods for the study of fast chemical processes of interest to the physical biochemist or cell biologist. These processes range from simple chemical reactions involving small organic molecules to those encompassing the dynamic behavior of biopolymers in supramolecular structures including intact cells. The principal aim of this work is to present a logical exposition of basic principles and of technical aspects of the various methods described rather than to provide a comprehensive survey of recent work in each field. Nevertheless, every chapter is amply illustrated with biological applications of current interest.

The term "fast methods" as used in the present context includes a diverse group of techniques with resolution ranging from the second to the picosecond time scale. With some exceptions, the book generally progresses along the lines of time resolution. Thus, classical flow methods with millisecond time resolution are presented first, followed by relaxation methods such as temperature-jump and pressure-jump and other perturbation techniques such as flash photolysis and pulse radiolysis which possess resolution in the microsecond range. The discussion of flow and relaxation methods is complemented by a chapter on rapid quench techniques. Whereas the former are employed to identify intermediate states of a reaction by rapid observation in real time, the latter permit the detection of transient species by slower methods. The book then moves into the nanosecond time scale with a discussion of time-resolved fluorescence measurements and finally progresses into the picosecond range with a chapter on picosecond spectroscopy. A number of other methods which have only been developed within the last decade are also included; namely, electrophoretic light scattering, time-resolved X-ray scattering and fluorescence photobleaching. Nuclear magnetic resonance and electron spin resonance techniques have been omitted from this work since recent developments in these areas have been adequately covered in the recent literature.

An effort has been made, whenever possible, to discuss applications of the various techniques to problems in cell biology. The inclusion of such material in a book on fast methods is timely for several reasons. First, the vast amount of background information acquired from purified biochemical preparations and model systems now serves to guide the design of meaningful experiments and provides an ever growing foundation on which to base the interpretation of data obtained from the more complex situations encountered in an intact cell. Second, recent technological advances make feasible the application of sophisticated biophysical methods to the study of cellular processes. The advent of a number of fast optical techniques such as absorption photometry, fluorimetry, and the measurement of optical anisotropy and light scattering provide a methodological link for the interchange of information between the physical biochemist and cell biologist. Examples of this are found in the chapters on electrophoretic light scattering, time-resolved fluorescence spectroscopy, flash photolysis and fluorescence photobleaching.

While we hope that this work may prove useful to the expert, we have strived to make the material readily intelligible to the newcomer in the field. To this end we have included one chapter (Ch. 1) in which we bring together selected topics in signal analysis which are of general applicability to many of the methods included in the book. Cross-references to this material are found where appropriate throughout the text. It may also be noted here that lasers have become ubiquitous tools in the field of fast methods. The reader who is unfamiliar with their operation will find a discussion of this topic in Chapter 10.

Finally, this book represents an international cooperation among authors of different countries and disciplines. The editors have made every effort in coordinating chapters so as to avoid inconsistencies, excessive overlap and errors. The editors are grateful to the contributors of this book who have written their contributions on time and have gracefully accepted modifications and suggestions of the editors.

Farmington, CT (U.S.A.)   R.I. Sha'afi
                                                                                S.M. Fernandez

Contributors

David BALLOU
Department of Biological Chemistry
University of Michigan
Ann Arbor, Michigan, U.S.A. 48109

Joan BORDAS
European Molecular Biology Laboratory
Hamburg Outstation
Notkestrasse 52, D-2000
Hamburg, FRG

H.B. DUNFORD
Department of Chemistry
University of Alberta
Edmonton, Alberta, Canada T6G 2G2

Yves ENGELBORGHS
Laboratorium voor Chemische en Biologische
Dynamica
Katholieke Universiteit te Leuven
Celestijnenlaan 200D
B-3030 Heverlee, Belgium

Salvador M. FERNANDEZ
Department of Physiology
University of Connecticut Health Center
Farmington, CT U.S.A. 06032

Daniel D. HAAS
Kodak Research Laboratories
Eastman Kodak Company
Rochester, New York U.S.A. 14650

Koichiro HAYASHI
The Institute of Scientific and Industrial Re-
search
Yamadakami, Suita, Osaka 565
Japan

Karel HEREMANS
Laboratorium voor Chemische en Biologische
Dynamica
Katholieke Universiteit te Leuven
Celestijnenlaan 200D
B-3030 Heverlee, Belgium

Dewey HOLTEN
Department of Chemistry
Washington University
St. Louis, Missouri, U.S.A. 63130

Toshihisa ISHIKAWA
Research Institute of Applied Electricity
Hokkaido University
Sapporo 060, Japan

Kazuo KOBAYASHI
The Institute of Scientific and Industrial Re-
search
Yamadakami, Suita, Osaka 565
Japan

Dennis KOPPEL
Biochemistry Department
University of Connecticut Health Center
Farmington, CT. U.S.A. 06032

Eckhard MANDELKOW
Max Planck Institute for Medical Research
Jahnstrasse 29
D-6900 Heidelberg, FRG

Ramadan I. SHA'AFI
Department of Physiology
University of Connecticut Health Center
Farmington, CT. U.S.A. 06032

Mamoru TAMURA
Research Institute of Applied Electricity
Hokkaido University
Sapporo 060, Japan

Susumo TSUBOTA
Research Institute of Applied Electricity
Hokkaido University
Sapporo 060, Japan

B.R. WARE
Chemistry Department
Syracuse University
Syracuse, New York, U.S.A. 13210

Isao YAMAZAKI
Research Institute of Applied Electricity
Hokkaido University
Sapporo 060, Japan

Contents

Preface   vii

Contributors   ix

Chapter 1.   Some general aspects of signal analysis
               by Salvador M. Fernandez and Ramadan I. Sha'afi   1

Chapter 2.   Stopped-flow, temperature-jump and flash photolysis
               techniques: effect of temperature, dielectric constant and
               viscosity
               by H.B. Dunford   11

Chapter 3.   Recent applications of the stopped-flow and pressure-jump
               relaxation techniques in the biological sciences
               by Yves Engelborghs and Karel Heremans   39

Chapter 4.   Rapid-quench methods in fast biochemical processes
               by David P. Ballou   63

Chapter 5.   Application of pulse radiolysis to biochemistry
               by Kazuo Kobayashi and Koichiro Hayashi   87

Chapter 6.   Flash photolysis studies in heterogeneous systems
               by Mamoru Tamura, Toshihisa Ishikawa, Susumu Ysubo-
               ta and Isao Yamazaki   113

Chapter 7.   Time-resolved X-ray scattering from solutions using syn-
               chroton radiation
               by Joan Bordas and Eckhard Mandelkow   137

Chapter 8.   Electrophoretic light scattering
               by B.R. Ware and Daniel D. Haas   173

Chapter 9. Time-resolved fluorescence spectroscopy
by Salvador M. Fernandez   221

Chapter 10. Picosecond spectroscopy
by Dewey Holten   281

Chapter 11. Fluorescence photobleaching as a probe of translational
and rotational motions
by D.E. Koppel   339

Subject Index   369

Some general aspects of signal analysis

SALVADOR M. FERNANDEZ
and RAMADAN I. SHA'AFI

University of Connecticut Health Center, Farmington, CT 06032, U.S.A.

Contents

1. Introduction 1
2. Periodic signals 1
   2.1. Power spectrum 2
   2.2. Discrete sampling: aliasing 3
3. Correlation functions 3
4. Deconvolution 6
References 9

1. Introduction

This chapter presents a brief introduction to some problems in signal analysis which are rather universal in the world of experimental science and which are shared in common by many of the methods included in this volume. More specifically, the important concepts of convolution, correlation, and power spectrum are discussed, as well as problems associated with the discrete sampling of wave forms. While these concepts are frequently encountered by the physical scientist or engineer, they may be less familiar to the biochemist or cell biologist. The objective here is not to pursue a complete or rigorous treatment of the subject of signal analysis but rather to develop an intuitive understanding of selected concepts which arise in the discussion of various fast methods. This chapter, therefore, is probably best approached in response to a reference to it in the rest of the text where the reader feels that a supplementary discussion on the particular topic may be helpful. References to this material have been included throughout the text where appropriate.

2. Periodic signals

A signal S that repeats itself after a time T (the independent variable need not be restricted to time) is said to be periodic; i.e.,

R.I. Sha'afi and S.M. Fernandez (Eds.), Fast Methods in Physical Biochemistry and Cell Biology
© 1983 Elsevier Science Publishers

$$S(t + T) = S(t) \tag{1}$$

The best known examples of periodic signals are the sine and cosine functions, such as

$$S(t) = A \sin(\omega_0 t - \varphi) \tag{2}$$

where A is the amplitude, ω_0 is the angular frequency and φ is the phase of the signal. Instead of using phase, the signal from eqn. 2 can be written as a sum of a sine term and a cosine term

$$S(t) = B \sin \omega_0 t - C \cos \omega_0 t \tag{3}$$

where $B = A \cos \varphi$ and $C = A \sin \varphi$. *Harmonics* of the fundamental frequency ω_0 such as $\cos n\omega_0 t$ or $\sin n\omega_0 t$ also have period T (although they also have a shorter period). It can be shown (see, for example, Tolstov, 1962) that any continuous periodic function of arbitrary shape can be made up of sine and cosine waves of harmonically related frequencies (Fourier series):

$$S(t) = a_0 + \sum_{n=1}^{N} (a_n \cos n\omega_0 t + b_n \sin n\omega_0 t) \tag{4}$$

where

$$a_n = \frac{2}{T} \int_0^T S(t) \cos n\omega_0 t \, dt$$

$$b_n = \frac{2}{T} \int_0^T S(t) \sin n\omega_0 t \, dt \tag{5}$$

$$a_0 = \frac{1}{T} \int_0^T S(t) dt$$

Thus, it is possible to speak of the "frequency spectrum" of a signal.

2.1. Power spectrum

By analogy to the fact that the power dissipated in a resistor is given by v^2/R (v, voltage; R, resistance), the square of any signal is often called the power. For a periodic signal then, the average "power" is defined as

$$\langle S^2(t) \rangle = \frac{1}{T} \int_0^T S^2(t) dt \tag{6}$$

It can be shown that substitution of Eqn. 4 into Eqn. 6 gives

$$\langle S^2(t)\rangle = a_0^2 + \tfrac{1}{2}\sum_{n=1}^{N}(a_n^2 + b_n^2) \tag{7}$$

where each term in the sum represents the average power at each frequency.

2.2. Discrete sampling: aliasing

Almost everyone is familiar with stroboscopic effects such as the backward motion of the stagecoach wheels or the apparent failure of the vertical hold in a television set when viewed in motion pictures. In signal analysis this phenomenon is called *aliasing* and it is of general importance because experimental data is often obtained by sampling a signal at discrete intervals. Whenever a frequency component is present whose period is less than twice the sampling interval, it will appear in the analysis as a much lower frequency (the movie frames in the examples above are not made rapidly enough). This phenomenon is intuitively illustrated in Fig. 1

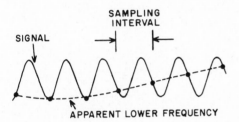

Fig. 1. An illustration of aliasing. When a periodic signal is discretely sampled at intervals which are longer than half a period, the resulting sampled wave form appears to have a much lower frequency.

which shows a sine wave sampled at regularly spaced intervals which are longer than half a period. If the sampling interval is chosen equal to one-half the reciprocal of the highest frequency component (f_c) aliasing will not occur; the frequency $2f_c$ is known as the Nyquist sampling rate. This is an extremely important concept in many fields of scientific application (see, for example, Chapter 8 on electrophoretic light scattering).

3. Correlation functions

Correlation functions are useful in determining whether two variables are correlated, that is, whether a change in one is accompanied by a change in the other. Correlation functions also constitute an important tool for the recovery of periodic signals buried in noise.

Consider two variables $S_1(t)$ and $S_2(t)$. The cross-correlation, C_{12}, of these two variables is defined as

4

$$C_{12}(\tau) = \lim_{T \to \infty} \frac{1}{2T} \int_{-T}^{T} S_1(t)S_2(t + \tau)dt \tag{8}$$

Eqn. 8 is sometimes abbreviated as

$$C_{12}(\tau) = \langle S_1(t)S_2(T + \tau) \rangle \qquad \text{or} \qquad C_{12}(\tau) = S_1(t) * S_2(t) \tag{9}$$

The cross-correlation depends only on the relative shift between the two signals; it does not matter whether signal 2 is advanced by an amount τ or signal 1 is delayed by the same amount. Furthermore, the integral of the correlation function has a positive contribution only if $S_1(t)$ and $S_2(t)$ are both positive or negative at the same time; it will have a negative contribution when one is positive and the other one is negative. The mathematical operations represented by Eqn. 8 are graphically illustrated in Fig. 2.

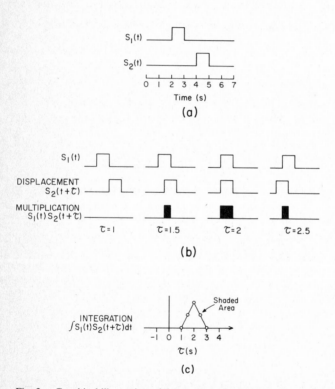

Fig. 2. Graphical illustration of the process of calculating the cross-correlation between two signals. In this case, both signals $S_1(t)$ and $S_2(t)$ are square pulses one of which $(S_1(t))$ leads the other $(S_2(t))$ by 2 sec. These are shown in a. b depicts the intermediate steps in the calculation of the cross-correlation for several values of the lag variable τ. The final result is shown in c. The value of the cross-correlation at each value of τ is given by the shaded areas in b.

The autocorrelation function is the correlation of a signal with itself:

$$C_{11}(\tau) = \langle S_1(t)S_1(t + \tau) \rangle \tag{10}$$

It can be shown that the autocorrelation and the power spectrum of a signal are related to each other through a Fourier transform. In other words, the power spectrum of a periodic signal can be obtained from either the squares of the Fourier coefficients of the signal, or from the Fourier coefficients of the autocorrelation function. This can, perhaps, be seen intuitively since Eqn. 6 is the same as Eqn. 10 when $\tau = 0$.

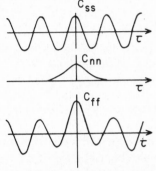

Fig. 3.   The upper graph shows the autocorrelation of a periodic signal; it is also periodic with the same period as the original signal. The middle panel illustrates one possible shape for the autocorrelation of noise; other shapes are possible depending on the nature of the noise, but the general feature of noise autocorrelation is that it approaches zero for large values of τ. The bottom panel represents the autocorrelation of a periodic signal which contains noise. Note the presence of a periodic component for long values of τ.

The autocorrelation can be useful for detecting a weak periodic signal in the presence of noise. Consider a periodic signal $S(t)$ containing noise $n(t)$ and assume that their average is zero. The combination of signal and noise is given by

$$F(t) = S(t) + n(t) \tag{11}$$

The autocorrelation of the combination is

$$C_{ff}(\tau) = \langle [S(t) + n(t)][S(t + \tau) + n(t + \tau)] \rangle \tag{12}$$

$$= \langle S(t)S(t + \tau) \rangle + \langle S(t)n(t + \tau) \rangle + \langle n(t)S(t + \tau) \rangle + \langle n(t)n(t + \tau) \rangle \tag{13}$$

$$= C_{ss}(\tau) + C_{sn}(\tau) + C_{ns}(\tau) + C_{nn}(\tau) \tag{14}$$

since the noise is completely random, the cross-correlations C_{ns} and C_{sn} of the signal with the noise should be zero provided that the average is taken over a sufficiently long time. Therefore,

$$C_{ff}(\tau) = C_{ss}(\tau) + C_{nn}(\tau) \tag{15}$$

The second term in Eqn. 15 approaches zero as the shift time τ is made long. This is intuitively plausible since for a completely random signal there is no correlation for large shifts. The first term, $C_{ss}(\tau)$, is the autocorrelation of the signal, which is periodic with period equal to that of the signal. Thus, it is seen that when a periodic signal is masked by noise its presence can be revealed by the autocorrelation function: if the autocorrelation shows periodicity for long delays, a periodic signal is present. These concepts are illustrated in Fig. 3.

4. Deconvolution

Another useful method of signal recovery which is frequently encountered is that of deconvolution. Experimental science relies on instruments to measure values of observable quantities as functions of some other variable. For the sake of concreteness we will take this other variable to be time, which is not inappropriate for a discussion of fast methods, although the arguments that follow are general and need not be so restricted.

Every measuring instrument distorts to some extent the actual time-dependent values of the desired variables. These values may or may not be negligible depending on the nature of the instrument and of the information sought. Elimination, at least in part, of the instrumental distortion of the measured variable is the aim of

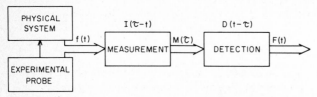

Fig. 4. Block diagram representing the distortion of a measured variable by the measurement and detection process. $f(t)$ is the "true" signal that is sought and $F(t)$ is the experimentally determined signal which results from the convolution of $f(t)$ with the instrumental response.

deconvolution. The process of instrumental distortion can be represented, in general terms, by a diagram such as the one shown in Fig. 4. The behavior of the physical system is investigated with a suitable probe which may take the form of a beam of photons or electrons, a magnetic field, etc. The interaction of the probe with the system is then sensed with some sort of detector. Theoretical considerations will often characterize the signal that is expected. This is called the "true signal" and presumably represents the signal that would be measured by a "perfect" instrument. This true signal is represented as $f(t)$ in Fig. 4; the actual output of the measurement system, however, $M(\tau)$, is related to the input $f(t)$ by the following *convolution* integral

$$M(\tau) = \int I(\tau - t)f(t)dt \qquad (16)$$

often represented in shorthand notation as

$$M(t) = I(t) * f(t) \qquad (19)$$

$I(\tau - t)$ completely represents the distortion of the true signal $f(t)$ by the measuring system and is called the system transfer function. Often $M(\tau)$ is not the final output of the experimental system but must be detected, amplified and possibly processed in other ways before yielding the final output signal $F(t)$. The effect of the detector and subsequent processing can be represented by a second convolution

$$F(t) = \int D(t - \tau)M(\tau)d\tau = D(t) * M(t) \qquad (18)$$

The total effect of the experimental system on $f(t)$ is then given by

$$F(t) = D(t) * I(t) * f(t) \qquad (19)$$

$$= T(t) * f(t) \qquad (20)$$

where $D(t) * I(t)$ has been replaced by $T(t)$ which is called the overall system transfer function.

The function $f(t)$ is the desired data but the measurement process yields $F(t)$. Deconvolution is a method which allows one to obtain an estimator of $f(t)$ when $T(t)$ can be measured or assumed. The accuracy with which deconvolution can be performed depends critically on the amount of noise present in the data. In fact, deconvolution can be viewed as the opposite of noise filtering. The latter involves the selective attenuation of high spectral frequencies, whereas deconvolution entails restoration of high spectral frequencies attenuated in the measurement process (Blass and Halsey, 1981). Therefore, a crucial prerequisite for successful deconvolution is a high signal-to-noise ratio in the data.

An impediment to the understanding of convolution and deconvolution stems from the fact that it is extremely difficult to visualize the mathematical operation of Eqn. 18. Therefore, it might be helpful to illustrate the process of convolution by intuitively considering in greater detail one specific example. For this purpose a problem which arises in time-resolved fluorimetry is examined (see Chapter 9). In this type of experiment one may be interested in the response of a homogeneous ensemble of fluorescent molecules to excitation by a short pulse of light. Physical theory predicts that when such an ensemble is excited by an infinitely short pulse of light (a delta function), the fluorescence decay $f(t)$ is given by

$$f(t) = f_0 e^{-t/\tau} \qquad (21)$$

where τ is the fluorescence lifetime. $f(t)$, in this case, is the true signal that is

8

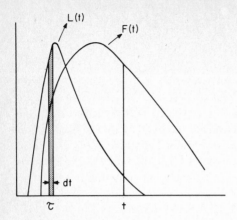

Fig. 5. Response of an ensemble of fluorescent molecules to excitation by a short pulse of light, $L(t)$, of finite width. The fluorescence intensity $F(t)$ emitted by the system at an arbitrary time t is given by the convolution of Eqn. 23.

sought. In practice, however, the light pulse, $L(t)$, possesses a finite width; therefore, the result of its interaction with the physical system does not yield $f(t)$ directly but a distorted function $F(t)$. This process is illustrated in Fig. 5.

To obtain an expression for $F(t)$ note that the exciting light pulse $L(t)$ can be viewed as made up of a series of infinitesimal "slices" of width dt (Fig. 5). The number of photons produced by the light pulse in one such infinitesimal interval centered at time τ is $L(\tau)dt$. In the limit as dt approaches zero, $L(\tau)dt$ becomes a delta function. Thus, the observed fluorescence intensity $F(t)$ at some later time t due to excitation by this infinitesimal segment of $L(t)$ is given by

$$F_\tau(t) = L(\tau)dtf(t - \tau), \qquad t > \tau \tag{22}$$

where a constant of proportionality has been incorporated into the function $L(\tau)$. To obtain the total observed fluorescence signal $F(t)$ at time t, the contributions of all the infinitesimal segments which make up $L(t)$ up to time t must be summed (assuming that the fluorescence system is linear with respect to the exciting light). It may be intuitively clear that in the limit as $dt \rightarrow 0$, the sum becomes an integral and $F(t)$ is given by

$$F(t) = \int_0^t L(\tau)f(t - \tau)d\tau \tag{23}$$

It can be shown (Brigham, 1974) that Eqn. 23 is equivalent to

$$F(t) = \int_0^t L(t - \tau)F(\tau)d\tau \tag{24}$$

which is analogous to Eqn. 18. This represents the distortion of the true signal by the measuring probe. In practice, one would need to additionally consider the distortion of $F(t)$ by the detector system which would result in a second convolution.

With this specific example as a guide it is easy to generalize to other situations. Thus, in an X-ray scattering experiment the observed diffraction pattern (continuing to ignore the detector response for the moment) would be the "true" signal only if the probe beam were infinitesimally small (a two-dimensional delta function). However, if the beam possesses finite dimensions the scattering pattern would be given by a convolution similar to that of Eqn. 24 (Chapter 7). A similar situation arises in fluorescence photobleaching experiments (Chapter 11). Several deconvolution methods are presented in Chapter 9; for other techniques the reader is referred to Blass and Halsey (1981).

References

Blass, E.W. and Halsey, G.W. (1981) *Deconvolution of Absorption Spectra*, Academic Press, New York.
Brigham, E.O. (1974) *The Fast Fourier Transform*, Prentice Hall, Englewood Cliffs, NJ.
Tolstov, G.P. (1962) *Fourier Series*, Prentice Hall, Englewood Cliffs, NJ.

CHAPTER 2

Stopped-flow, temperature-jump and flash photolysis techniques: effect of temperature, dielectric constant and viscosity

H.B. DUNFORD

Department of Chemistry, University of Alberta, Edmonton, Alberta, T6G 2G2, Canada

Contents

1. Introduction 11
2. Theory of reaction kinetics 12
3. Rapid reaction techniques 14
   3.1. Steady-state enzyme kinetics 14
   3.2. Stopped-flow kinetics 14
   3.3 Relaxation techniques 23
   3.4. The temperature-jump method 25
   3.5. Flash photolysis 27
4. Effect of varying solvent parameters 29
   4.1. Introduction 29
   4.2. Solvent dielectric constant 29
   4.3. Solvent viscosity 32
References 36

1. Introduction

This article is divided into four parts, including this brief introduction. An outline is given for the theory of reaction kinetics which includes the *raison d'être* for reducing complicated overall processes into their elementary steps or reactions. Some of the most widely used rapid reaction techniques for enzyme reactions are described: steady-state kinetics and three transient state methods: stopped-flow, temperature-jump and flash photolysis techniques. Steady-state kinetics are discussed mainly for comparative purposes. Finally, the effects of varying solvent parameters are examined, in particular temperature, dielectric constant and viscosity.

R.I. Sha'afi and S.M. Fernandez (Eds.), Fast Methods in Physical Biochemistry and Cell Biology
© *1983 Elsevier Science Publishers*

2. Theory of reaction kinetics

The basic theory of biochemical or chemical kinetics is valid for all time scales; it is only the techniques which must be changed in order to perform experiments in different time domains. To start our discussion let us define some basic terms. If one could isolate and observe the elementary reaction with the stoichiometry

$$ES \xrightarrow{k_{cat}} E + P \tag{1}$$

where ES is an enzyme–substrate complex, E the native enzyme and P the reaction product, then the rate "law" (the empirical mathematical relationship required to fit the experimental rate quantitatively for all concentrations of ES) would be

$$+ \frac{d(P)}{dt} = - \frac{d(ES)}{dt} = k_{cat}(ES)^1 \tag{2}$$

where the brackets indicate concentrations and k_{cat} is the rate constant (a catalytic rate constant, hence the subscript "cat" for an enzyme–substrate complex decomposing to yield product). Eqn. 2 is an example of a first-order rate equation, since the exponent is one for (ES) on the right side of the equation. (Normally this exponent is not shown if it is one.) The rate $d(P)/dt$ has units of $M \cdot s^{-1}$ (molar concentration per second) and since (ES) has M units, then k_{cat} has units of sec^{-1}.

In practice it is extremely difficult to isolate and study reaction 1. The simplest reaction scheme one finds in the literature is, of course, the Michaelis–Menten scheme.

$$E + S \underset{k_{-1}}{\overset{k_1}{\rightleftarrows}} ES \xrightarrow{k_{cat}} E + P \tag{3}$$

This consists of three elementary reactions, $E + S \rightarrow ES$, $ES \rightarrow E + S$ and $ES \rightarrow E + P$ and the overall process is called, logically, the overall reaction. The overall reaction can also be represented as

$$S \xrightarrow{E} P \tag{4}$$

the conversion of S to P catalyzed by E. It can be studied by using steady-state enzyme kinetics in which linear plots of (P) vs. t are obtained (Fig. 1). This proves that for the overall process 4 the rate law is

$$\frac{d(P)}{dt} = v \tag{5}$$

where v, the velocity, is constant for each experiment involving a given initial value

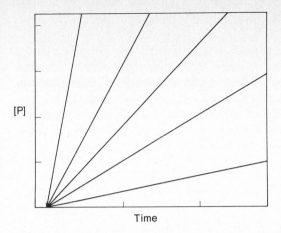

Fig. 1.   Plots of product concentration vs. time for steady-state (zero-order) enzyme kinetics. The greater the slope the greater the initial substrate concentration.

of S. One is observing a zero-order reaction and the units of v are $M \cdot sec^{-1}$.

Eqns. 4 and 5 illustrate one of the unique powers of the kinetic method. A naive inspection of Eqn. 4 might lead one to predict that first-order kinetics would be observed. Since the experimental observation is zero order kinetics, one can conclude immediately that there is more than one elementary reaction involved. We started with a three-step process, Eqn. 3; now we have proof that some such complicated scheme is necessary. These equations also illustrate a weakness of steady-state kinetics. Although reaction 3 is adequate to explain the observation of zero-order kinetics, we now know that it is inadequate to explain all of the features of any enzyme-catalyzed reaction. As a single illustration, since many enzyme-catalyzed reactions are reversible, then Eqns. 3 and 4 can be reversed. Since an initial enzyme–substrate complex must exist for $E + S \rightarrow ES$, it also must exist for $E + P \rightarrow EP$. Therefore, at least one more enzyme–substrate (or enzyme–product, EP) complex must be inserted into Eqn. 3.

A few final notes on definitions and units: for an elementary process such as $E + S \rightarrow ES$, the reaction is second order since there are two concentration terms on the right side of Eqn. 6, each to the first power.

$$\frac{d(ES)}{dt} = k_1(E)(S) \qquad (6)$$

The rate constant, k, has units of $M^{-1} \cdot sec^{-1}$.

If one could detect all of the elementary steps in an overall reaction and measure their rate constants, then one would attain a kineticist's definition of the reaction mechanism. This, of course, implies that the reactant(s) and product(s) are identified in every step. In principle every k can be measured by kinetic techniques and in principle every structure can be determined by various spectroscopic techniques.

Although the difficult and elusive goal of understanding the complete, unambiguous mechanism of any enzymatic reaction has not yet been attained, spectacular progress is being made on every front. In this article the primary concern will be the measurement of the elementary rate constants. Not only is the theory for elementary rate processes simple, but the conversion of experimental observation into deduction about mechanism is also much more simple than it is for steady-state results on an overall process. The conversion of experimental observations on an overall reaction into deductions about mechanism is often complicated, obscure and hence open to qualitative as well as quantitative errors.

3. Rapid reaction techniques

3.1. Steady-state enzyme kinetics

The steady-state technique for studying enzyme reactions derives its name from the fact that the total amount of enzyme is partitioned into various forms at constant concentration. For example, from Eqn. 3 the total amount of enzyme E_0 is partitioned between the form E (native enzyme) and ES (enzyme–substrate complex). The steady-state approximation states that $d(E)/dt = d(ES)/dt = 0$. This condition is valid provided $(S) \gg (E)_0$ and provided a small time interval has elapsed, which is much smaller than the "dead" time or time required to start observations of a steady-state rate.

The use of steady-state enzyme kinetics is a legitimate rapid reaction technique since it can be used for all known enzymatic reactions involving their natural substrates. These are all fast reactions. From any series of experiments such as those shown in Fig. 1 one can always extract two parameters: V_{max} the maximum velocity and K_m the Michaelis constant. If Eqn. 3 were valid, K_m would be equal to $(k_{-1} + k_{cat})/k_1$, a rough approximation of a dissociation constant with units of M. For more realistic equations, any resemblance between K_m and a dissociation constant, beyond possession of the correct units $(\sec^{-1})$, is tenuous. However, the relationship

$$V_{max} = k_{cat}(E)_0 \tag{7}$$

is always valid. The term V_{max} means the maximum attainable reaction velocity. This requires that the substrate concentration becomes infinitely large so that E is quantitatively converted into ES. We have already seen that k_{cat} has units of $\sec^{-1}$, but these simple units do not adequately describe the significance of k_{cat}. In words, k_{cat} means the maximum possible number of molecules of product formed per second per molecule of enzyme, and it is called the turnover number. (One could substitute molar concentration for molecules since the units for expressing the product and the enzyme are the same and cancel.) Thus k_{cat} expresses the theoretical upper limit for the overall catalytic rate or efficiency. For a detailed discussion of steady-state enzyme kinetics see, for example, Cornish-Bowden (1976), Laidler and Bunting (1973), Plowman (1972) and Wong (1975).

In order to perform successful steady-state experiments one uses enzyme concentrations typically in the range of $10^{-8} - 10^{-10}$ M and substrate concentrations perhaps in the range $10^{-2} - 10^{-5}$ M. Since formation of the initial enzyme–substrate complex is second order (Eqn. 6), decreasing the concentration of both reactants lowers the rate of the overall reaction. Typically the time scale in Fig. 1 is of the order of a minute or minutes.

A sterling attempt was made by Alberty and coworkers to extract all of the elementary rate constants for a single substrate reaction as a function of pH by use of steady-state kinetics (Peller and Alberty, 1961). The enzyme used in the study was fumarase. Although the ultimate goal was not achieved, upper or lower limits were defined for every constant. This work may be regarded as a definitive study of what may or may not be accomplished by developing and applying steady-state enzyme kinetic theory.

If one has a two-substrate reaction, life for the steady-state kineticist becomes even more complicated. In the so-called ping-pong mechanism as defined by Cleland (see Plowman, 1972) (see Eqn. 15) substrate A binds to E, product P departs leaving the enzyme in a modified state F, then substrate B binds and product Q departs. All other permutations involving A, B, P and Q are theoretically possible. Despite the complexity of the system the ping-pong mechanism is readily differentiated from other reaction schemes.

The power of the steady-state method is greatly extended by use of isotopic labeling (Boyer, 1978; Klinman, 1978; Rose, 1979), a topic which is beyond the scope of this treatise.

3.2. Stopped-flow kinetics

One of the problems with steady-state enzyme kinetics is that one is forced to use such small concentrations of enzyme that direct observations can only be made on the product or substrate; all information about the behavior of the enzyme is obtained by inference. In a typical steady-state experiment one has two identical cuvettes containing every reaction component, buffer, etc., but not the enzyme. These are thermostated in a dual-beam spectrophotometer. One adds a measured amount of enzyme to one cuvette (say 10^{-8} M final concentration), stirs, covers the cell compartment and switches on the spectrophotometer. Thus, absorbance of product at a fixed wavelength is measured as a function of time. The time from addition of the enzyme to start of the measurement is of the order of 5 sec. If one conducted a similar spectrophotometric experiment using 10^{-5} M enzyme at least one of the following events would probably happen: (a) the reaction would be over before measurements began; (b) the steady-state theory would be invalid; (c) one would be unable to interpret the data.

The problem of rapid mixing must be understood in order to appreciate the development of the stopped-flow apparatus described in this section. All kinds of mixing techniques have been tried and the shortest time in which complete mixing of reactants has been attained is of the order of 1 msec (one thousandth of a second). Suffice to say that forcing two different solutions through tangential jets into a mix-

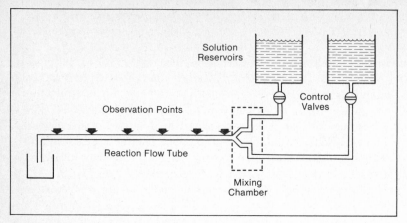

Fig. 2.   Schematic diagram of a continuous flow apparatus.

ing chamber is the most efficient mixing method. The flow rate must be sufficiently large to ensure turbulent flow. Problems concerning various types of mixers and detectors for stopped-flow kinetic studies have been recently reviewed by Berger (1978).

Chemists have had the luxury of many slow reactions to study, but this is not true for biochemists. It is no accident that the pioneering work for studying rapid reactions in solution was performed by biochemists. Hartridge and Roughton (1923) developed a continuous-flow apparatus in which they studied oxygen and carbon monoxide binding to hemoglobin (Fig. 2). Since observations are made at various positions along a flow tube, distance becomes interchangeable with time. The disadvantages of the technique are that large solution volumes are required and the sensitivity is low, since observations are made across a narrow flow tube. The latter difficulty has been removed by an ingenious modification of Roughton's apparatus by Holzwarth (1979a). In Holzwarth's continuous rapid-flow apparatus measurements are made by viewing the reaction right through the mixing chamber into the flow tube. Much faster reactions can be studied. However, if they are complex, difficulties arise, since an integration-with-respect-to-distance procedure is used to extract the rate constants.

Development of the modern stopped-flow apparatus (Fig. 3) was pioneered by Chance (1951a), who developed the accelerated- and stopped-flow methods, and Gibson (Gibson and Milnes, 1964) who put the mechanical stop into the stopped-flow method (see also Caldin (1964) for an extended discussion). Sensitivity was increased by lengthening the observation path. The total length of the observation chamber is viewed instead of its width. Solutions are conserved by stopping the flow after mixing the solutions. To do this reproducibly, a constant pressure head is required to drive the solutions through the mixer at a uniform rate, and a mechanical stop is required to interrupt the flow after a fixed time interval.

In a conventional stopped-flow apparatus the two reactant solutions are placed in

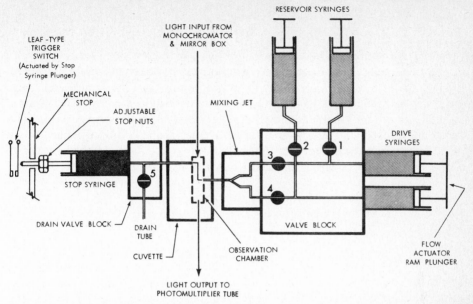

Fig. 3.   Schematic diagram of the Gibson–Durrum stopped-flow apparatus. Courtesy of the Dionex Corporation, Sunnyvale, CA.

storage syringes and transferred into the drive syringes as required (Fig. 3). Pneumatic pressure is applied to the drive syringes by opening a gas reservoir. This forces a piston forward against the drive syringes until the flow is stopped. With 2-ml drive syringes typically 5 – 10 experiments can be performed before the reactant solutions are used up. Observations are usually made spectrophotometrically using monochromatic light. These are single-light-beam experiments (in contrast to the dual-beam experiments performed in a spectrophotometer). Therefore, a constant-intensity light source combined with a stable photodetector tube and stable electronic circuitry is required for accurate measurements. These requirements are not difficult to meet since the time intervals are so short. If one can observe the formation or decay of a single-reactant species by measuring absorbance as a function of time, the absorbance, A, is directly proportional to concentration, c, according to Beer's law: $A = abc$, since "a", the molar absorptivity, is a constant, and b, the optical path length through the solution, is also constant.

If the reaction is first order, then

$$-\frac{dc}{dt} = kc \tag{8}$$

Eqn. 8 is Eqn. 2 rewritten in general terms involving the concentration of any reactant. Integration yields

$$\ln \frac{c_0}{c} = kt \tag{9}$$

where c_0 is the concentration of reactant at time $t = 0$. Plots of c vs. t and $\ln c$ vs. t are shown in Fig. 4. Note that a ratio of concentrations occurs in Eqn. 9. It is therefore possible to use any concentration units one wants since they cancel out. Since A is proportional to c it is possible to use absorbance units directly. Furthermore, the really important kinetic quantities, *change* of c (Δc) with respect to time, and *change* of absorbance (ΔA) with respect to time, are also proportional to each other. An important experimental implication is that one does not need to operate at a wavelength where one is observing only the absorbance of a single reactant.

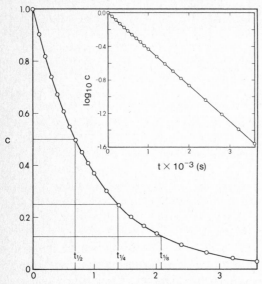

Fig. 4. Rate of disappearance of reactant in a first-order reaction. Concentration of reactant c is plotted on a relative scale. The independence of $t^{1/2}$ from c is illustrated. The inset shows a semi-logarithmic plot of the same data in which the negative value of the slope is equal to k, the first-order rate constant.

One could be monitoring a change in absorbance due to the simultaneous disappearance of reactant and appearance of product, but ΔA would still be proportional to Δc. Another implication of this fundamentally important property is that provided one is operating under first-order conditions, one does not need to know the quantities "a" or "b" in order to measure the rate constant, k.

It is only for first-order reactions that the half-life is independent of c_0. Substitution of $c = 0.5c_0$ and $t = t_{1/2}$ into Eqn. 9 yields

$$\ln 2 = kt_{1/2}$$

$$t_{1/2} = (\ln 2)/k \tag{10}$$

Therefore, except for an effect on the amplitude of the signal, the precision of determining a first-order rate constant is unaffected by a change in c_0. Stated another way, one can choose c_0 and zero time arbitrarily on a first-order decay curve and not affect the accuracy of the determinations of k (Fig. 4). The choice of c_0 only shifts the logarithmic plot up or down; it does not affect the slope.

None of the above advantageous properties of first-order kinetics is found in zero- or second-order reactions. Inherent properties of zero-, first- and second-order reactions are summarized in Table 1. For further details see, for example, Moore and Pearson (1981). Finally, one can force a reaction to occur under first-order conditions.

For example, the elementary reaction

$$E + S \xrightarrow{k'} ES \tag{11}$$

would obey the rate law

$$-\frac{d(E)}{dt} = k'(E)(S) \tag{12}$$

but if $(S) \gg (E)$ then (S) is effectively a constant and can be combined with k' to yield a pseudo-first-order rate constant k

$$-\frac{d(E)}{dt} = k(E) \tag{13}$$

TABLE 1
Summary of mathematical properties of zero-, first- and second-order reactions in terms of concentration of reactants) c

Order	Differential[a] rate equation	Integrated rate equation	Units of k	Half-life, $t_{1/2}$	To obtain k from a linear slope, plot:
Zero	$-\dfrac{dc}{dt} = k$	$c_0 - c = kt$	M sec^{-1}	$\dfrac{c_0}{2k}$	c vs. t, $k = -$ (slope)
First	$-\dfrac{dc}{dt} = kc$	$\ln \dfrac{c_0}{c} = kt$	sec^{-1}	$\dfrac{\ln 2}{k}$	$\ln c$ vs. t, $k = -$ (slope)
Second[b]	$-\dfrac{dc}{dt} = kc^2$	$\dfrac{1}{c} - \dfrac{1}{c_0} = kt$	M^{-1} sec^{-1}	$\dfrac{1}{c_0 k}$	$1/c$ vs. t, $k =$ slope

[a] To express the equations in terms of rate of formation of product substitute $c = a - x$ where a is initial concentration of reactant and x is amount of product at any time t.
[b] It is assumed that the concentrations of reactants are equal.

Since

$$k = k'(S) \qquad (14)$$

a plot of k vs. (S) would be linear with slope k'. Therefore, the pseudo-first-order constant k (sec^{-1}) can be converted into the second-order rate constant, k' (M^{-1} sec^{-1}). Thus, one gains all of the inherent advantages of studying a reaction under first-order conditions.

We have seen that if one studies a two-substrate enzyme reaction instead of a one-substrate reaction life becomes much more difficult for a steady-state enzyme kineticist. This is not necessarily true for a stopped-flow kineticist. A single-substrate single-enzyme reaction is difficult to control since once it is started it is difficult if not impossible to stop. One might use the analogy of locking a cat and a dog in a closet. Upon opening the door after a certain time interval one knows that the action has been fast and furious, but one has absolutely no detailed knowledge of it. A blow-by-blow description is lacking.

However, consider the ping-pong mechanism described at the end of the previous section of this chapter. The reaction may be written as a series of elementary steps

$$E + A \underset{k_{-1}}{\overset{k_1}{\rightleftarrows}} EA$$

$$EA \xrightarrow{k_2} F + P \qquad (15)$$

$$F + B \underset{k_{-3}}{\overset{k_3}{\rightleftarrows}} FB$$

$$FB \xrightarrow{k_4} E + Q$$

The steady-state kineticist cannot isolate individual steps since the enzyme must be "turning over" in order to gain any kinetic information. However, a stopped-flow kineticist can isolate the first two steps by reacting only E + A. Typically, $(E)_0$ would be $\sim 10^{-5}$ M and (A) 10^{-2} M to ensure pseudo-first-order conditions. If k_1 and $k_{-1} \ll k_2$ then one would observe a process which appears to be simply

$$E + A \xrightarrow{k_{app}} F + P \qquad (16)$$

and the apparent rate constant k_{app} (M^{-1} sec^{-1}) could be obtained as illustrated in Eqns. 11–14. The formation of the intermediate complex, EA, can be inferred, but one has no direct evidence for it (contrary to some early reports in the transient-state enzyme kinetics literature). On the other hand, if $k_{-1} \geq k_2$ then k_1, k_{-1} and k_2 may be determined depending upon whether pre-equilibrium is maintained (Strickland et al., 1975).

Also one could take the product mixture from Eqn. 16 and perform more experiments reacting F + B with similar results for the second half of the reaction. This type of experiment involving a reactive intermediate F is greatly simplified using pseudo-first-order conditions. Thus the requirement of two substrates for the overall reaction provides the transient state kineticist with a handle whereby individual elementary steps can be isolated and studied. Examples of stopped-flow reaction traces are shown in Fig. 5.

So far we have only discussed optical detection. Although this is the most commonly employed method, other non-optical techniques such as calorimetry, determinations of pH or oxidation-reduction potential are currently used in conjunction with stopped-flow experiments. More recently other interesting detection methods have been introduced such as the nuclear magnetic resonance flow system of Manuck et al. (1973); the circular dichroism system of Luchins (1977), used for the study of hemoglobin denaturation, and the combined stopped-flow laser-flash photolysis of Sawicki and Gibson (1978) for the study of carbon monoxide and oxyhemoglobin. For additional details on alternative detection techniques see Hiromi (1979).

A potential difficulty with absorbance measurements may be introduced by chemiluminescence from the sample. If this occurs as the result of the reaction under observation, then one simply shuts off the exciting light beam and measures the chemiluminescence. If one of the reactants is fluorescent, an additional method for monitoring the reaction kinetics becomes possible. In this case, the sample is excited at the appropriate wavelength and the fluorescence is detected with the phototube placed perpendicular to the exciting beam. Commercial equipment is available for this purpose.

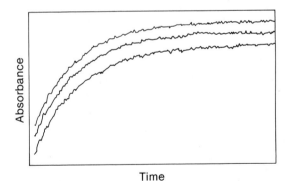

Fig. 5. Repetitive stopped-flow measurements at 437 nm on the reaction of cyanide with chloroperoxidase. Temperature 25°C, 0.05 M citrate buffer pH 5.60, total ionic strength 0.11. $[KCN]_{total}$ = 1 × 10^{-3} M, [ClPO] = 0.80 mM. The full time scale is equal to 100 msec. $k_{obs} = 56 \pm 3$ sec^{-1}. Experiments performed in the author's laboratory by Dr. A.M. Lambeir on a Union Giken Rapid Reaction Analyzer Model 601.

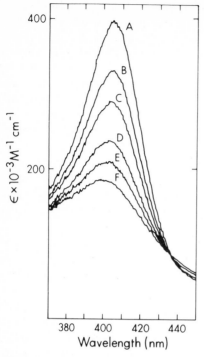

Fig. 6. Rapid-scan spectrophotometric measurement of the reaction of human erythrocyte catalase with methyl hydroperoxide at pH 7.1 and 25°C. A spectrum of native catalase. B–F show progression of compound I formation at ~1, 2, 7, 12 and 47 msec after start of the reaction. [Enzyme] = 1.5 μM, [Methyl hydroperoxide] = 100 mM. Reprinted by permission from *J. Biol. Chem.* (Palcic and Dunford, 1980).

Two further extensions of the stopped-flow technique are worthy of note. If the solutions are turbid, then light transmission is impaired by the turbidity. However, by use of a dual-wavelength instrument it is still possible to extract accurate rate data. This clever modification was devised by Chance (1951b). Finally, rapid scan equipment is becoming commercially available. An example of use of this technique is shown in Fig. 6. There are many advantages to simultaneous detection at many wavelengths using a white light source over measurement at a single wavelength using a monochromatic light source.

Stopped-flow manufacturers are listed by Hiromi (1979). To this list should be added Cantech Scientific Limited, 35 Purdue Bay, Winnipeg, Manitoba, Canada R3T 3C6. Among the best stopped-flow equipment for accurate thermostating is that of Nortech Laboratories. The apparatus sold by the Union Giken Company is among the best for shortest dead time (time from the start of mixing to the beginning of accurate rate measurements). In one of the latter company's designs the drive syringes are eliminated and pressure is applied directly to the reactant solutions.

Use of stopped-flow apparatus is an example of transient-state kinetics. The change in concentration of the native enzyme or of a reactive intermediate is followed as a function of time, in contrast with steady-state studies where, by definition, the concentrations are invariant with respect to time. Two more transient-state techniques are described below.

3.3. Relaxation techniques

So far we have seen that the minimum time required to start a steady-state experiment is of the order of 5 sec. The dead time of a stopped-flow apparatus is 1–5 msec, depending upon the apparatus. If the half-life of a reaction is appreciably shorter than 1 msec, then the reaction is too fast to be studied by the stopped-flow method. A further improvement in dead time is attained in Holzwarth's apparatus (Holzwarth, 1979a), but so far this equipment requires too much solution for most biological studies. The next major breakthrough in time scales has been attained by use of relaxation techniques.

Any parameter that influences the position of a chemical equilibrium can be used to create a step perturbation of the equilibrium. The relaxation of the system to the new equilibrium state can then be followed with a suitable detection system. The most common parameters that have been used so far for rapid perturbation are temperature, pressure, electric field and concentration. The way a relaxation kinetic equation is derived from an equilibrium expression is extensively described in several good text books (Bernasconi, 1976; Strehlow and Knoche, 1977). The simplification that is obtained with small perturbations is not restricted to fast reactions.

Consider a system already at equilibrium

$$E + I \underset{k_{-1}}{\overset{k_1}{\rightleftharpoons}} EI \tag{17}$$

where I is an enzyme inhibitor. If a small perturbation is applied to the system (for our present purposes the type of perturbation is not important) then the equilibrium is upset. In other words, for a time interval the concentrations of all the species E, I and EI become different from their equilibrium concentrations. The system then relaxes to its new equilibrium position — hence the term a relaxation process.

The net rate of production of EI is given by

$$\frac{d(EI)}{dt} = k_1(E)(I) - k_{-1}(EI) \tag{18}$$

At equilibrium

$$k_1(E)_{eq}(I)_{eq} = k_{-1}(EI)_{eq} \tag{19}$$

where the subscript, eq, indicates final equilibrium concentration. The net rate of production of EI is zero at equilibrium. The conservation relation

$$(E)_0 = (E) + (EI) \tag{20}$$

must also be valid at any time t, as must the relations

$$(E) = (E)_{eq} + \Delta(E)$$

$$(I) = (I)_{eq} + \Delta(I) \tag{21}$$

$$(EI) = (EI)_{eq} + \Delta(EI)$$

$$\Delta(EI) = -\Delta(E) = -\Delta(I) \tag{22}$$

The delta terms indicate how far the concentrations of each species is from its equilibrium value at any time t. These terms are very small. If the right sides of Eqn. 21 are substituted into Eqn. 18 one obtains

$$\frac{d(EI)}{dt} = \frac{d\Delta(EI)}{dt} = k_1[(E)_{eq} + \Delta(E)][(I)_{eq} + \Delta(I)] - k_{-1}[(EI)_{eq} + \Delta(EI)] \tag{23}$$

which by use of Eqn. 22 becomes

$$\frac{d\Delta(EI)}{dt} = k_1(E)_{eq}(I)_{eq} - k_{-1}(EI)_{eq} - k_1\Delta(EI)[(E)_{eq} + (I)_{eq}] + \\ + k_1\Delta^2(EI) - k_{-1}\Delta(EI) \tag{24}$$

The first two terms on the right side of Eqn. 24 cancel (Eqn. 19) and the Δ^2 term obtained by multiplying out the quadratic term is negligible compared to any Δ term to the first power. Therefore,

$$-\frac{d\Delta(EI)}{dt} = \{k_1[(E)_{eq} + (I)_{eq}] + k_{-1}\}\Delta(EI) \tag{25}$$

All of the terms inside the { } brackets are constants. Therefore, one has a first-order equation in which

$$\frac{1}{\tau} = k = k_1[(E)_{eq} + (I)_{eq}] + k_{-1} \tag{26}$$

The relaxation time, τ, is simply the reciprocal of the first order rate constant. If one determines τ for a series of concentrations of reactants, then a plot of $1/\tau$ vs. $[(E)_{eq} + (I)_{eq}]$ gives a straight line with the slope equal to k_1 (M^{-1} sec^{-1}) and the intercept equal to k_{-1} (sec^{-1}). Furthermore, if $(I)_0$, the initial concentration of inhibitor, is considerably greater than $(E)_{eq}$, then $(I)_0 = (I)_{eq}$ and

$$\frac{1}{\tau} = k_1(I)_0 + k_{-1} \tag{27}$$

Therefore, an even simpler relation is obtained.

It turns out that the equations for all slightly perturbed equilibria can be reduced to first-order conditions which is the way all kinetics should be conducted. Beautiful! The Nobel prize in chemistry in 1966 was shared by Manfred Eigen for the development of a whole battery of relaxation techniques. In the next section we shall describe in more detail the most popular of these, the temperature-jump method.

3.4. The temperature-jump method

A schematic representation of a conventional temperature-jump (T jump) apparatus is shown in Fig. 7. The solution to be studied is placed in a small cell, capac-

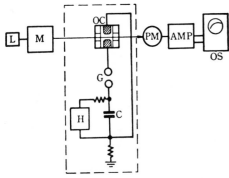

Fig. 7. Schematic representation of a temperature-jump apparatus. L, light source; M, monochromator; PM, photomultiplier; AMP, amplifier; OS, oscilloscope; H, high-voltage power supply; C, capacitor; G, spark gap; OC, observation cell. Often G is replaced by an electronic triggering device. Reprinted with permission from Hiromi (1979), p. 129.

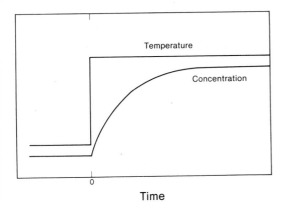

Fig. 8. Schematic representation of a temperature-jump experiment. The discharge of a capacitor through a conduction solution provides a step-wise temperature jump and the concentration of a reactant relaxes to its new equilibrium concentration at the higher temperature.

ity 1 or 2 ml, which contains electrodes connected to a high-voltage capacitor. Upon discharge of the capacitor through the solution, the solution is heated because of its resistance to electric flow. (The ions in the solution provide the electrical conduction.) The heating effect may be regarded as a step function (Fig. 8). Provided the chemical reaction has a finite enthalpy change ΔH^0, according to the equation

$$\frac{\mathrm{d}\ln K}{\mathrm{d}T} = \frac{\Delta H^0}{RT^2} \tag{28}$$

then a change in temperature changes the value of the equilibrium constant, K. Suddenly the system finds itself out of equilibrium and the concentrations change until a new equilibrium is attained (Fig. 9). From this relaxation process the forward and reverse rate constants can be measured as outlined in section 3.3.

The solution must have an appreciable dielectric constant. If it is a protic solvent (protons are conducting) and if an inert electrolyte is added, then the solution has sufficient ability to conduct the electric current, but sufficiently high resistance that heating occurs. Typically a temperature rise of 4 or 5°C is most ideal. The resultant perturbation to the equilibrium is sufficiently large to produce a readily measurable signal. However, it is not so large that linearization of the differential equations (neglect of the Δ^2 terms) is an inaccurate approximation (see section 3.3).

Typically the dead time of a T-jump apparatus is of the order of 2–10 μsec (2 × 10^{-6} to 10 × 10^{-6} sec). Therefore, the time range for rate constant measurements is greatly extended from that for a stopped-flow apparatus. An ultra-fast T jump with dead time of the order nsec (10^{-9} sec) can be achieved in one of two ways. Either the discharge of the high-voltage capacitor can be cut short by use of appropriate electronic circuitry, or a coaxial cable with impedance matched to that of the reaction cell can be used in place of the capacitor. Conventional T jumps are available commercially, but not ultra-fast T jumps (Hiromi, 1979).

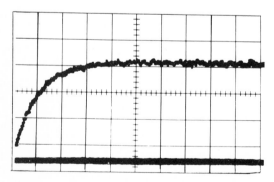

Fig. 9. Oscilloscope trace of voltage vs. time in a temperature-jump experiment performed at 413 nm on the reaction of hemin with imidazole in aqueous ethanol at 25°C. The horizontal scale is 1 msec/large-scale division. Reprinted by permission from *Can. J. Chem.* (Hasinoff et al., 1969).

T jumps may be used to study enzyme–inhibitor reactions as described in section 3.3, antigen–antibody reactions, and ligand binding, including O_2 binding. If the substrate–product equilibrium is not shifted too far in one direction, then the reaction of an enzyme with its normal substrate can be studied. If one has non-aqueous non-conducting solutions then a laser may be used to heat the solution by absorption of light by a dye. Water may be heated by absorption of light quanta from a laser of appropriate frequency (Holzwarth, 1979b).

3.5. *Flash photolysis*

The other half of the Nobel prize shared by Eigen was awarded to Norrish and Porter for the development of the flash photolysis technique. In this technique a short pulse of light is used to displace an equilibrium or to disturb a steady state in a time interval which is short compared to the time course of the subsequent dark relaxation. The dark reaction is usually monitored by measuring an absorbance change as a function of time. For example, light quanta absorbed by the molecules present in solution can result in photodissociation; the subsequent recombination process can then be investigated. Application of this technique, for instance, to the study of myoglobin–CO complexes, which can be dissociated with a quantum efficiency of unity (one CO molecule is released for each light quantum absorbed) are discussed in Chapter 6 of this book.

Flash photolysis provides an alternative to mixing techniques for the initiation of a reaction and it is particularly useful when the reaction under study is diffusion limited. Examples of such cases are also presented in Chapter 6. A schematic representation of a flash photolysis apparatus is shown in Fig. 10.

As lasers were being developed in the 1960's and 1970's the time scales for light pulses were drastically reduced. For Q-switched lasers the time scale approached nsec (10^{-9} sec) and for mode-locked lasers psec (10^{-12} sec). Thus, instead of a flash-lamp source, light of sufficient intensity but in much shorter bursts can be supplied by lasers.

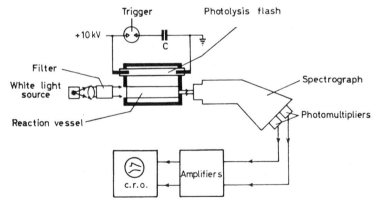

Fig. 10. Schematic diagram of a flash photolysis apparatus. Reproduced by permission from *Proc. Roy. Soc.* (Bridge and Porter, 1958).

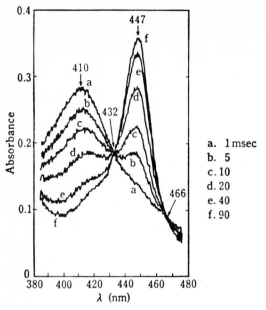

Fig. 11. Spectral changes caused by flash photolysis of the cytochrome P-450 carbon monoxide complex. Reproduced by permission from Hiromi (1979), p. 158. Work of Y. Ishimura and T. Iizuka.

An example of the use of flash photolysis and rapid scan method using a commercial instrument is shown in Fig. 11. Workers in this author's laboratory have used the same commercial instrument to carry out stopped-flow, rapid-scan, T-jump and flash photolysis experiments. With the flip of a switch the stopped-flow instrument is converted into a combined stopped-flow rapid-scan apparatus. A solid-state photodiode array is used to collect light over a 96-nm wavelength region. The stopped-flow cuvettes can be replaced by a flash photolysis unit. Two high-voltage circuits identical in every respect, except capacitance, supply the current for the flash and detection lamps. Again, either monochromatic or rapid-scan spectrophotometric detection is possible. Finally, a T-jump cell can be inserted into the light detection path. It utilizes the same high-voltage circuitry used for the flash detection lamp. Thus, this compact self-contained apparatus possesses the most popular of the transient-state kinetic techniques.

The capital investment is generally much smaller for transient-state kinetic techniques in comparison to spectroscopic techniques used to determine structure. It appears, however, that transient-state kinetics (theory, experiment and interpretation) is of sufficient sophistication to require a full-time specialist. Verification would appear to be provided by the number of stopped-flow apparatuses which this author has seen sitting in a state of disrepair in various laboratories. It would appear that for little more capital investment but with sufficient priority to hire a professional transient-state kineticist, many laboratories could have an entire battery of powerful kinetic techniques at their disposal.

4. Effect of varying solvent parameters

4.1. Introduction

In sections 1–3 of this chapter some of the relevant theory and techniques were described from which an overall enzyme-catalyzed process can be studied in terms of its elementary reactions. Let us now turn our attention to the additional information that can be obtained by varying various parameters which affect properties of the solvent.

One of the most widely varied parameters is pH. Fundamental information about the importance of Brønsted acid–base catalysis can be obtained from the experimental results. The pK_a values of catalytic groups can be measured with precision. Thus, thermodynamic parameters can be derived from kinetic data. The converse statement is not true. Since classical thermodynamics is concerned solely with equilibrium, by definition information about kinetics and mechanism cannot be obtained from classical thermodynamic studies. Horseradish peroxidase is an outstanding example of an enzyme with a very wide range of pH stability. This stability has been exploited to demonstrate kinetically the importance of acid–base catalysis in each of the elementary reaction steps (Dunford and Stillman, 1976). The relevant pK_a measurements have been confirmed by delicate spectrophotometric and pH-stat measurements (Hayashi and Yamazaki, 1978).

Temperature is also a potentially valuable variable. From the variation of a rate constant with temperature one can deduce the energy barrier (activation energy) of a reaction:

$$k = \frac{\ell T}{h} e^{\Delta S^{+}/R} e^{-\Delta H^{+}/RT} \tag{29}$$

In Eqn. 29 k is the rate constant, $\ell T/h$ a frequency factor, ΔS^{+} the entropy of activation, ΔH^{+} the enthalpy of activation or activation energy, R the universal gas constant and T temperature on the Kelvin scale (see for example Frost and Pearson, 1961). From a linear plot of $\ln(k/T)$ vs. $1/T$ the slope is equal to minus the value of $\Delta H^{+}/R$. Eqn. 29 is derived from transition-state theory. In this theory properties are described for the critical complex which occurs at the maximum in the potential energy barrier. A combination has been made of Brønsted acid–base catalytic theory and transition-state theory for enzyme reactions (Dunford, 1974, 1975). In the remainder of this section the significance of the dielectric constant and viscosity of solvents and their variation will be discussed.

4.2. Solvent dielectric constant (Table 2)

The force f between two ions is given by Coulomb's law

$$f = \frac{Z_1 Z_2 e^2}{\varepsilon r^2} \tag{30}$$

TABLE 2

Values of dielectric constants, ε, at 25°C

Solvent	ε
Vacuum	1
Carbon tetrachloride	2.2
Dioxane	2.1
Methanol	33
Water	80
HCN	140

where Z_1 and Z_2 are the ionic valences $(+1, -1,$ etc.$)$, e the charge of the electron, ε the dielectric constant and r the distance between ions. The units of Eqn. 30 need not concern us in this discussion. Eqn. 30 must be applied many times over to describe all of the electrostatic interactions in a protein. According to Eqn. 30 the force between ions (either attractive or repulsive) is inversely proportional to the dielectric constant. Thus, a medium of high dielectric constant helps to screen or protect two ions from each other. Dielectric constant and solvent polarity or dipole moment tend to go together. Thus water is a polar molecule because it has a dipole moment. It has the high dielectric constant value of 80. Dioxane is an interesting molecule because it has two dipole moments pointing in opposite directions which therefore cancel. Hence dioxane has a low dielectric constant. However, because it has polar components dioxane is miscible with water in all proportions. Liquid hydrogen cyanide has a dielectric constant of 140 but it is a solvent system shunned by most experimentalists!

If the temperature of a solvent is lowered its dielectric constant increases. This is because of a decrease in the random motion of the molecular dipoles. It is thus possible, by the addition of a second solvent component of lower dielectric constant, to compensate for increased dielectric constant at lower temperature.

Douzou (1977) has developed this technique to an art form, whereby solvent dielectric constant is maintained exactly at a constant value as solvent temperature is lowered from room temperature to −50°C or lower. Thus enzyme solutions can be cooled over a large temperature interval in "anti-freeze" solvent systems of constant dielectric constant. There is no doubt that the dielectric constant of the medium is intimately connected with protein integrity. One can readily denature a protein by trying to dissolve it in a nonpolar solvent such as chloroform. However, other factors are also important. A change in pH changes the charge on a protein and pH changes both with temperature and solvent composition (Douzou, 1977; Bates, 1964). Specific interactions of solvent and protein may occur. A more pragmatic approach to the use of "anti-freeze" solvents has been taken by Fink (1977). The relative importance of dielectric constant compared to other solvent parameters would appear to be a valuable area of research.

In Figs. 12 and 13 dielectric constant data of Åkerlöf (1932) and Åkerlöf and Short (1936) have been rearranged into a form whereby ε is maintained constant despite changing temperature. Thus, one can use pure solvent at one temperature as a reference and conduct experiments at lower temperatures in solvent systems with identical values of ε. It is important to test for specific effects of the second solvent component. For example ethanol will bind in the active site of peroxidase (Dunford and Hewson, 1977). From a practical standpoint glycerol is a notoriously impure solvent. It may be necessary to vacuum distill it in order to attain purity. Similarly,

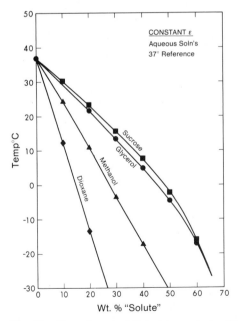

Fig. 12. Plots of constant dielectric constant ε. With decrease in temperature increase in the amount of organic component keeps ε constant. Pure water at 37°C is used as the reference. Data interpolated from Åkerlöf (1932) and Åkerlöf and Short (1936).

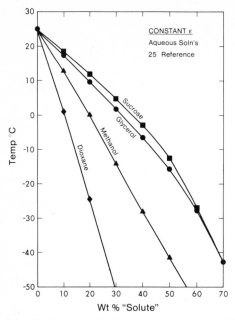

Fig. 13. Same as Fig. 12 except pure water at 25°C is used as the reference.

spectral-grade methanol should be used; otherwise "reagent-grade" methanol will probably require considerable purification.

The objective in the use of "anti-freeze" solvents is to increase the lifetime of transient-state species so that their properties can be studied more readily. Spectacular results have been achieved (Douzou, 1977; Fink, 1977). According to Eqn. 29, the lower the temperature, the slower a reaction, and the greater the activation energy, the greater the effect of lowered temperature. The objective of isolating reactive intermediates is more likely to be attained for enzymatic reactions with two substrates. The single-substrate reaction is still like a cat and a dog in a closet and the fact that the closet is cold may not have discernible benefits. Other complications may occur. For example an enzymatic reaction which has chemically controlled rate-limiting steps at room temperature may be converted to one that is diffusion-controlled at low temperature. Thus the comprehensive study of ligand recombination with myoglobin after flash photolysis (Austin et al., 1975) has been shown to correlate with solvent viscosity. Therefore, the reaction inside the protein appears to be diffusion-controlled at high enough solvent viscosity and/or low enough temperature (Hasinoff, 1977, 1978, 1981; Beece et al., 1980).

4.3. Solvent viscosity

The viscosity of the solvent is important because it determines the diffusion coefficients of solute molecules. According to the Stokes–Einstein equation

$$D = \frac{kT}{6\pi\eta r} \tag{31}$$

the diffusion coefficient of a solute molecule D is inversely proportional to its radius r and the solvent viscosity η (Caldin, 1964). In a diffusion-controlled reaction the rate of the reaction is determined by how fast the two reactant molecules can diffuse together (Smoluchowski, 1917)

$$k_f = \frac{4\pi N}{1000}(D_E + D_S)(r_E + r_S)M^{-1}\,sec^{-1} \tag{32}$$

where k_f is the diffusion-controlled second-order rate constant, N is Avogadro's number and the subscripts E and S refer to the two reactant molecules, enzyme and substrate. According to Eqns. 31 and 32 there is a compensation between the magnitudes of the diffusion coefficient and molecular radius. A large molecule will diffuse slowly but it presents a large target for the other reactant molecule. Eqn. 32 does not take into account that the active site of an enzyme occupies only a small fraction of the total accessible surface area of the enzyme; and corrections for this and other omissions form a lively area of research (Hill, 1975; Knowles and Albery, 1977; Nakatani and Dunford, 1979). Nevertheless, Eqn. 32 leads to a prediction of the right magnitude ($k_f = 10^{10}\,M^{-1}\,sec^{-1}$ at room temperature). For enzyme reactions a rough general rule may be that $k_D = 10^8 - 10^9\,M^{-1}\,sec^{-1}$ at room temperature. The Smoluchowski equation has been available since 1919 so that it is quite remarkable that a comparable equation for how fast reactants diffuse apart was not published until the work of Eigen (1954). According to Eigen

$$k_r = \frac{3(D_E + D_S)(r_E + r_S)\,sec^{-1}}{a^3} \tag{33}$$

where k_r is the diffusion-controlled rate constant for the dissociation of a complex E–S and "a" is the radius of the E–S complex. The magnitude of k_r is of the order of $10^9\,sec^{-1}$. A proper discussion of a diffusion-controlled reaction must take the dissociation reaction into account. Thus the equation

$$E + S \underset{k_r}{\overset{k_f}{\rightleftarrows}} E - S \xrightarrow{k} Products \tag{34}$$

introduces the proper perspective. A diffusion-controlled reaction does not necessarily occur simply because the reactants combine or collide at the diffusion-controlled rate — that is always occurring amongst all solute molecules. The essential criterion is that $k \gg k_r$, in other words, the complex AB must react chemically faster than it dissociates into its components. If $k_r \gg k$, then one simply has formation of an E–S complex which immediately dissociates. An equilibrium association constant can be written for this non-productive association

$$K = \frac{(E - S)}{(E)(S)} = \frac{k_f}{k_r} = \frac{4\pi Na^3}{3000} \; M^{-1} \tag{35}$$

in which the magnitude is calculated by dividing Eqn. 32 by Eqn. 33. If the overall chemical reaction rate k is finite but still much smaller than k_r, then the equilibrium $E + S \rightleftharpoons E\text{-}S$ is maintained despite the relatively slow chemical reaction which is occurring. Thus the ratio k/k_r is of crucial importance. If it is much greater than one the reaction is diffusion-controlled; much less than one and it is chemically controlled. A complete spectrum of intermediate cases is possible. If the ratio k/k_r is sufficiently large then the steady-state approximation applies to E-S and the apparent rate constant, k_{app}, is given by

$$k_{app} = \frac{k_f k}{k_r + k} \; M^{-1} \, sec^{-1} \tag{36}$$

It would appear that the only readily accessible way to test for diffusion control is to vary the viscosity of the solvent. Thus for reactions in two different solvent media

$$k_f^\circ \eta^\circ = k_f \eta; \qquad k_r^\circ \eta^\circ = k_r \eta \tag{37}$$

(Nakatani and Dunford, 1979; Loo and Erman, 1977). Therefore a plot of $1/k_{app}$ vs. η/η^0 should yield a straight line with the slope equal to $1/k_f^0$ and the intercept equal to the ratio $k_r^0/k_f^0 \, k$. Such a plot was obtained by Nakatani and Dunford (1979) for the reaction of horseradish peroxidase with p-nitroperbenzoic acid at 25°C. The aqueous diffusion-controlled association rate constant k_f^0 was found to be 1.3×10^8 $M^{-1} \, sec^{-1}$ and $k_r^0/k = 2.2$. Therefore for every E-S complex which reacted to form horseradish peroxidase compound I there was an average of 2.2 which dissociated to initial reactants. The results may be regarded as evidence for the existence of the E-S encounter complex.

The viscosities of some solvent mixtures pass through a maximum as a function of solvent composition as illustrated in Figs. 14 and 15 for dioxane–water and methanol–water systems.

If the dielectric constant is proven to be of such importance that small variations have an important influence on protein integrity then it is possible to design solvent systems in which ε is held constant but η is varied at a constant temperature. The compositions of one such set of solvents is shown in Table 3. A three-component system is used consisting of water, dioxane and glycerol. The glycerol provides the major change in η with a small amount of dioxane used to maintain ε at a constant value as η (and glycerol concentration) is decreased. It would be important to demonstrate that none of the solvent systems exhibit specific interactions with the enzyme whose reactions are being tested for diffusion control.

Two enzymes which appear not to be affected by small changes in ε, at least in more viscous solvents, are horseradish peroxidase and yeast cytochrome c

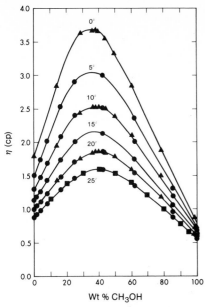

Fig. 14. Plot of viscosity in centipoise vs. aqueous methanol composition. Landolt-Börnstein (1969), data of the author.

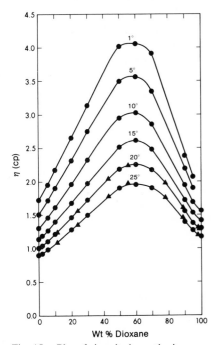

Fig. 15. Plot of viscosity in centipoise vs. aqueous dioxane composition. Geddes (1933), data of the author.

36

TABLE 3[a]

Viscosities of aqueous solutions containing both dioxane and glycerol at 16°C, which have the same dielectric constant as water at 37°C

Dioxane (wt. %)	Glycerol (%)	Viscosity (cp)
11	0	1.61
5.5	17	2.31
4.4	20.4	2.49
3.3	23.8	2.70
2.2	27.2	2.90
1.1	30.6	3.14
0	34	3.64

[a] I am indebted to Dr. I.M. Ralston who compiled the data in Table 3. It is assumed that a linear extrapolation in the dielectric constant ε occurs between the 11% dioxane solution and the 34% glycerol solution which appears to be justified (Åkerlöf, 1932).

peroxidase (Dunford and Hewson, 1977; Loo and Erman, 1977). More information on other enzyme systems is required before generalizations may be made.

References

Åkerlöf, G. (1932) Dielectric constants of some organic solvent–water mixtures at various temperatures, J. Am. Chem. Soc., 54, 4123–4139.
Åkerlöf, G. and Short, O.A. (1936) The dielectric constant of dioxane–water mixtures between 0 and 80°, J. Am. Chem. Soc., 58, 1241–1243.
Austin, H.A., Beeson, K.W., Eisenstein, L., Fraunfelder, H. and Gunsalus, I.C. (1975) Dynamics of ligand binding to myoglobin, Biochemistry, 14, 5355–5373.
Bates, R.G. (1964) Determination of pH, Theory and Practice, Wiley, New York, pp. 172–230.
Beece, D., Eisenstein, L., Fraunfelder, H., Good, D., Marden, M.C., Reinisch, L., Reynolds, A.H., Sorensen, L.B. and Yue, K.T. (1980) Solvent viscosity and protein dynamics, Biochemistry, 19, 5147–5157.
Berger, R.L. (1978) Some problems concerning mixer and detectors for stopped-flow kinetics studies, Biophys. J., 24, 2–24.
Bernasconi, C.F. (1976) Relaxation Kinetics, Academic Press, New York.
Boyer, P.D. (1978) Isotope exchange probes and enzyme mechanisms, Acts. Chem. Res., 11, 218–224.
Bridge, N.K. and Porter, G. (1958) Primary processes in quinones and dyes, II. Kinetic studies, Proc. Roy. Soc. (London), A244, 276–288.
Caldin, E.F. (1964) Fast Reactions in Solution, Blackwell, Oxford, pp. 41–48, 10–11.
Chance, B. (1951a) Rapid and sensitive spectrophotometry, I. The accelerated and stopped-flow methods for the measurement of the reaction kinetics and spectra of unstable compounds in the visible region of the spectrum, Rev. Sci. Instr., 22, 619–627.
Chance, B. (1951b) Rapid and sensitive spectrophotometry, III. A double beam apparatus, Rev. Sci. Instr., 22, 634–638.
Cornish-Bowden, A. (1976) Principles of Enzyme Kinetics, Butterworths, London.
Douzou, P. (1977) Cryobiochemistry: an Introduction, Academic Press, New York.
Dunford, H.B. (1974) The advantages of transition state and group acid dissociation constants for pH-dependent enzyme kinetics, J. Theor. Biol., 48, 283–298.

Dunford, H.B. (1975) Collision and transition state theory approaches to acid-base catalysis, J. Chem. Ed., 52, 578–580.

Dunford, H.B. and Hewson, W.D. (1977) Effect of mixed solvents on the formation of horseradish peroxidase compound I, The importance of diffusion-controlled reactions, Biochemistry, 16, 2949–2957.

Dunford, H.B. and Stillman, J.S. (1976) On the function and mechanism of action of peroxidases, Coord. Chem. Rev., 19, 187–251.

Eigen, M. (1954) Über die Kinetik sehr schnell verlaufender Ionenreaktionen in wässeriger Lösung, Z. Physik. Chem., NF1, 176–200.

Fink, A.L. (1977) Cryoenzymology: the study of enzyme mechanisms at subzero temperatures, Accts. Chem. Res., 10, 233–239.

Geddes, J.A. (1933) The fluidity of dioxane–water mixtures, J. Am. Chem. Soc., 55, 4832–4837.

Gibson, Q.H. and Milnes, L. (1964) Apparatus for rapid and sensitive spectrophotometry, Biochem. J., 91, 161–171.

Hartridge, H. and Roughton, F.J.W. (1923) A method of measuring the velocity of very rapid chemical reactions, Proc. Roy. Soc. (London), A104, 376–394.

Hasinoff, B.B. (1977) The diffusion-controlled reaction kinetics of the binding of CO and O_2 to myoglobin in glycerol–water mixtures of high viscosity, Arch. Biochem. Biophys., 183, 176–188.

Hasinoff, B.B. (1978) Diffusion controlled transient kinetics of the reaction of carbon monoxide with myoglobin in a supercooled very high viscosity glycerol–water solvent, J. Phys. Chem., 82, 2630–2631.

Hasinoff, B.B. (1981) Flash photolysis reactions of myoglobin and hemoglobin with carbon monoxide and oxygen at low temperatures, Evidence for a transient diffusion-controlled reaction in super-cooled solvents, J. Phys. Chem., 85, 526–531.

Hasinoff, B.B., Dunford, H.B. and Horne, D.G. (1969) Temperature jump kinetics of the binding of imidazole to ferriprotoporphyrin IX, Can. J. Chem., 47, 3225–3232.

Hayashi, Y. and Yamazaki, I. (1978) Heme-linked ionization in compounds I and II of horseradish peroxidases A_2 and C, Arch. Biochem. Biophys., 190, 446–453.

Hill, T.L. (1975) Effect of rotation on the diffusion-controlled rate of ligand-protein association, Proc. Natl. Acad. Sci. (U.S.A.), 75, 72, 4918–4922.

Hiromi, K. (1979) Kinetics of Fast Enzyme Reactions, Halsted, New York, pp. 159, 167–184.

Holzwarth, J.F. (1979a) Fast continuous flow, in: W.J. Gettins and E. Wyn-Jones (Eds.), Techniques and Application of Fast Reactions in Solution, Reidel, Dordrecht, pp. 13–24.

Holzwarth, J.F. (1979b) Laser temperature jump, in: W.J. Gettins and E. Wyn-Jones (Eds.), Techniques and Applications of Fast Reactions in Solution, Reidel, Dordrecht, pp. 47–59.

Klinman, J.P. (1978) Kinetic isotope effects in enzymology, Adv. Enzymol., 46, 415–494.

Knowles, J.R. and Albery, W.J. (1977) Perfection in enzyme catalysis: the energetics of triosephosphate isomerase, Acct. Chem. Res., 10, 105–111.

Laidler, K.J. and Bunting, P.S. (1973) The Chemical Kinetics of Enzyme Action, 2nd Edn., Clarendon Press, Oxford.

Landolt-Börnstein (1969) 5-Teil, II Band, Springer, New York, pp. 366–378.

Loo, S. and Erman, J. (1977) The rate of reaction between cytochrome-c peroxidase and hydrogen peroxide is not diffusion limited, Biochim. Biophys. Acta, 481, 279–282.

Luchins, J.I. (1977) Stopped-Flow Circular Dichroism, Ph. Thesis, Columbia University, New York, University Microfilms, Inc., Ann Arbor, MI.

Manuck, B.A., Maloney Jr., J.G. and Sykes, B.D. (1973) Kinetics of the interaction of methyl isonitrile with hemoglobin chains: measurement by NMR, J. Mol. Biol., 81, 199–205.

Moore, J.W. and Pearson, R.G. (1981) Kinetics and Mechanism, Wiley, New York, pp. 12–82.

Nakatani, H. and Dunford, H.B. (1979) Meaning of diffusion-controlled association rate constants in enzymology, J. Phys. Chem., 83, 2662–2665.

Palcic, M.M. and Dunford, H.B. (1980) The reaction of human erythrocyte catalase with hydroperoxides to form compound I, J. Biol. Chem., 255, 6128–6132.

38

Peller, L. and Alberty, R.A. (1961) Physical chemical aspects of enzyme kinetics, in: G. Porter, (Ed.), Progress in Reaction Kinetics, Vol. 1, Pergamon, Oxford, 237–260.

Plowman, K.M. (1972) Enzyme Kinetics, McGraw-Hill, New York.

Rose, I.A. (1979) Positional isotope exchange studies of enzyme mechanisms, Adv. Enzymol., 50, 361–395.

Sawicki, C.A. and Gibson, Q.H. (1978) The relation between carbon monoxide binding and the conformational change of hemoglobin, Biophys. J., 24, 21–28.

Smoluchowski, M.V. (1917) Versuch einer mathematischen Theorie der Koagulations Kinetik kolloider Lösungen, Z. Physik. Chem., 92, 129–168.

Strehlow, H. and Knoche, W. (1977) Fundamentals of Chemical Relaxation, Verlag Chemie, Weinheim, New York.

Strickland, S., Palmer, G. and Massey, V. (1975) Determination of dissociation constants and specific rate constants of enzyme-substrate (or protein ligand), Interactions from rapid reaction kinetic data, J. Biol. Chem., 250, 4048–4052.

Wong, J.T.-F. (1975) Kinetics of Enzyme Mechanisms, Academic Press, New York.

CHAPTER 3

Recent applications of the stopped-flow and pressure-jump relaxation techniques in the biological sciences

YVES ENGELBORGHS* and KAREL HEREMANS

Laboratory of Chemical and Biological Dynamics, Katholieke Universiteit te Leuven, Celestijnenlaan 200 D, B-3030 Leuven, Belgium

Contents

1. Introduction 40
2. Biological applications of stopped-flow techniques 40
   2.1. Ligand binding to macromolecules 40
      2.1.1. Detection of the ligand response 41
      2.1.2. Detection of the protein response 42
   2.2. Protein–protein interactions 43
   2.3. Conformational studies 44
      2.3.1. Circular dichroism stopped flow 44
      2.3.2. Chemical probing 45
      2.3.3. Hydrogen–deuterium exchange studies 45
   2.4. Permeability of membranes to rapidly permeating molecules 45
3. Special applications 49
   3.1. High-pressure studies 49
   3.2. Optical scanning stopped flow 50
   3.3. Injection into a fixed volume 50
4. Non-optical detection systems 50
   4.1. Stopped-flow calorimetry 50
   4.2. Stopped-flow with glass electrode detector 51
   4.3. Dielectric relaxation detection 51
   4.4. NMR and EPR stopped flow 51
5. Pressure-jump relaxation technique 51
   5.1. Introduction 51
   5.2. Instrumentation 53

---

* Research Associate of the National Fund for Scientific Research, Belgium.

R.I. Sha'afi and S.M. Fernandez (Eds.), Fast Methods in Physical Biochemistry and Cell Biology
© 1983 Elsevier Science Publishers

40

5.3. Applications 55
    5.3.1. Enzymology 55
    5.3.2. Subunit assembly in ribosomes 56
    5.3.3. Metarhodopsin equilibrium 56
    5.3.4. Myosin assembly 56
    5.3.5. Polymerization of glutamate dehydrogenase 57
    5.3.6. Tubulin rings 58
References 58

1. Introduction

Hartridge and Roughton (1923) introduced the technique of rapid mixing for the study of fast reactions in solution. In those early years, time resolution was only possible because in the continuous flow method, time is proportional to distance, through the flow velocity. The stopped-flow approach was introduced by Chance (1940) when the electronic techniques necessary for this method became available. Since then, many modifications have been made to improve the performance of the technique and to adapt it to a variety of detection systems. Important progress has also been made in data acquisition, the most substantial being the introduction of the transient recorder, which allowed replacement of photographic data recording from a storage oscilloscope, by a direct link to the digital computer. Stopped-flow instrumentation has been discussed by Chance (1974) and Hiromi (1979). The requirements for complete automation and the different constructions described in the literature have been reviewed extensively by Crouch et al. (1977). Determination of the dead time, quality control and possible problems of the technique are discussed by Hiromi (1979). Details about instrumentation, theory and experimental arrangements of the stopped-flow and relaxation methods have been dealt with in Chapter 2 of this volume. Therefore, here we concentrate on recent applications to a variety of biochemical systems. Applications to enzyme reactions are fully developed in a book by Gutfreund (1972). The proceedings of three recent symposia on fast reactions contain a number of good reviews and many valuable references to the literature (Parsegian, 1978, 1980; Getting and Wyn-Jones, 1978).

2. Biological applications of stopped-flow techniques

2.1 Ligand binding to macromolecules

Many biological processes can be classified as an interaction between a small molecule and a biological macromolecule. Substrate and product binding to enzymes, drug and hormone interactions with their receptors, metal ion complexation, etc. can be considered as such. Equilibrium and kinetic considerations of protein–ligand interactions are reviewed by Gutfreund (1974). Generally, ligand binding is studied by mixing the macromolecule with a large excess of ligand, to reduce the problem to pseudo-first-order conditions. In this way it is also possible to prove the existence of a rate-limiting conformational change of the whole complex, when the observed rate constant is no longer dependent on the ligand concentration. The results depend, however, on the entity that is detected in the reaction.

2.1.1. Detection of the ligand response

To increase the signal-to-noise ratio, one should preferentially monitor changes in the spectrum of the macromolecule where the contributions of the excess-free ligand are minimal. This is, however, not always possible. We were confronted with this problem in the case of colchicine binding to tubulin (Lambeir and Engelborghs, 1981). It is the colchicine molecule itself which becomes fluorescent upon binding to the protein, but a combination of low-fluorescence quantum yield and very high absorbance of the excess-free ligand (inner filter effect) renders the measurements quite difficult. The inner filter effect was minimized using a laboratory-built stopped-flow. Excitation was done radially with a light beam of 2 mm by 8 mm, obtained from a light guide close to the measuring chamber, with an optical pathlength of only 2 mm. Fluorescence was collected over a wide angle. In this way binding kinetics could be measured up to 8 mM colchicine. At these concentrations, corrections for dimerization of colchicine had to be applied (Engelborghs, 1981). Even then the rate-limiting conformational change could not be reached. A strong deviation from the linear pseudo-first-order conditions allowed, however, the calculation of the equilibrium constant for fast initial binding and the rate constant of the slow conformational change leading to the final fluorescent complex.

Dissociation kinetics can be studied separately, by preventing reassociation. This can be accomplished by strongly diluting the complex, which is, however, impractical. Better is the addition of complexing agents or of a large excess of a competitively binding ligand in so-called displacement reactions. This latter approach was employed to study the dissociation of the drugs warfarin and phenylbutazone from their complex with the protein human serum albumin (Maes et al., 1982). The two ligands can be differentiated because warfarin fluorescence is increased upon binding, while phenylbutazone quenches the tryptophan fluorescence of serum albumin. In displacement reactions the dissociation constants of both ligands were determined very accurately as well as the activation energies.

A ligand that very frequently occurs in biological systems is the nucleotide. Unfortunately nucleotides only absorb in the UV range of the spectrum, where proteins also contribute considerably. Therefore a number of colored and fluorescent nucleotide analogues have been developed. Their spectral properties are reviewed by Trentham et al. (1976). Some of the chromophoric analogues have absorption spectra in the range of 310–330 nm and therefore overlap quite well with the tryptophan emission of proteins. They can thus act as acceptors for radiationless energy transfer upon ligand binding. Binding of the fluorescent ATP analogue ε-ATP to actin was elegantly studied by Waechter and Engel (1977). Confronted with an extreme instability of the nucleotide-free protein, a multimixing stopped flow was developed to produce a transient concentration of free actin by rapid mixing with EDTA. Subsequently a large excess of Ca^{2+} was added to allow recombination of the protein with the nucleotide–metal complex. Here it is clear that only kinetic methods can yield an accurate value for the binding constant.

Special considerations in relation to hormone binding to receptors spread over a cellular surface are reviewed by Delisi (1980). The non-covalently bound reporter ethidium bromide was used to detect ligand-induced changes in the acetylcholine receptor upon binding of agonists, antagonists and different toxins (Quast et al., 1979).

The kinetics of pyrene exchange between human high-density lipoproteins, and between vesicles were studied by Charlton et al. (1976), and by Almgren (1980) respectively. Exchange of fluorescent lipids between vesicles was studied by Doody et al. (1980) and by Nichols and Pagano (1981). As with the lipoproteins, it was shown that transfer occurs via a rate-limiting dissociation into the water phase, and not by vesicle collision. This dissociation rate shows a sharp maximum at the phase transition, proving the importance of irregularities in the bilayer.

2.1.2. Detection of the protein response

Although fast reaction kinetics are quite useful to elucidate reaction mechanisms, they achieve their full biological importance in the study of fast biological processes, such as muscle contraction. The mechanism of ATP hydrolysis and its relationship to force generation has been extensively studied and is reviewed by Trentham et al. (1976).

In fast skeletal muscle, maximum tension develops about 10 msec after excitation by depolarization of the cell membrane. Tension then relaxes back in about 50 msec. During these first 10 msec, Ca^{2+} is released from the sarcoplasmic reticulum and binds to troponin C, leading in this way to a chain of conformational changes that finally allow actin to interact with myosin. In the following period of 50 msec, Ca^{2+} is sequestered again and the different conformational changes are reversed. The question therefore arises as to which binding sites are involved in the control of these fast processes. Johnson et al. (1979) studied the binding of Ca^{2+} to troponin C, labeled with dansylaziridine, using a fluorescence stopped-flow apparatus. The protein has 4 Ca^{2+}-binding sites; 2 high-affinity sites, which also bind Mg^{2+} competitively, and 2 low-affinity, exclusively Ca^{2+}, binding sites. Stopped-flow experiments show that only the Ca^{2+}-specific sites can dissociate fast enough to show an appreciable change in saturation during a single contraction cycle. A simulation study from the same group (Robertson et al., 1981) reveals that due to their very slow dissociation, the Ca^{2+} content of the mixed sites will depend on the immediate history of the muscle, i.e. the frequency and intensity of the train of contraction cycles. Similar considerations were also made for other Ca^{2+}-binding proteins in muscle: calmodulin, parvalbumin and myosin, both for fast and slow muscles.

Cox and Stein (1981) studied the binding of Ca^{2+} and Mg^{2+} to a sarcoplasmic calcium-binding protein from the sandworm (*Nereis diversicolor*). This protein has 3 mixed binding sites. The protein is particularly interesting because the fluorescence spectrum of the apoprotein and the two metal complexes, all differ considerably. A fluorescence stopped-flow study was made by Cox and Engelborghs (1983). The association rates were too fast to be measured, and were probably diffusion or dehydration controlled. In the study of the dissociation steps, the fluorescence changes observed in complexation reactions were much slower than in displacement reactions. The explanation is that in complexation reactions what is observed is a change of conformation to that of the apoprotein and not the off rate of the metal ion. This off rate, however, is rate limiting for the displacement reactions which directly lead to the conformation of the final complex without passing through the state of the apoprotein.

Other mechanisms that lead to discrepancies between off rates determined with complexing agents and displacement reactions are cooperative phenomena (Gutfreund, 1974) or the presence of a large number of ligand-binding sites close together, such as on a cell surface (Delisi, 1980). In cooperative systems it is very often the off rate that depends on the degree of saturation, which makes the difference between complexing and displacement reactions understandable.

2.2. Protein – protein interactions

For the study of protein–protein interactions, several ways are open. Covalently linked reported groups can be sensitive to direct association or to induced conformational changes. Competition for the interfaces with fluorescent probes such as ANS could also be used. A more direct monitoring of the associations is however possible by using light-scattering or turbidity measurements. Specially the use of small laser light sources such as He-Ne source (633 nm) or an argon ion laser (488 nm) has made the use of a light-scattering stopped flow more practical. A thorough discussion of the signal-to-noise problems, and precautions for filtering the solution is made by Liddle et al. (1977). Unlike a fluorescence stopped flow where an emission filter can be used, special attention has to be given to the reduction of reflections in the scattering stopped flow. Systems studied are quaternary structural changes in hemoglobin, hemocyanin, ribosomes, glutamate dehydrogenase, etc. A stopped flow with turbidity detection was used by Deranleau et al. (1980) to study shape changes in the red blood cell upon mixing with ADP.

Because the rotational freedom of a small ligand or a protein is reduced when it is rigidly bound in a large complex, fluorescence polarization can also be used to detect association reactions. Goss et al. (1980) studied in this way the binding of initiation factor IF3 to the ribosomal subunits.

Fibrin polymerization has been studied by Hantgan and Hermans (1979), using a laboratory-built light-scattering stopped flow. The instrument was especially designed to allow rapid disassembly after each experiment to remove the cloth that had been formed. The study is generally made difficult by the necessity for initial proteolytic steps, a rather fast reaction and the appearance of an enormous increase in scattering upon cloth formation. In the normal system the proteolytic enzyme thrombin removes two fibrinopeptides A from the A chains and fibrinopeptide B from the B chain. After gelation, covalent linkages are made by another factor. Hantgan and Hermans (1979) simplified the system by using reptilase. This enzyme removes only the fibrinopeptide A, with the consequence that gelation is strongly retarded. In this way it was possible to study a first and rapid transient increase in light scattering, which was attributed to the formation of linear associations of fibrin monomers into the so-called protofibrils. This polymerization did not show any indications for a nucleation process and was assumed to proceed according to the random association model:

$$f_i + f_j \leftrightarrow f_{i+j}$$

where each association step was assumed to be identical. The subsequent large increase in light scattering was shown to be due to lateral association of these protofibrils. The removal of fibrinopeptide B by thrombin is not an absolute necessity but speeds up this lateral association. It was also deduced that polymerization into protofibrils is probably necessary for fibrin to be a substrate for this proteolytic reaction. The kinetics of the association between myosin subfragment S1 and actin were studied by Geeves and Gutfreund (1982) using turbidity detection. A multistep mechanism was deduced from the observation that the association rate constant is much smaller and the activation energy much larger than diffusion controlled.

Recently the spectrum of detection possibilities was extended by the introduction of the X-ray-scattering stopped flow. Thanks to the availability of synchroton sources and position-sensitive detectors, time resolution became possible. Moody et al. (1980) have studied the dissociation of aspartate transcarbamylase on the second time scale. Continuing improvements of this technique are expected and will certainly have very interesting possibilities.

2.3. Conformational studies

2.3.1. Circular dichroism stopped flow

This technique has been recently reviewed by Bailey (1981). An interesting combination between fluorescence and circular dichroism (CD) stopped flow was applied to the acid unfolding of apomyoglobin by Kihara et al. (1980). While at neutral pH this protein has 55% of its amino acids in the α-helix structure. In acid medium this is reduced to 35%. When subjected to a rapid pH jump in the stopped flow, a first change was observed within the mixing time, and a second phase in the tens of milliseconds range. The change in the CD spectrum was also within the mixing time of the instrument. These data clearly show that both types of spectroscopy report on different aspects of the structure. Fluorescence probes the local environment of tryptophan and/or tyrosine which were correlated with the rupture of the heme pocket. The CD changes show that the overall secondary structure is lost much faster.

2.3.2. Chemical probing

Chemical modification of specific residues in proteins has been used for a long time to identify those residues that are involved in the specific functions of the protein under study. These methods suffer, however, from an inherent weakness because strongly linked allosteric effects can never be excluded until a complete structure is known from X-ray analysis. A kinetic analysis including a correlation with loss of function is somewhat more specific, but still cannot avoid this ambiguity. Chemical modification of tryptophan residues with N-bromosuccinamide has been employed to study the exposure of some residues to the solvent. Peterman and Laidler (1979) studied the modification of N-acetyltryptophanamide: the tripeptide Gly-Trp-Gly: apocytochrome c, and a-chymotrypsin. The highest rate constant was found with the simple tryptophan derivatives: $8 \times 10^5 \, \mathrm{M}^{-1} \, \mathrm{sec}^{-1}$. Apocytochrome c showed a small reduction to about half this value. This protein is quite small and has a random coil configuration. In α-chymotrypsin several classes of tryptophan residues could be differentiated, in agreement with the X-ray data. The results are comparable to the exposure data obtained with acrylamide quenching.

An interesting application is the search for protection against modification of specific residues by binding of ligands such as substrates or drugs, suggesting the involvement of the residue in the binding site (with the restriction mentioned above). Hiromi et al. (1977) applied this method to lysozyme and its binding of N-acetylglucosamine. Maes and Engelborghs (unpublished results) studied the kinetics of the modification of the single tryptophan residue in human serum albumin, and the protection by the binding of warfarin or phenylbutazone. With both ligands a rather limited protection was found, which was in agreement with a limited screening of the acrylamide quenching. Due to the small screening of the tryptophan by the drugs, it is not possible to decide whether the tryptophan is part of the binding site or is influenced by a limited allosteric effect.

2.3.3. Hydrogen–deuterium exchange studies

This method has been used for a long time and has revealed the existence of conformational fluctuations. Exchange of peptide hydrogens is studied with NMR or Raman spectroscopy. Only recently the exchange properties of side chains was studied using UV absorption measurements.

Exchange at the indole NH of tryptophan and at the phenol OH of tyrosine was studied by Nakanishi et al. (1978), and Nakanishi and Tsuboi (1978) respectively. Upon mixing with D_2O a very fast initial change is observed which is attributed to a direct solvent perturbation effect. Further slower changes are observed, which are dependent on the conformation of the protein around the residue. Exchange at nucleic acids has also been studied (Nakaniski et al., 1977).

2.4. Permeability of membranes to rapidly permeating molecules

The principle of this method is to use light scattering or transmission to measure rapid changes in cell volume. The volume changes can then be related to permeability

changes through theoretical considerations as discussed below. In practice, two solutions, one containing a dilute suspension of cells and the other containing the permeating molecule, are rapidly mixed. When steady flow is achieved, the flow is
stopped abruptly and the fluid is isolated in an observation tube through which light
passes. The time course of cell volume changes which occur on a millisecond time
scale can be measured indirectly from changes in the intensity of either 90°-scatte-
red light or 180°-transmitted light. In the application of this method, various approaches have been developed for rapid mixing and for recording (Sha'afi, 1981). It
must be pointed out that when employing this technique the experimental conditions must be carefully chosen so that the light intensity is linearly related to cell volume. Usually this relationship holds true when the changes in cell volume are relatively small. Failure to take this into consideration could lead to erroneous conclusions.

When red cells or any other cells are placed in a medium containing an iso-osmo-
lal concentration of impermeant solute together with a suitable concentration of a
permeant non-electrolyte which enters the cell less rapidly than water, the cell initially shrinks and then returns to its initial volume after passing through a well defined minimum. In the shrinking phase, water moves out of the cell owing to the excess of osmotically active material externally, while the non-electrolyte moves
down its concentration gradient into the cell. At the minimum point, the inward
flow of the solute is exactly balanced by the outward volume flow of water. Subsequently, the osmotic pressure gradient reverses its direction, in part because the impermeant solute is now more concentrated in the cell than in the medium: water enters the cell along with the permeant solute, and the volume reaches a new steady-
state value when the permeating non-electrolyte is equally distributed between the
cells and the medium. The cell volume will respond to changes in medium osmolali-
ty (hyperosmolar) in one of four ways: (a) it will decrease to a new steady-state value; (b) it will decrease and then increase to the initial equilibrium; (c) no change in
volume takes place; or (d) the cell swells and then shrinks. A mirror image situation
is obtained when the solution is hypo-osmolar. Condition (a) is obtained when the
cell membrane is completely impermeable to the solute, (b) when the solute is permeable but less so than water, (c) when both the solute and water are equally permeable, and condition (d) when the solute is more permeable than water (usually
called anomalous osmosis). We will consider only the first two conditions since the
last two are less informative.

In analysis of osmotic flow and net solute movement, the two basic equations
which are invariably used are those given by Katchalsky and Curran (1965):

$$J_v = -L_p \Delta \pi_i + L_{pd} \Delta \pi_s \tag{1}$$

$$J_d = -L_{dp} \Delta \pi_i + L_d \Delta \pi_s \tag{2}$$

Flow into the cell is considered to be in the positive direction. J_v is the volume flow
expressed as volume per unit area per unit time, and L_p is the hydraulic water con-

ductivity expressed as $cm^3/dyne$, sec. The osmotic pressure due to the permeant solute is denoted by $\Delta\pi_s$, which is defined as $\Delta\pi_s = RT(C_s^o - C_s^{\Delta x})$; $\Delta\pi$ has units of $dyne/cm^2$; R is the gas constant and T is the absolute temperature. C^o and $C^{\Delta\pi}$ denote the concentration outside and inside the cell respectively. In these equations we speak of differences in concentrations in bulk phases since the partition coefficient which relates the concentration in the membrane phase to that of the bulk phase is incorporated in the permeability coefficients. The subscripts i and s refer to impermeant and permeant solute respectively. L_{pd} is the cross-coefficient for the volume flow arising from differences in the osmotic pressure of the permeant solute, $\Delta\pi_s$, when there is no difference of either *hydrostatic* or osmotic pressure produced by impermeant solutes ($\Delta\pi_i = 0$). L_{dp} is the relative diffusional solute mobility per unit *hydrostatic* (or impermeant solute) pressure difference when $\Delta\pi_s = 0$. Although L_p is always positive, L_{pd} and L_{dp} are both negative and have the same units as L_p. If the Onsager reciprocal relation holds, then $L_{pd} = L_{dp}$. J_d is the diffusional flow and is a measure of the relative velocity of solute to solvent flux. L_d is a phenomenological coefficient, related to the permeability coefficient, and describes solute movement down its own concentration gradient in the absence of hydrostatic or osmotic pressure differences due to impermeant solutes. Rather than to measure L_d directly, it is much more convenient experimentally to measure ω_s, the permeability coefficient. From Eqns. 1 and 2, one can derive the following equations (Sha'afi, 1981):

$$J_s = (1 - \sigma)\overline{C_s}J_v + \omega_s\Delta\pi_s \tag{3}$$

ω_s is related to the phenomenological coefficients by

$$\omega_s = (L_dL_p - L_{pd}L_{dp})\overline{C_s}/L_p \tag{4}$$

and $\overline{C_s}$ is defined by

$$\overline{C_s} = (C_s^o - C_s^{\Delta x})\ln C_s^o/C_s^{\Delta x} \tag{5}$$

in which $C_s^{\Delta x} = n_s/V'$ and $C_i^{\Delta x} = C_{i,o}^{\Delta x}V_o'/V'$ in which n_s is the amount of permeant solute in the cell water whose volume is V'. V is cell volume and b is the sum of the volumes of fixed framework and solute dry weight so that $V' = V - b$. $C_{i,o}^{\Delta x}$ is the intracellular concentration of the impermeant solute at $t = 0$. Now, J_v can be expressed as $J_v = (1/A)(dV'/dt)$, and $J_s = (1/A)(dn_s/dt)$. A is the cell area which in the case of red cells is assumed to remain constant. Rewriting Eqns. 1 and 3 one obtains

$$v(dv/d\tau) = -(1 + \alpha)v + (\beta + \alpha s) \tag{6}$$

$$(ds/d\tau) = (1 - \sigma)(\overline{C_s}/C_s^o)(dv/d\tau) + r(v - s)/v \tag{7}$$

where

$$v = V'/V_o', \quad \tau = AL_pRTC_i^ot/V_o', \quad \alpha = \sigma C_s^o/C_i^o, \quad \beta = C_{i,o}^{\Delta x}/C_i^o$$

$$\sigma \equiv - L_{pd}/L_p \equiv - L_{dp}/L_p, \quad s = n_s/C_s^oV_o', \quad r = \omega_s/L_pC_i^o$$

When the membrane is completely impermeable to the solute (case a), Eqn. 7 reduces to zero. In Eqn. 6, s will be equal to 0 and $\alpha = C_s^o/C_i^o$, since by definition the reflection coefficient, σ, is equal to unity when the membrane is completely impermeable to the substance. The solution of Eqn. 6 for the boundary conditions $\tau = 0$, $v = 1.0$ is given by Eqn. 8:

$$\tau = (1 + \alpha)^{-2}\ln \frac{(1 + \alpha) - \beta}{(1 + \alpha)v - \beta} + (1 - \alpha)^{-1}(1 - v) \tag{8}$$

The hydraulic conductivity, L_p, can be calculated by Eqn. 8 from the experimentally measured time course of cell volume change. In this solution, the assumption is made that the concentration of the impermeant solute in the bathing medium remains constant. This assumption is justified since in actual experimental conditions the volume occupied by the red cells is only 1% of the total volume of the suspension. Furthermore, L_p is assumed to remain constant during the time course of volume change, but it is not necessarily independent of medium osmolarity or direction of water movement. On the basis of perturbation analysis an approximate solution for Eqn. 6 was worked out (Sha'afi, 1981). In this solution, which has the advantage of being easier to use, a normalized time constant is inversely proportional to the square of a normalized external pressure. The proportionality constant is directly related to L_p. This approach is quite satisfactory when the total change in cell volume is not too large.

When the membrane is permeable to the solute under study, the reflection coefficient σ can be calculated by the zero time method developed by Goldstein and Solomon (see Sha'afi, 1981). In this method, use is made of volume change at $\tau = 0$, when $v \to 1.0$, and $s \to 0$ so that

$$\lim_{\tau \to 0} (dv/d\tau) = \beta - (1 + \alpha).$$

Solving for s from Eqn. 6, differentiating with respect to τ to obtain $ds/d\tau$ and then equating $ds/d\tau$ with Eqn. 7, one obtains the following differential equation which describes the changes in cell volume when the membrane is permeable to the solute under study (for simplicity $\beta = 1.0$):

$$(v^2)(d^2v/d\tau^2) + (v)(dv/d\tau)^2 + (1 + \alpha + r)(v)(dv/d\tau) \tag{9}$$

$$- (\alpha)(v)(1 - \sigma)(\overline{C_s}/C_s^o)(dv/d\tau) + r(v - 1) = 0$$

An exact solution for this differential equation cannot be found. The presence of the time-dependent $\overline{C_s}$, and of $(dv/d\tau)^2$, contributes to the difficulty of integrating this equation. Fortunately, the permeability coefficient, ω_s, of the solute under study can be obtained from this equation by the minimum volume method (Sha'afi, 1981). At the minimum $\tau = \tau_m$, $v = v_m$ and $(dv/d\tau)_m = 0$. Under these conditions, one obtains

$$r = (\omega_s/L_pC_i^0) = (v_m^2)(d^2v/d\tau^2)_m/(1 - v_m) \tag{10}$$

Eqn. 10 has been used to calculate the permeability coefficients of mammalian red cell membranes to various solutes. Using second-order perturbation analysis an approximate solution for the differential Eqn. 9 has been worked out (Sha'afi, 1981).

The experimental technique outlined here can also be used to measure water and other non-electrolyte movements across the membrane of other cells as well as liposomes (Bangham et al., 1974). The surface area A of mammalian red cells remains constant for small changes in cell volume. Because of this condition the mathematical analysis is considerably simplified. In the case of other cells, the variation of the surface area during the time course of cell volume change must be taken into consideration in the theoretical analysis. Normally, the cell under study is assumed to be spherical and the surface area A is expressed in terms of the cell volume V.

3. Special applications

3.1. High-pressure studies

The study of the effect of pressure on the rates of chemical reactions in solution has proved to be of great value for an understanding of mechanisms of reactions in solution (Heremans, 1982). In our group a stopped-flow instrument was developed to study the effect of pressure on the rates and equilibria of fast conformational changes of proteins (Heremans et al., 1980). Here we refer only to our study on the alkaline isomerization of horse heart cytochrome c (Heremans and Wauters, 1980). It is known that the redox potential of cytochrome c drops at alkaline pH. This is due to a change in conformation of the oxidized form. The unprotonated isomer of the protein can be reduced by ascorbate. The kinetic profile of the reduction at alkaline pH reveals two steps: a fast one due to the reduction of the active conformation and a slow one (in the range of seconds) which is due to transformation of the inactive species to the active species. The process can be followed conveniently through the change of the porphyrin absorption bands of the protein. The process is a substitution of the lysine-79 for the met-80 in the coordination sphere of the porphyrin iron. From equilibrium studies we find that the reducible form is 45 ml/mole smaller than the irreducible form. The activation volume for the formation of the reducible form is 20 ml/mole. These volume changes are of the same order of magnitude as those observed for the reversible unfolding of proteins under pressure. Similar studies with α-chymotrypsin reveal the importance of the salt bridge for the activity of the enzyme. Since electrostatic interactions are very pressure sensitive,

50

the salt bridge is broken at high pressure. This result is a loss of activity of the protein under pressure. The process is fully reversible.

Ralston et al. (1981, 1982) have studied the effect of pressure on the interaction of ligands with horseradish peroxidase. From the activation volumes a clear distinction can be made between ligands which bind with proton transfer to the protein and ligands which bind with proton and electron transfer.

3.2. Optical scanning stopped flow

It is clear that a scanning stopped-flow instrument can drastically reduce the amount of experiments needed, provided that enough different intermediates can be detected in fairly different wavelength regions. Previously, scanning instruments employed a moving dispersion element. However, with the advent of the vidicon tubes, electronic scanning is now possible, making the construction much simpler (Ridder and Margerum, 1977).

3.3. Injection into a fixed volume

An interesting variant of the stopped flow has been suggested by Klingenberg (1964). In this instrument, a small sample is diluted in a large volume by pushing the solution into the mixing chamber which simultaneously moves down a spectrophotometric cell. A slightly different cell for the same purpose was designed by Karr and Purich (1980) and used in the study of microtubule dissociation. Here a strong dilution is necessary to reduce the protein concentration below the critical concentration, while a long optical pathway is necessary to have a measurable signal.

4. Non-optical detection systems

4.1. Stopped-flow calorimetry

A great deal of experience about the possibilities of stopped-flow instruments with thermal detection has been gained by the group of Berger (Bowen et al., 1980). This method allows the detection of fast reactions which exhibit no spectral changes. Special precautions have to be taken in relation to thermostatization and reduction of friction. These authors also suggested the use of a mechanical coupling between the reagent and stop syringes to reduce the heating due to adiabatic compression upon stopping the flow. A simple modification of the Aminco stopped flow for thermal detection was employed to study the binding of CO to deoxyhemoglobin by Malyi et al. (1980).

4.2.   Stopped flow with glass electrode detector

A stopped flow with direct pH detection was developed by Berger (1978) and by Crandall et al. (1978). It was applied by the latter authors to the study of bicarbonate–chloride exchange in erythrocyte suspensions.

4.3.   Dielectric relaxation detection

Some binding processes are not accompanied by the usual light spectroscopic changes. Therefore Scheider (1979) has developed a stopped flow with dielectric relaxation detection possibilities. This was used for the study of the binding of long-chain fatty acids to human serum albumin. The contribution of the protein to the dielectric constant of the solution is strongly dependent on its saturation with fatty acids. In this instrument the detection of this change is done by subjecting the mixed solutions to a sequence of short electric square pulses. The response of the system to every pulse is Fourier analyzed to yield the dielectric constant, and in this way its time dependency after mixing is constructed. As the dielectric increment for bovine serum albumin is quite different from that of human albumin, even dissociation of the complex could be studied using bovine albumin as a complexing agent.

4.4.   NMR and EPR stopped flow

These methods are less popular in view of the rather sophisticated methodology. Basically the principle used is the same as for dielectric detection. After the flow is stopped, the Pulse Fourier Transform System is triggered so that a single scan of an NMR can be made in about 2 sec (Sykes and Grimaldi, 1978). A computer-controlled stopped-flow technique for electron spin resonance has been described by Gascoyne (1981).

5.   *Pressure-jump relaxation technique*

5.1.   Introduction

For the study of reactions that are too fast to be accessible by the stopped-flow method, relaxation techniques have been developed (Eigen and De Maeyer, 1974). Any parameter that influences the position of a chemical equilibrium can be used to create a step perturbation of the equilibrium. The relaxation of the system to the new equilibrium state can then be followed with a suitable detection system. The parameters that have been used so far for rapid perturbation are temperature, pressure, electric field and concentration. The way a relaxation kinetic equation is derived from an equilibrium expression is extensively described in several good text books (Bernasconi, 1976; Strehlow and Knoche, 1977). The simplification that is obtained with small perturbations is not restricted to fast reactions.

Pressure can be used quite advantageously in biological systems. Almost all protein–protein association processes occur with a positive volume change. Moreover, for most proteins the onset of irreversible denaturation is only observed at pressures higher than 1000 atm (Heremans, 1982).

The effect of pressure on a chemical equilibrium is described as follows:

$$\left(\frac{\delta \ln k}{\delta P}\right)_T = -\frac{\Delta V_0}{RT} \tag{11}$$

However a fast pressure jump is too fast to allow for thermal equilibration. The pressure drop is therefore adiabatic, or at constant entropy (s) instead of constant temperature. Therefore we have to write:

$$\left(\frac{\delta \ln K}{\delta P}\right)_S = \left(\frac{\delta \ln K}{\delta P}\right)_T + \left(\frac{\delta \ln k}{\delta T}\right)_P \cdot \left(\frac{\delta T}{\delta P}\right)_S \tag{12}$$

$$\left(\frac{\delta \ln K}{\delta P}\right)_S = -\frac{\Delta V_0}{RT} + \frac{\Delta H_0}{RT} \cdot \frac{\alpha_T}{C_p \cdot \varrho} \tag{13}$$

herein is C_p the specific heat of the solution; ϱ the density of the solution; and α_T the thermal expansion coefficient. ΔH_0 is the change in molar enthalpy of the reaction.

Table 1 gives the values of the different parameters involved and the calculated factor of the second term, in the case of pure water. As is clear in most cases the second term can be neglected in aqueous media; however, in organic solvents it can be very important.

The pK of the buffers used is also changed by pressure. As ions cause electrostriction of water, all ionization reactions will be strongly favored by increasing pressure. Those buffers that dissociate without creating more ions, however, will practically be pressure insensitive. Volume changes for some acids commonly used in buffers are listed below (Neuman et al., 1973):

Phosphoric acid (1st ionization): −15.5 ml/mole
Phosphoric acid (2nd ionization): −24.0 ml/mole
Acetic acid: −11.2 ml/mole
Tris H$^+$:   + 1   ml/mole
Cacodylic acid: −13.2 ml/mole

The pressure-jump technique has some distinct advantages over the more classical temperature-jump (T-jump) relaxation method. Pressure at the final state can be stable for very long periods (at least in some instruments), in contrast to a fairly rapid decay to the original temperature in the T-jump instruments. Therefore systems with both fast and slow responses can be studied. The solution is static as compared to a stopped flow, where considerable shear forces develop which can destroy fragile biopolymers such as microtubules or actin filaments. As compared to

TABLE 1
Temperature change of water due to adiabatic pressure changes

T (°C)	$\alpha \times 10^6$ (K^{-1})	ϱ (g/ml)	C_p (cal/g/K)	$dT/dp \times 10^3$ (K/atm)
0	− 63.14	0.999868	1.00738	−0.44
5	15.98	0.999992	1.00368	0.10
10	87.90	0.999728	1.00129	0.60
15	150.73	0.999129	0.99976	1.05
20	206.61	0.998234	0.99883	1.46
25	257.05	0.997075	0.99828	1.83
30	303.14	0.995678	0.99802	2.23
35	345.71	0.994063	0.99795	2.59

α, thermal expansion coefficient of H_2O; ϱ, density of H_2O; C_p, specific heat of H_2O.

Data from: *Handbook of Chemistry and Physics*, R.D. Weast (Ed.), The Chemical Rubber Co., OH, pp. F-5, D-124.

the temperature jump, using Joule heating, no restrictions are made on the ionic composition of the solution.

Pressure-jump instruments were initially built for conductivity detection, but subsequent modifications were introduced to allow for a variety of detection techniques including light absorbance, scattering, fluorescence, CD and ORD.

J. Davis and Gutfreund (1976) reviewed the scope and applicability of the technique in the biological sciences. Since then the different subjects mentioned have been elaborated and a number of new applications have been published.

5.2. Instrumentation

The pressure-jump technique has been reviewed up to 1974 by Knoche (1974). In most instruments a fast pressure drop is obtained using a bursting membrane: the disc is in fact ruptured as pressure is gradually built up. The disadvantage of the rupturing disc is however that it is not possible to equilibrate the system at high pressure, and therefore it is not possible to obtain accurate amplitude data. Also the replacement of the disc after each experiment takes some time. The latter problem was solved in the instrument of Knoche and Wiese (1976) by using a brass ribbon instead of individual discs. To obtain easy repetition and stable high pressures, several alternatives are possible. A mechanical release valve was designed by Vanhorebeek (1970) and applied in the pressure jump described by J. Davis and Gutfreund (1976) (see Fig. 1). This release valve consists of a small piston which closes the oil compartment, and a trigger system to remove the piston rapidly. The exit of the oil compartment is a small hole, which can be closed by the piston. On top of the piston stem a small ball can be located with the aid of a needle. Piston and ball are kept in place by tightening the valve. Hydrostatic pressure is then applied from

54

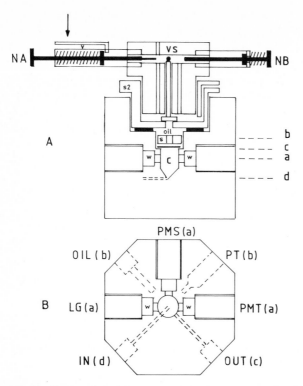

Fig. 1. Construction of the pressure-jump instrument as described by J. Davis and Gutfreund (1976). The application of the release valve allows the establishment of stable high pressures. (A) Vertical section of the pressure compartment and the release valve. Pushing the lever at the arrow releases the needle NA which hits the ball on top of the piston. The piston is pushed upwards by the oil which escapes and pressure drops to 1 atm. Needle NB is used to replace the ball on top of the piston stem and the oil compartment is closed again by tightening the valve screw VS. w, sapphire window; c, solution compartment covered with teflon membrane; oil, oil compartment; s, screw which holds the teflon membrane; s2, screw which covers the oil compartment except for a small opening which is to be closed by the piston of the release valve. a, b, c and d indicate vertical levels at which different connections are situated (cf. B). (B) Horizontal section at level a. The other connections are projected in the same plane. Their vertical position is however indicated in parentheses. LG, light guide; PMT, photomultiplier for transmission; PMS, photomultiplier for scattering or fluorescence; IN, inlet of solution; OUT, exit of solution. The instrument is tilted in such a way that this exit point is the hight point of the sample compartment. In this way trapping of air bubbles is avoided. PT, pressure transducer in connection with oil compartment provides trigger signal; OIL, oil inlet from pump, in connection with oil compartment.

a hand pump, and the system is equilibrated at high pressure. Pressure release is obtained by releasing a spring-loaded trigger needle that hits the ball with high speed. The ball is pushed away and the piston is lifted under the influence of the oil pressure and oil can escape to 1 atm. The pressure drop takes between 50 and 200 μsec/100 atm. The experiment can easily be repeated by loosening the valve, replacing the ball and fastening again. Although it generally takes less than a minute to repeat the experiment, this cannot be considered a real repetitive pressure jump. The instrument is very well suited for biological applications. In our laboratory this instrument was adopted for use with light-scattering detection. Diaphragms were introduced to minimize reflection and a high pressure Hg-Xe lamp was used as a light source. Particularly the high intensity Hg line at 365 nm is useful for light scattering. The transmitted light is used as a reference signal, after appropriate attenuation.

An instrument especially designed for light-scattering detection of protein–protein interactions was made by Kegeles and Ke (1975). Pressure build-up takes, however, 2 msec using He gas and a fast solenoid valve. With this instrument Kegeles studied the assembly of hemocyanin and of ribosomes.

A repetitive pressure-jump instrument was designed by Yasunaga et al. (1978) also using gas pressure and an electric solenoid valve. A more elegant system was developed by Clegg and Maxfield (1976) and Halvorson (1979). Here a repetitive pulse train is obtained from a stack of piezoelectric plates. Pressure jumps obtained are very small but the repetition rate is high and accumulation of a large number of transients is no problem. This instrument should be particularly useful for cooperative systems.

Conductometric detection is extensively reviewed by Knoche (1974) but is of little use in biological systems. The special problems with high-pressure window effects in relation to CD and ORD detection are described by Grunewald and Knoche (1978) and in relation to fluorescence polarization by Paladini and Weber (1981).

High-pressure jumps are also of biological interest, since many deep-sea animals have to bear pressures as high as 1000 atm. With these instruments it is also possible to determine activation volumes which are often much easier to interpret than activation energies. An instrument has been described by Heremans et al. (1980).

The performance of a pressure-jump instrument can simply be controlled using the same indicator solution as for a temperature jump: 2×10^{-5} M cresol red in 0.1 M Tris–HCl buffer, 0.1 M NaCl, pH 8, following absorbance changes at 570 nm.

5.3. Applications

5.3.1. Enzymology

A large number of stopped-flow and quenched-flow studies have shown that the catalytic step in an enzyme-catalyzed reaction is very often preceded and followed by several conformational changes of the complex. These conformational changes can also be studied using relaxation techniques on the equilibrium mixture of substrates and products in the presence of large concentrations of enzyme. Because the binding constants for substrates and products are usually different, the equilibrium on the enzyme will be different from that in solution. The dehydrogenases are studied in this way by Gutfreund and his coworkers (Coates et al., 1977; Hardman et al., 1978). These are particularly suited for this purpose in view of the many signals available, and the possibility to distinguish between free and bound ligands.

5.3.2. Subunit assembly in ribosomes

Ribosomes are association products of two subunits, the 30 S and 50 S subunits, which assemble into the 70 S ribosomes. Two types of association products seem to be formed in apparent equilibrium: the so-called loose and tight couples. Although procedures are described to isolate these association products, e.g. tight couples, they seem to interconvert gradually into the mixture. When 70 S ribosomes are subjected to a pressure jump, two relaxation processes are observed, one of which is too fast for the pressure jump but which can be analyzed with the temperature-jump technique where a rate constant of $6 \times 10^9 \, M^{-1}sec^{-1}$ was determined (Chaires et al., 1979). Both relaxation times were concentration dependent. The final picture that emerges is a two-step mechanism. In the first step a diffusion-controlled association takes place leading to the almost immediate establishment of the first equilibrium which is situated far to the left. In a second, much slower step, the final stable ribosome is formed. Although this step is not accompanied by a change in light scattering, it shows up because it displaces the first equilibrium. Tight and loose couples show a similar behavior. The relaxation amplitudes are dependent on the concentration of Mg^{2+} ions, and both types show a maximal amplitude at different Mg^{2+} concentrations. In this way they can be distinguished.

5.3.3. Metarhodopsin equilibrium

Pressure-relaxation experiments have been applied to such complex biological systems as the rod outer segments. When rhodopsin of the rod outer segments is bleached by an intense light pulse, metarhodopsin I is formed, which on a time scale of seconds, equilibrates with metarhodopsin II. Further reactions to metarhodopsin III or hydrolysis to opsin and retinaldehyde are much slower. The equilibrium between metarhodopsin I and II is pressure sensitive (Lamola et al., 1974) and its kinetics could be studied using the pressure-jump technique (Attwood and Gutfreund, 1980). The conformational change can be followed by absorbance measurements at 380 nm. From the pressure effect on the equilibrium, a volume change of 60 ml/mole was calculated.

5.3.4. Myosin assembly

J.S. Davis made extensive use of the pressure-jump technique to study the assembly of myosin into thick filaments (J.S. Davis, 1981a, b). Both equilibrium and kinetic studies were made. In the solution conditions used, i.e., pH 8.1, 5°C, 5 mM bicine buffer, 0.15 M KCl, filaments were formed of a uniform length of 630 nm (as compared to 1550 nm for the native thick filaments). Structurally both types of filaments are completely comparable, with a thickness of 12.5 nm and a bare zone of 180 nm. It had been shown previously by Josephs and Harrington (1966), that assembly into thick filaments is very strongly dependent on KCl concentration and on pH.

First, J.S. Davis et al. (1982) showed, by a technique of fixation under pressure, that partial dissociation induced by pressure leads to uniform shortening of the filaments without destroying them. Correlation with the change of turbidity showed that turbidity can be used to measure the length of the filaments within a single series of experiments. The specific turbidity was determined. A critical concentration was observed and as this is the reciprocal of the propagation equilibrium constant, the volume change for propagation was calculated from the pressure dependence of this critical concentration. A value of 240 ml/mole was obtained. It should be realized that this is deduced from the effect of pressure on the propagation equilibrium constant, determined at a length close to the bare zone. From the analysis of the relaxation curves it was further deduced that the active entity in propagation is a myosin dimer. This was confirmed by fixation and electron microscopy. Davis also showed that the on-rate constant is independent of the length of the filament, but the off-rate constant is strongly dependent, and increases considerably with increasing length. Simulation studies showed that the narrow length distribution experimentally found could be explained on the basis of this length dependency of the off rate.

5.3.5. Polymerization of glutamate dehydrogenase
Glutamate dehydrogenase is the enzyme that catalyzes the oxidative deamination of glutamate. It is subject to regulation by many ligands. It is a hexameric structure, but can polymerize into long linear aggregates. Its polymerization has been studied by dynamic light scattering (Cohen et al., 1976). Whether this polymerization is of any physiological relevance is a matter of dispute (Thusius, 1977; Cohen and Benedek, 1979). A kinetic analysis was done by Thusius et al. (1975) using the temperature-jump technique. It was found that a random association model was most adequate to describe the data. All polymers are able to associate with each other and there is no special role for the monomer. The general reaction equation therefore reduces to

$$P_i + P_j \rightleftharpoons P_{i+j}$$

As a consequence a single relaxation time is observed, which is dependent on the square of the total protein concentration. The polymerization reaction was also studied using the pressure jump by Engelborghs (unpublished results). The system turns out to be very well suited to check the possibilities of the instrument in the case of light-scattering detection. The polymerization was studied at low temperatures and also the amplitudes were analyzed. The concentration dependence was in agreement with the model proposed by Thusius. However, at temperatures lower than about 7°C, a negative activation energy was found for the association step. This implies the existence of a fast pre-equilibrium. So far we have no independent evidence for a conformational change of the enzyme. The pressure dependence of the amplitudes allowed the calculation of the reaction volume. This reaction volume was temperature dependent and showed a maximum of 54 ml/mole at 7°C. Reaction volumes were also determined by static measurements by Heremans (1974). Halvorson (1979) studied the polymerization of GDH, using a piezoelectric pressure jump as described by Clegg and Maxfield (1976). Very remarkably, his finding of

58

an additional very fast step was attributed to fast phosphate exchange. We have no evidence for such an additional process.

5.3.6. Tubulin rings

When microtubules from brain tissues are isolated and cooled to 4°C, dissociation products are formed which consist of pure tubulin and some ring-shaped aggregates. Tubulin itself is a dimeric protein with a molecular weight of 120 000 dalton. The rings are built from tubulin dimers and the so-called microtubule-associated proteins. The structure and composition of the rings is dependent on the solvent conditions, e.g. pH, ionic strength, etc. A detailed description of ring formation is found in a series of articles from the group of Borisy (Marcum and Borisy, 1978a, b). Whether these rings have any physiological significance is not yet clear. The fact is that they are never seen in cell preparations, but only as in vitro dissociation products. Evidence thus far points to these rings as being side products on the pathway of assembly that only form when the main pathway of microtubule formation is blocked (Engelborghs et al., 1977). When a solution of cooled microtubules at pH 6.5, where only one type of rings is present, is subjected to pressure, substantial dissociation occurs. When pressure is released, light scattering increases again with a rather complex relaxation spectrum (Engelborghs et al., 1980). A fast step is observed in the millisecond range and two other steps in the seconds time scale. The first fast steps do not seem to be concentration dependent. This phenomenon was interpreted as a conformational change accompanied by a change in light scattering, e.g. a ring opening. The two slower steps are both concentration dependent. In the presence of an excess of monomers, the fastest step in the second range showed an increased relaxation rate. The interpretation of these different steps is of course difficult. Exchange between radioactive tubulin dimers and the rings was studied by Zeeberg et al. (1980) and by Simon et al. (1980). It was shown that exchange had a half-life at 35°C of less than 2 min, but about 30 min at 4°C. As our fast steps were observed at about 0°C, the slow exchange can only be explained by a sequential exchange mechanism.

Pressure-jump experiments have been performed on whole cells (Inoue et al., 1975) to study the (slow) reappearance of microtubules in the spindle figure. A special miniature pressure chamber was developed for microscopic observations. In combination with microinjection of fluorescent proteins, this technique seems very promising for the study of the dynamics of the cytoskeleton.

References

Almgren, M. (1980) Migration of pyrene between lipid vesicles in aqueous solution, A stopped-flow study, Chem. Phys. Lett., 71, 539–543.

Attwood, P.V. and Gutfreund, H. (1980) The application of pressure relaxation to the study of the equilibrium between metarhodopsin I and II from bovine retinas, FEBS Lett., 119, 323–326.

Bailey, P. (1981) Fast kinetic studies with chiroptical techniques: Stopped flow circular dichroism and related methods, Prog. Biophys. Mol. Biol., 37, 149–280.

Bangham, A.D., Hill, M.W. and Miller, N.G.A. (1974) Preparation and use of liposomes as models of biological membranes, in: E.D. Korn (Ed.), Methods in Membrane Biology, Vol. 1, Plenum, New York, pp. 1–68.

Berger, R.L. (1978) Some problems concerning mixers and detectors for stopped flow kinetic studies, Biophys. J., 24, 2–20.

Bernasconi, C.F. (1976) Relaxation Kinetics, Academic Press, New York.

Bowen, P., Balko, B., Blevins, K., Berger, R.L. and Hopkins, H.P. (1980) Stopped-flow microcalorimetry without adiabatic compression: application to reactions with half-lives between 3 and 50 ms, Anal. Biochem., 102, 434–440.

Chaires, J.B., Kegeles, G. and Wahba, A.J. (1979) Relaxation kinetics of E. coli ribosomes: evidence for the reaction of 30 S IF3 complex with 50 S ribosomal subunits, Biophys. Chem., 9, 405–412.

Chance, B. (1940) The accelerated flow method for rapid reactions, J. Franklin Inst., 229, 455–476, 613–640, 737–766.

Chance, B. (1974) Rapid flow methods, in: G.G. Hammes (Ed.), Investigation of Rates and Mechanisms of Reactions, Techniques of Chemistry, Vol. VI, Part II, 3rd edn., Wiley, New York, pp. 5–62.

Clegg, R.M. and Maxfield, B.W. (1976) Chemical kinetic studies by a new small pressure perturbation method, Rev. Sci. Instrum., 47, 1383–1393.

Coates, J.H., Hardman, M.J., Shore, J.D. and Gutfreund, H. (1977) Pressure relaxation studies of isomerizations of horse liver alcohol dehydrogenase linked to NAD^+ binding, FEBS Lett., 84, 15–28.

Cohen, R.J. and Benedek, G.B. (1976) The functional relationship between the polymerization and catalytic activity of beef liver glutamate dehydrogenase, I. Theory, J. Mol. Biol., 198, 151–178.

Cohen, R.J. and Benedek, G.B. (1979) The functional relationship between the polymerization and catalytic activity of beef liver glutamate dehydrogenase, III. Analysis of Thusius' critique, J. Mol. Biol., 129, 37–44.

Cohen, R.J., Jedziniak, J.A. and Benedek, G.B. (1976) The functional relationship between the polymerization and catalytic activity of beef liver glutamate dehydrogenase, II. Experiments, J. Mol. Biol., 108, 179–199.

Cox, J.A. and Engelborghs, Y. (1983) Fast kinetics of divalent metal ion exchange in nereis sarcoplasmic calcium binding protein, submitted.

Cox, J.A. and Stein, E.A. (1981) Characterization of a new sarcoplasmic calcium-binding protein with magnesium-induced cooperativity in the binding of calcium, Biochemistry, 20, 5430–5436.

Crandall, E.D., Obaid, A.L. and Forster, R.E. (1978) Bicarbonate-chloride exchange in erythrocyte suspensions: stopped-flow pH electrode measurements, Biophys. J., 24, 35–47.

Crouch, S.R., Holler, F.J., Notz, P.K. and Beckwith, P.M. (1977) Automated stopped-flow systems for fast reaction rate methods, Appl. Spectr. Rev., 13, 165–259.

Davis, J. and Gutfreund, H. (1976) The scope of moderate pressure changes for kinetic and equilibrium studies of biochemical systems, FEBS Lett., 72, 199–207.

Davis, J.S. (1981a) The influence of pressure on the self-assembly of the thick filament from the myosin of vertebrate skeletal muscle, Biochem. J., 197, 301–308.

Davis, J.S. (1981b) Pressure-jump studies on the length-regulation kinetics of the self-assembly of myosin from vertebrate skeletal muscle into thick filaments, Biochem., J., 197, 309–314.

Davis, J.S., Buck, J. and Greene, E.P. (1982) FEBS Lett., 140, 219–222.

Delisi, C. (1980) The biophysics of ligand-receptor interactions, Quart. Rev. Biophys., 13, 201–230.

Deranleau, D.A., Rothen, C., Streit, M., Dudler, D. and Luescher, E.F. (1980) A stopped-flow laser turbidimeter for studying changes in the shape of cell stimulated by external agents, Anal. Biochem., 102, 288–290.

Doody, M.C., Pownall, H.J., Koa, Y.J. and Smith, L. (1980) Mechanism and kinetics of transfer of a fluorescent fatty acid between single-walled phosphatidylcholine vesicles, Biochemistry, 19, 108–116.

Eigen, M. and DeMaeyer, L. (1974) Theoretical basis of relaxation spectrometry, in: G.G. Hames (Ed.), Investigation of Rates and Mechanisms of Reactions, Techniques of Chemistry, Vol. VI, part II, 3rd edn., Wiley, New York, pp. 63–146.

60

Engelborghs, Y. (1980) A thermodynamic study of colchicine and colcemid dimerization, J. Biol. Chem., 256, 3276–3278.

Engelborghs, Y., Demaeyer, L.C.M. and Overbergh, N. (1977) A kinetic analysis of the assembly of microtubules in vitro, FEBS Lett., 80, 81–85.

Engelborghs, Y., Robinson, J. and Ide, G. (1980) A pressure relaxation study of tubulin oligomer formation, Biophys. J., 32, 440–443.

Gascoyne, P.R.C. (1981) A reagent-efficient computer-controlled stopped-flow technique for electron spin resonance spectroscopy, J. Phys. E: Sc. Instrum., 14, 62–64.

Geeves, M.A. and Gutfreund, H. (1982) The use of pressure perturbations to investigate the interaction of rabbit muscle myosin subfragment 1 with actin in the presence of MgADP, FEBS Lett., 140, 11–15.

Getting, W.J. and Wyn-Jones, E. (Eds.) (1978) Techniques and Applications of Fast Reactions in Solution, D. Reidel, Dordrecht.

Goss, D.L., Parkhurst, L.J. and Waba, A.J. (1980) Kinetic studies of the rates and mechanism of assembly of the protein synthesis initiation complex, Biophys. J., 32, 283–294.

Grunewald, B. and Knoche, W. (1978) Pressure jump method with detection of optical rotation and circular dichroism, Rev. Sci. Instrum., 49, 797–802.

Gutfreund, H. (1972) Enzymes: Physical Principles, Wiley Interscience, London.

Gutfreund, H. (1974) Equilibria and kinetics of protein–ligand interaction, in: H. Gutfreund (Ed.), Chemistry of Macromolecules, MTP International Review of Science, Biochemistry Series 1, Butterworths, London, University Park Press, Baltimore, MD, pp. 261–286.

Halvorson, H. (1979) Relaxation kinetics of glutamate dehydrogenase self-association by pressure perturbation, Biochemistry, 18, 2480–2487.

Hantgan, R.R. and Hermans, J. (1979) Assembly of fibrin — a light scattering study, J. Biol. Chem., 254, 11272–11281.

Hardman, M.J., Coates, J.H. and Gutfreund, H. (1978) Pressure relaxation of the equilibrium of the pig heart lactate dehydrogenase system, Biochem. J., 171, 215–223.

Hartridge, H. and Roughton, F.J.W. (1923) Proc. Roy. Soc. (London), A104, 376.

Heremans, K. (1974) The self-association of biological macromolecules, high pressure light scattering studies on glutamate dehydrogenase, in: Proceedings of the 4th International Conference on High Pressure, Kyoto, pp. 627–630.

Heremans, K. (1982) High pressure effects on proteins and other biomolecules, Annu. Rev. Biophys. Bioeng., 11, 1–21.

Heremans, K. and Wauters, J. (1980) High pressure stopped flow studies of fast conformational changes in proteins, in: B. Vodar and P. Marteau (Eds.), High Pressure Science Technology, Pergamon Press, Oxford, pp. 845–847.

Heremans, K., Snauwaert, J. and Rijkenberg, J. (1980) Stopped-flow apparatus for the study of fast reactions in solution under high pressure, Rev. Sci. Instrum., 51, 806–808.

Hiromi, K. (1979) Kinetics of Fast Enzyme Reactions – Theory and Practice, Kodansha, Tokyo, Halsted Press, Wiley, New York.

Hiromi, K., Kawagishi, T. and Ohnishi, M. (1977) Kinetic studies on the chemical modification of lysozyme by n-bromosuccinimide and its protection by substrates and analogs, J. Biochem., 81, 1583–1586.

Inoue, S., Fuseler, J., Salmon, E.D. and Ellis, G.W. (1975) Functional organization of mitotic microtubules, Physical chemistry of the in vivo system, Biophys. J., 15, 725–743.

Johnson, J.D., Charlton, S.C. and Potter, J.D. (1919) A fluorescence stopped flow analysis of Ca^{2+} exchange with troponin C, J. Biol. Chem., 254, 3497–3502.

Josephs, R. and Harrington, W.F. (1966) Studies on the formation and physical chemical properties of synthetic myosin filaments, Biochemistry, 5, 3474–3487.

Karr, T.L. and Purich, D.L. (1980) A rapid dilution cuvette for kinetic studies of microtubule disassembly, Anal. Biochem., 104, 311–314.

Katchalsky, A. and Curran, P.F. (1965) Nonequilibrium Thermodynamics in Biophysics, Harvard University Press, Cambridge, MA, p. 67.

Kegeles, G. and Ke, C. (1975) A light-scattering pressure-jump kinetics apparatus, Anal. Biochem., 68, 138–147.

Kihara, H., Yakamura, K. and Tabushi, I. (1980) Kinetic study on the unfolding of apomyoglobin, detected by fluorescence and circular dichroism stopped-flow, Biochem. Biophys, Res. Commun., 95, 1687–1693.

Klingenberg, M. (1964) The moving mixing chamber, in: B. Chance, Q.H. Gibson, R.H. Eisenhardt and K.K. Lonberg-Holm (Eds.), Rapid Mixing and Sampling Techniques in Biochemistry, Academic Press, New York, pp. 61–66.

Knoche, W. (1974) Pressure-jump methods, in: G.G. Hammes (Ed.), Investigation of Rates and Mechanisms of Reactions, Techniques of Chemistry, Vol. VI, Part II, 3rd edn., Wiley, New York, pp. 187–210.

Knoche, W. and Wiese, G. (1976) Pressure-jump relaxation technique with optical detection, Rev. Sci. Instrum., 47, 220–221.

Lambeir, A. and Engelborghs, Y. (1981) A fluorescence stopped flow study of colchicine binding to tubulin, J. Biol. Chem., 256, 3279–3282.

Lamola, A.A., Yamane, T. and Zipp, A. (1974) Effects of detergents and high pressure upon the metarhodopsin I – metarhodopsin II equilibrium, Biochemistry, 13, 738–745.

Liddle, P.F., Jacobs, D.J. and Kellet, G.L. (1977) A stopped-flow laser light-scattering photometer for the study of the kinetics of macromolecular association–dissociation reactions, Anal. Biochem., 79, 276–290.

Maes, V., Engelborghs, Y., Hoebeke, J., Maras, I. and Vercruysse, A. (1982) Fluorimetric analysis of the binding of warfarin to human serum albumin: equilibrium and kinetic study, Mol. Pharmacol., 21, 100–107.

Malyi, M., Smith, P.D., Balko, B. and Berger, R.L. (1980) Thermal kinetics using a modified commercial stopped flow apparatus, Rev. Sci. Instrum., 51, 896–899.

Marcum, J.M. and Borisy, G.G. (1978a) Characterization of microtubule protein oligomers by analytical ultracentrifugation, J. Biol. Chem., 253, 2825–2833.

Marcum, J.M. and Borisy, G.G. (1978b) Sedimentation velocity analysis of the effect of hydrostatic pressure on the 30 S microtubule protein oligomer, J. Biol. Chem., 253, 2852–2857.

Moody, M.F., Vachette, P., Foote, A.M., Tardieu, A., Koch, M.H.J. and Bordas, J. (1980) Stopped-flow X-ray scattering: the dissociation of aspartate transcarbamylase, Proc. Natl. Acad. Sci. (U.S.A.), 77, 4040–4043.

Nakanishi, M. and Tsuboi, M. (1978) Measurement of hydrogen exchange at the tyrosine residues in ribonuclease A by stopped-flow and ultraviolet spectroscopy, J. Am. Chem. Soc., 100, 1273–1275.

Nakanishi, M., Tsuboi, M., Salio, Y. and Nagamura, T. (1977) FEBS Lett., 81, 61–64.

Nakanishi, M. Nakamura, H., Hirakawa, A.Y., Tsuboi, M., Nagamura, T. and Sijo, Y. (1978) Measurement of hydrogen exchange at the tryptophan residues of a protein by stopped-flow and ultraviolet spectroscopy, J. Am. Chem. Soc., 100, 272–276.

Neuman, R.C., Kauzmann, W. and Zipp, A. (1973) Pressure dependence of weak acid ionization in aqueous buffers, J. Phys. Chem., 77, 2687–2691.

Nichols, J.W. and Pagano, R.W. (1981) Kinetics of soluble lipid monomer diffusion between vesicles, Biochemistry, 20, 2783–2789.

Paladini, A.A. and Weber, G. (1981) Absolute measurements of fluorescence polarization at high pressures, Rev. Sci. Instrum., 52, 419–427.

Parsegian, V.A. (Ed.) (1978) Biophysical discussions: fast biochemical reactions in solutions, membranes and cells, Biophys., J., 24 (1).

Parsegian, V.A. (Ed.) (1980) Biophysical discussions: proteins and nucleoproteins, structure, dynamics and assembly, Biophys. J., 32 (1).

Peterman, B.F. and Laidler, K.J. (1979) The reactivity of tryptophan residues in proteins: stopped-flow kinetics of fluorescence quenching, Biochim. Biophys. Acta, 577, 314–323.

Quast, U., Schimerlik, M.I. and Raftery, M.A. (1979) Ligand-induced changes in membrane-bound acetylcholine receptor observed by ethidium fluorescence, 2. Stopped-flow studies with agonists and antagonists, Biochemistry, 18, 1891–1901.

62

Ralston, I.M., Wauters, J., Heremans, K. and Dunford, H.B. (1981) Effects of pressure and temperature on the reactions of horseradish peroxidase with hydrogen cyanide and hydrogen peroxide, Biophys. J., 36, 311.

Ralston, I.M., Wauters, J., Heremans, K. and Dunford, H.B. (1982) Activation volumes for horseradish peroxidase compound II reactions, Biophys. Chem., 15, 15–18.

Ridder, G.M. and Margerum, D.W. (1977) Simultaneous kinetic and spectral analysis with a vidicon rapid-scanning stopped-flow spectrometer, Anal. Chem., 49, 2098–2108.

Roberston, S.P., Johnson, J.D. and Potter, J.D. (1981) The time-course of Ca^{2+} exchange with calmodulin, troponin, pervalbumin, and myosin in response to transient increases in Ca^{2+}, Biophys. J., 34, 559–569.

Scheider, W. (1979) The rate of access to the organic ligand-binding region of serum albumin is entropy controlled, Proc. Natl. Acad. Sci. (U.S.A.), 76, 2283–2287.

Sha'afi, R.I. (1981) Permeability in water and other polar molecules, in: S.L. Bonting and J.J.H.H.M. de Pont (Eds.), Membrane Transport, Vol. 2, Elsevier Biomedical Press, Amsterdam, pp. 29–60.

Simon, C., Carlier, M.F. and Pantaloni, D. (1980) Exchange between tubulin and rings at equilibrium: participation of both species in microtubule assembly, Eur. J. Cell Biol., 22, 295.

Strehlow, H. and Knoche, W. (1977) Fundamentals of chemical relaxation, Verlag Chemie, Weinheim.

Sykes, B.D. and Grimaldi, J.J. (1978) Stopped-flow nuclear magnetic resonance spectroscopy, Methods Enzymol., 49, 295–321.

Thusius, D. (1977) Does a functional relationship exist between the polymerization and catalytic activity of beef liver glutamate dehydrogenase?, J. Mol. Biol., 115, 267–274.

Thusius, D., Dessen, P. and Jallon, J.M. (1975) Mechanism of bovine glutamate dehydrogenase self-association, Kinetic evidence for a random association of polymer units, J. Mol. Biol., 92, 413–432.

Trentham, D.R., Eccleston, J.F. and Bagshaw, C.R. (1976) Kinetic analysis of ATPase mechanisms, Quart. Rev. Biophys., 9, 217–282.

Vanhorebeek, R. (1970) Construction of a Pressure-Jump Relaxation Spectrometer, Thesis, University of Leuven.

Waechter, F. and Engel, J. (1977) Association kinetics and binding constants of nucleotide triphosphates with G-actin, Eur. J. Biochem., 74, 227–232.

Yasunaga, T., Tatsumoto, N., Harada, S. and Hiraishi, M. (1978) Pressure-jump apparatus suitable for repetitive experiments, Rev. Sci. Instrum., 49, 1747–1748.

Zeeberg, B., Cheek, J. and Caplow, M. (1980) Exchange of tubulin dimer into rings in microtubule assembly–disassembly, Biochemistry, 19, 5078–5086.

Rapid-quench methods in fast biochemical processes

DAVID P. BALLOU

*Department of Biological Chemistry, University of Michigan,
Ann Arbor, MI 48109, U.S.A.*

Contents

1. Introduction 63
2. Requirements of rapid-quenching techniques 66
    2.1. Mixing 66
    2.2. Flow velocity 67
    2.3. Quenching 68
    2.4. Other parameters 69
3. Design of apparatus for quenching methods 70
    3.1. Ram design 70
    3.2. Bath-and-flow system design 72
    3.3. Quenching design – chemical and cryogenic 77
    3.4. Other considerations in design and use 79
4. Examples of use 80
    4.1. Calibration 80
    4.2. Chemical quenching 81
         4.2.1. Determination of a mechanistic path 81
         4.2.2. Determination of fraction of active substrate 82
         4.2.3. Identification of an intermediate in a mechanism 82
         4.2.4. Isotope-trapping techniques 83
         4.2.5. Hormone-receptor-mediated transmembrane permeability
                changes 84
    4.3. Rapid freezing 84
5. Conclusions 85
Acknowledgements 85
References 85

1. Introduction

The reactions, interactions, and conformational changes of enzymes, proteins, and supramolecular structures are very dynamic processes. Therefore, to understand

R.I. Sha'afi and S.M. Fernandez (Eds.), Fast Methods in Physical Biochemistry and Cell Biology
© *1983 Elsevier Science Publishers*

the chemistry of these processes it is necessary to use fast kinetic methods which identify intermediate states. The previous chapters have dealt with techniques that require rapid observation of such events as they are occurring, usually by photometric methods. However, many of the chemists' most powerful probes of structures are not always suited to rapid observations. Examples include electron paramagnetic resonance (EPR) and nuclear magnetic resonance (NMR) spectroscopies, radioisotope measurements, and mass spectrometry. In this chapter I will describe some rapid quenching techniques which fall into the rapid-mixing category of fast reaction techniques and which permit the measurement of trapped transient species by slower methods. This writing reiterates some of the points made in other excellent reviews (Gibson, 1969; Gutfreund, 1969; Chance, 1974) while delineating the salient features of continuous-flow and chemical quenching techniques. In addition, some recent advances in instrumentation and applications in this area will be discussed.

The principles of flow methods were first enunciated by Hartridge and Roughton (1923). They discussed many of the experimental problems associated with making rapid reaction measurements in continuous-flow devices and studied the mechanics of thoroughly mixing two reactants in a short period of time (1 msec) (Hartridge and Roughton, 1923). Two reactants were placed into separate large reservoir bottles. The two bottles were connected by tubing and a suitable mixer, and gas pressure was used to drive the solutions through this tubing at a constant rate of flow. At a short distance from the mixer, observations of the reacting mixture could be made by thermal or spectroscopic means. Fig. 1 schematically shows one example of their method which uses the Hartridge reversion spectroscope (RS) (Hartridge, 1923) to measure spectral characteristics of hemoproteins at various points along the observation tube (OT). The time (t) that had passed between mixing and observation (the observation time) was related to the flow velocity and the distance from the mixer by:

$$t = S/U \tag{1}$$

S = distance from mixer to point of observation; U = flow velocity. In this continuous-flow mode, a constant flow rate is maintained and several observations are made at various distances from the mixer, each of which corresponds to a "time clamp" on the reaction. Alternatively, various observation times can be elicited by varying the flow rate (U) while maintaining a constant point of observation. A series of such discharges with appropriate observation enables the construction of a kinetic curve. In the rapid-quenching variation of this method, rather than observing the mixture at a point downstream from mixing, another mixer is employed to introduce a reagent which will quickly quench the reaction. This quenched mixture is collected and analysis is performed at a later time. In contrast, in the stopped-flow technique the reaction mixture is injected into an observation tube as rapidly as possible, and observation of that volume of reactants continues for as long as desired (usually, a few seconds or less).

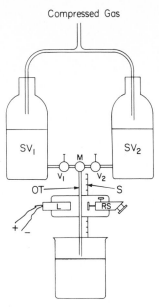

Compressed Gas

Fig. 1.   Schematic of Hartridge–Roughton apparatus. SV_1 and SV_2 are storage reservoirs, V_1 and V_2 are valves, M is the mixer, OT is the observation tube, S is a scale for measuring the distance from the mixer, RS is the reversion spectroscope, and L is the lamp source.

In spite of the fact that many ingenious flow devices have since been constructed, there has been very little improvement in the time resolution that Hartridge and Roughton were able to attain (1 msec). However, the methods to be described below do show major improvements in economy of materials required (Hartridge and Roughton required 3–20 liters of reactants for the determination of a single kinetic curve), and in the variety and sensitivity of the techniques useful for analyses.

The stopped-flow technique (the reader is referred to the article by Dunford (Ch. 2) in this volume for an excellent description of the stopped-flow technique) has the advantage that the entire kinetic curve can be obtained with one sample mixture (or "shot"). On the other hand, since measurements must be in the time frame of the kinetic events, certain analytical methods are not accessible to this technique. In the continuous-flow methods, and in the rapid-quenching variations, the analytical approaches can be optimized for sensitivity and/or resolution. However, since the kinetic curve is sampled only at discrete points rather than as a continuous function, details in the kinetic progress of the reaction are more difficult to extract.

A typical rapid-quenching system is shown in schematic form in Fig. 2. The reactants are placed into syringes A and B and the quenching agent is placed into syringe C. A ram drives the syringe plungers at a constant velocity causing the reactants to mix in mixer A. As the mixture traverses reaction tube DE, it ages (Eqn. 1). At mixer B the quenching agent stops the reaction and the quenched sample is

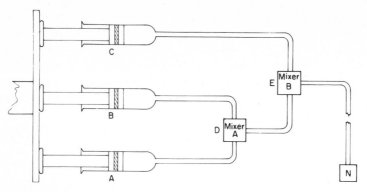

Fig. 2. Schematic of a generalized quenching device. A, B, and C are reactant/quenching solution driving syringes. The tube, DE, is the reaction tube, N is a nozzle for the rapid-freezing technique. Reservoir vessels and valves are not shown.

collected at N. By varying the volume of DE, different quenching times can be obtained.

Although in principle one can increase the length (and volume) of the reaction tube, this is only practical up to a certain limit, since long times imply large volumes of sample. The alternative is to fill reaction tube DE with a first push, pause for the desired time, and then drive the solutions out with a second push. Tube DE is chosen to hold just the requisite amount for analysis. This "push–push" mode uses twice the amount needed for analysis, but very long times can be obtained. For short times the single push is the most practical. The "break-even" point between "push–push" and single push will depend upon the amount of material needed for analysis and the rate of flow chosen for the experiment.

2. Requirements of rapid-quenching techniques

2.1. Mixing

Thorough mixing of reagents in a time short compared with the actual dynamic events of the reaction is an obvious requirement for any of the rapid-mixing techniques. Many designs of successful mixers have been constructed in the past. Chance (1974) has discussed many of these, showing results of detailed tests and demonstrating some of the unique requirements for mixing both unequal volumes and solutions of high viscosities. The essential principles (see Chance, 1974) for suitable mixers are: (1) that flow rates be sufficient to promote turbulent flow; (2) that the two solutions are broken up and intermingled into microscopic "packets" so that turbulence and diffusion can promote final mixing; (3) that sufficient energy is imparted to the system to promote (1) and (2); and (4) that cavitation is avoided both for optical reasons and for the simple fact that it effectively dilutes the mass in the reactant solution. For chemical reactants with low viscosities, and reactivities represented by rate constants of less than 300 sec, rapid-mixing systems perform quite well when fairly simple 4-jet or 8-jet mixers are employed which force small jets to impinge upon each other either directly or tangentially with turbulent flow. When conditions become more extreme, mixers with more sophisticated designs are required. The mixers which will be described in this chapter are not designed for these latter cases. For more complete discussions of mixer design and testing procedures, consult Chance (1974) and Gutfreund (1969).

2.2. Flow velocity

Ideally, in the continuous-flow method, all volume elements of the reacting mixture should pass from the mixing chamber to the observation point (or quenching point) in the same amount of time. This requires that there be a constant flow velocity for the entire period of flow and that there be no intermixing of volume elements of fluid which were mixed at different times. The fluid should travel along the reaction tube as if it were a solid plug, a condition referred to as plug or bulk flow. In reality, it is found that in tubes the flow velocity is always higher at the center than at the periphery. Therefore, without rapid radial intermixing, the fluid along the walls will arrive at the observation point later than that at the center of the tube. The net effect of the non-uniform velocities is longitudinal dispersion of the solution and a blurring of the time scale. This problem becomes acute in laminar flow where very little radial mixing occurs. However, with turbulent flow the radial volume elements are intermixed with the result that all of the solution arrives at any given point downstream with essentially the same age, i.e. plug flow is approximated. The problem is that the requirements for obtaining turbulence are not precisely known. A commonly accepted notion is that if the Reynolds number (RN) exceeds 2000 (Eqn. 2), the solution is likely to be flowing turbulently. A critical discussion of some of the difficulties with this idea has been published (Hansen, 1977).

$$RN = \frac{U\varrho d}{\eta} \tag{2}$$

U = flow velocity (cm/sec); ϱ = density of solution (g/cm^3); d = diameter of tube (cm); η = viscosity (poise).

Table 1 gives the flow velocities required to attain a RN of 2000 for some commonly available tubing utilized for continuous-flow instruments. Although it is probable that turbulence will not be achieved in all cases by utilizing these conditions, calculations show that the effects that laminar flow actually imposes upon the determination of first order reaction profiles are somewhat smaller than one's intuition might predict (Ballou, 1971). Nevertheless, to obtain reliable results, the flow velocities given in Table 1 should be considered minimum values for continuous-flow experiments.

Although the methods of Hartridge and Roughton (1923) give excellent time resolution, the quantities of reagents required for kinetic determinations are, for

TABLE 1
Flow velocity required for achieving Reynold's numbers of 2000 in
various tubes[a]

Diameter		Flow velocity
In.	mm	(ml/sec)
0.01	0.254	0.40
0.02	0.508	0.80
0.03	0.762	1.20
0.04	1.016	1.60
0.05	1.27	2.00

[a] This assumes a viscosity of 1 centipoise and a density of 1 g/cm^3.

most purposes, exorbitant. However, if one can produce a pulsed flow, wherein an
extremely rapid acceleration of flow precedes the period of constant flow velocity
during which measurement takes place, and then this constant flow velocity is suc-
ceeded by a rapid stopping of flow, very little reagent will be wasted. The constant
period should be adjusted so that the particular type of measurement required can
be optimized, or in the case of quenching experiments, the requisite amount of
material for analysis can be collected. In quenching experiments, since the entire
sample is usually collected and analyzed, it is imperative that the flow velocity have
the form of a step function with respect to time. This ensures that the entire
quenched sample is of the same reaction age. A slow acceleration and/or decelera-
tion of flow would contribute sample of indeterminate reaction age and would thus
constitute uncertainty in the measurement. This requirement places severe demands upon the devices which produce this flow since in the limit, instantaneous
acceleration implies infinite force. By using appropriate remotely actuated valves,
it is possible to discard that solution which flows during the early and late stages of
the pulse (Lymn et al., 1971). This relaxes the demands upon the drive system, but
increases the complexity of the flow system and wastes material. It also obviates the
possibility of using the "push–push" technique described above. Most quenching
devices do not use this valved method, but rely on a more robust driving ram.

2.3. Quenching

It is clear that a suitable means of quenching must be available for this technique to
be viable. In practice, this implies that quenching must be complete in a time that is
short compared with the half-life of the reaction. Furthermore, it is important that
the method used for quenching adversely affect neither the reaction being studied
nor the analytical methods utilized for quantifying the results.

Several methods are commonly used. For biological reactions a rapid change in pH will often effectively stop the reaction. Since pH changes can be made at diffusion-controlled rates, such methods, when applicable, are ideal. Alternatively, one might introduce another substrate at a very high concentration to effectively dilute the substrate being studied. This ploy is very effective when the first substrate is labeled and the second is not. Later I shall discuss this "pulse–chase" method more fully. It is sometimes possible to react the substrate (or intermediate) with a reagent which traps a particular species in a state that can be identified and quantitated. This clearly would be valuable in trapping species such as Schiff bases or aldehydes. Another method (although slower) would be to differentially extract the mixture into a solvent which by its selective extraction quenches the reaction. This requires excellent mixing at the point of quenching.

An interesting approach to quenching is the rapid freezing technique (Bray, 1961; Ballou and Palmer, 1974; Bray et al., 1979). This takes advantage of the Arrhenius activation energy which dictates that reactions proceed more slowly at lower temperatures. In this method, rather than observing or chemically quenching the reaction mixture at a specific distance from the mixer (the observation point), the reaction mixture is sprayed into a special EPR tube containing cold isopentane. This produces a sample frozen into tiny particles of ice at about $-140°C$, a temperature at which many reactions are essentially dormant. The frozen particles settle into the neck of the tube and are then packed into the measuring portion of the tube for subsequent measurement. Samples produced by this method can be analyzed by EPR, a technique that frequently requires low temperatures. Low temperatures would otherwise preclude rapid-flow devices from being used with aqueous solutions.

Since the quenching method chosen is selected to individually suit the process being studied, it is clear that the adequacy of quenching must be carefully evaluated in each case. If the quenching procedure prescribes chemical means, a suitable method is to show that a 3–5-fold variation in concentration of the quenching agent causes no (or minor) effects on the results. Chemical quenching of a reaction is expected to be a second-order process, implying that the above should be a reliable test.

Chemical quenching has usually been accomplished by one of two methods. In one, the reactants are expelled from the reaction tube leading from the mixer into a vessel containing quenching solution. This simple method places no demands upon the design of the flow system. However, a more reliable method employs a second mixer into which the reacting solution and the quenching reagent are introduced. Quenching can be accomplished more quickly (Froelich et al., 1976), and the reactants are collected in constant ratios so that incomplete extraction, spattering, or other such experimental difficulties will be minimized.

2.4. Other parameters

As in other kinetic methods, temperature control is of utmost importance. A variation of 1°C can easily produce a 5–10% variation in rate. Temperature control is most easily implemented by means of a water bath in which the entire flow system is submerged, coupled with an external temperature-controlled bath and circulator. An alternative thermoelectrically cooled system which is more compact and less expensive will be described in this chapter.

It is also important to be able to readily change reactant solutions so that a full range of experimental conditions can be employed. This is usually met by having replaceable reservoir vessels (syringes, tonometers, etc.) and 3-way valves leading to driving syringes. Stopped-flow instruments routinely use this method.

When performing chemical quenching experiments, it is necessary to have at least three driving syringes: two for the reactants and one for the quenching reagent. Frequently it is convenient to have a fourth syringe; thus two reactants can be premixed for a prescribed period before reacting with a third reagent, and finally the quenching reagent can be introduced from the fourth syringe. My experience is that four driving syringes will cover most applications.

3. Design of apparatus for quenching methods

The approach taken in this chapter will be to discuss the various parts of rapid-quenching devices with reference to published and commercial designs. The technical features will be detailed when they are new to the field or are particularly relevant to the understanding of the method. The goal is that the enterprising experimentalist should be able to use these ideas and the designs from the literature references given, to evaluate available quenching instruments or construct his/her own. Engineering drawings will not generally be included here, but any scientist should feel free to consult with the author about such details.

3.1. Ram design

The production of pulsed-flow for rapid-quench and related techniques requires a well-designed ram for driving the reactant syringe plungers. To closely approximate a step-function velocity profile, the ram must be able to provide a force which is very large at both the initiation and termination of flow. During the constant-flow period, the force will usually be considerably reduced. The original design of Hartridge and Roughton (1923), which employed hydraulic or gas pressure to drive the solutions, did not produce a sharp step function. Obtaining a period of constant flow for the measurement required ignoring the early and late flowing material, resulting in consumption of large amounts of material. One commercial instrument (Durrum D-133, Dionex Corp., Sunnyvale, CA) uses gas pressure as a means of driving the syringes (just as in their stopped-flow instrument), but partially avoids the above problem by using the "push–push" mode as the general procedure. Thus the solution is rapidly pushed into a reaction tube, a delay of desired period occurs, and then the solution is rapidly pushed out of the tube with the second push. The reaction time is primarily determined by the delay period rather than the flow period. This works reasonably well for times longer than about 25 msec, but has difficulty at shorter times. Moreover, the lack of control of flow prevents it from being useful for the rapid-freezing technique. The major advantage of this instrument is that it can be used for conventional and multiple-mixing stopped-flow as well as for chemical quenching. Many of the parts of the standard Durrum D-110 stopped-flow instrument are used in the D-133, and therefore existing stopped-flow devices may be converted to quenching systems at a moderate price.

A second method employs a 1/2 HP, SCR-controlled DC motor to drive an automobile flywheel at approximately 1500 rpm (Ballou and Palmer, 1974). This combination stores a large amount of energy in the flywheel for driving the syringes. The flywheel axle is led directly into a 100 : 1 reducing gearbox with output (15 rpm) to a spiral cam (Fig. 3) and follower, thus affording linear displacement to the syringe plungers. The motor and flywheel are maintained at constant rotational velocity, whereas the cam is activated only when the syringes are to be driven. This is effected by placing a pin through both the cam and the output plate of the gearbox to directly connect them together. When the cam follower reaches the end of the cam, flow stops; no clutches or brakes are necessary. For achieving longer reaction times (greater than about 250 msec), a "push–push" cam was devised (Fig. 3B) which was constructed of two continuous equal sections of the spiral. Section II causes a reaction tube (DE of Fig. 2) to become filled with the volume required for measurement, and section I, which is offset angularly relative to the first, pushes an equal amount after a suitable delay, thus forcing the aged mixture into a collection vessel while refilling the reaction tube with new material. The latter material is, of course, wasted. This ram has been tested extensively (Ballou, 1971) and it produces excellent pulsed flow. Its main disadvantages are that in the "push–push" mode it only is capable of driving fixed distances and thereby, fixed volumes and, in addition, the range of long times is somewhat limited by the rotational period of the cam (3–4 sec). Its simplicity and cost (*ca.* U.S. $ 1500), however, are distinct advantages over other systems.

A third method of driving syringes employs a powerful motor which is electronically controlled. Angular motion is converted to linear motion through efficient ball-screw assemblies. Various combinations of motors and clutches have been

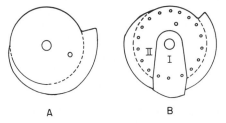

A   B

Fig. 3. Cams used for the Ballou ram. A is the cam used for single push experiments. B is the cam used for "push–push" experiments. Part II is the section of cam which fills the reaction tube, and part I, which can be rotated and then fixed to part II (by inserting a pin through the indicated holes), causes ejection of the reacted material at a delayed time.

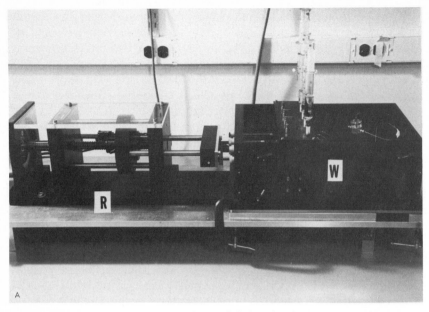

Fig. 4. The Ballou-Update quenching device. (A) Overall side view showing the Update Instruments ram (R) and the water bath (W). (B) A view of the driving syringes (S), the modified Hamilton valves (V), and the union adaptors (U). See text for details. Note that there are bubbles in these driving syringes. For proper results, these must be flushed out by forcing them into the reservoir vessels. (C) View of the reaction tubes (narrow stainless steel tubes), Ballou mixer (M), and just behind the foreground Ballou mixer, an HPLC "tee" (T). Note the union adaptor fittings leading into the mixers.

used for producing pulsed flow, but the most elegant and versatile system is sold by Update Instruments, Inc., Madison, WI. This system is driven by a printed circuit motor with an electronic power supply. An intelligent controller allows the user to establish velocity, displacement, and delay times for single or multiple pushes. A photograph (Fig. 4) shows this system in combination with a bath-and-flow system developed in the author's laboratory to be described in the next section. The primary disadvantage of the Update Instruments ram is its cost (more than U.S. $ 10 000), but the quality of construction and ease of use may be worth the cost. Another apparatus which uses a stepping motor drive (Froelich et al., 1976) also appears to perform quite well, although not at the level of either the Update Instruments or the cam-driven rams. Its cost is expected to be intermediate between those of the latter two.

3.2. Bath-and-flow system design

The flow system pictured in Fig. 4 is a new design incorporating many of the improved features developed in the past 5 years. It consists of a temperature-regulated water bath (W) with 4 driving syringes (usually Hamilton Gas-tight, Hamilton

73

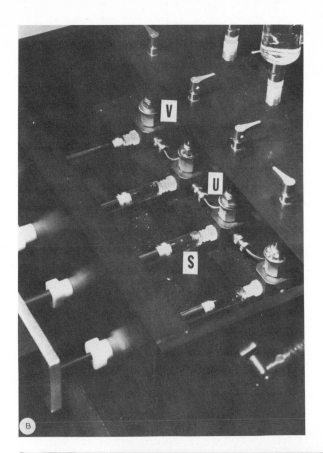

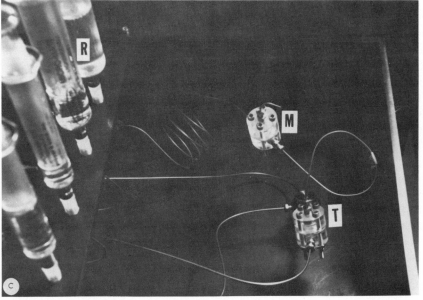

Co. Reno, NV), 4 Hamilton miniature 3-way valves (V), various lengths of HPLC tubing to connect the different components, and 2 or 3 mixers (M) for initiating and quenching the reactions.

The water bath is constructed of anodized aluminum and is large enough to permit convenient access to the components (10 in. wide, 15 in. long, 6 in. high). It is clamped to the baseplate of the Update Instruments ram (R) as shown. Temperature is controlled by circulating water from an external temperature-regulated bath or with an integral temperature controller (not shown in Fig. 4) which attaches to the bottom plate of the bath. This controller consists of a lucite plate (to thermally insulate the system from the ram) which has 4 wells machined in it symmetrically. In these wells are 4 Melcor (Trenton, NJ) CP1.4-127-06L ceramic thermoelectric modules which are pressed against the bottom of the aluminum bath with copper plates. The copper plates, in turn, are soldered to copper tubing which carries a slow stream of cooling tap water. This system is powered by an unregulated variable DC power source. The latter consists of a Staco model 291 0-132 VAC variable transformer, a Triad F-282 U 60 VAC, 6 A step-down transformer, and a 12 A bridge rectifier. The variable power supply makes it possible to provide the proper amount of cooling to meet the demands of the system at different temperatures, in contrast to refrigeration systems which are "on/off" devices. Fine temperature control is provided by a 250-W stainless steel heating element controlled by a RFL model 72A zero-crossing controller and sensor (RFL Industries, Boonton, NJ). Stirring by closed-loop pump or external stirrer assures uniform temperature. This system is very compact, requires no bench space, and is inexpensive (*ca.* U.S. $ 400 parts plus machining); it works very well, and it can easily be adapted to other similar devices.

The flow system is in many ways similar to other designs (Ballou and Palmer, 1974; Update Instrument System 1000). Certain new features will be detailed. The major new feature being incorporated into this flow device is the use of commercial HPLC tubing and fittings wherever possible. The ready availability of a variety of tubing and parts, the non-compressibility of tubing (which is important in producing true pulsed flow), the interchangeability of parts, and the qualifications for anaerobic work makes this approach extremely desirable for most rapid-flow work (including stopped-flow devices).

The Hamilton valves are modified somewhat. Longer handle stems have been added for convenience (Fig. 4B). The standard valves use a spring to load the rotating plug and dynamically correct for differential expansions and contractions in the valve due to changes in temperature. However, the hydraulic pressures in quenching devices can occasionally exceed the spring pressure and leaks will occur. Therefore, the springs have been replaced by sleeves which hold the plug tightly into the valve by means of the top nut. This arrangement does not allow any release as a result of sudden pressure changes occurring in quenching experiments. However, one must be careful to continually check that the top nut is finger-tight. These valves work quite well, but occasionally can still be sources of leaks in these experiments. An alternative (which would require an adaptor to connect the syringes — see below), is the use of SSI 3-way model 02-0183 HPLC valves (Scientific Systems, Inc., State College, PA). These valves are only slightly more expensive than the Hamilton valves, but can withstand fluid pressures to several hundred atmospheres. These valves are not of the switching type, and therefore all three channels are connected to each other when the valve is open. This makes it difficult to fill syringes while avoiding cross-contamination. Although more complicated, the use of HPLC "tee's" and two 2-way valves per syringe can avoid this problem. HPLC switching valves such as made by Valco or Rheodyne would be more suitable, but would also be more expensive (approximately U.S. $ 350 each).

The valves are connected to reservoir syringes or vessels with either HPLC tubing or plastic tubing of the LDC/Cheminert or Altex designs. Syringes, Hamilton valves, and other similar devices can be mated to HPLC parts by using various combinations of 1/4 in. × 28 UNF threaded fittings and modified HPLC 1/16 in. unions (U) (Fig. 4B). HPLC unions (Valco zero dead volume female unions) are cut into two equal lengths and the outside is threaded to 1/4 in. × 28 UNF. Standard HPLC tubing ends (male) can be used to fasten tubing to one side of the modified union while the other can be mated to any 1/4 in. × 28 UNF fitting via female unions. Figs. 4A and C show some applications. The reservoir syringes are mounted in 1/4 in. × 28 UNF luer fittings connected to plastic female unions and then to the modified HPLC unions. When using HPLC switching valves instead of the Hamilton valves shown in Fig. 4B, the syringes can be mated to HPLC tubing by using a Valco 1/4–1/16-in. zero dead volume reducing union (ZVRU.250/062). The 1/4-in. nut and ferrule are replaced with a Kel-F or Teflon nut with a tip shaped like the ferrule. The other end of the nut has a female Luer taper to receive the syringe and a small hole (1 mm) drilled down the center of the nut. The union can be clamped to a suitable holder to mount the syringe. This arrangement makes it possible to change syringes with minimal difficulty.

The mixers most successfully used for this technique are 4-jet tangential mixers as shown in Fig. 5. These mixers have jets which impinge tangentially under turbulent conditions. An important addition to this design is gasket 2 which has a small hole that produces further mixing (Ballou, 1971). These mixers are easy to construct, have a dead volume of only 1–2 μl, and can be disassembled easily for cleaning. Tests have shown that mixing is more than 95% complete within 3 msec under a variety of conditions of flow rate, viscosity, and mixing ratios appropriate for rapid quenching (Ballou, 1971). The above modified HPLC union adaptors, in addition to receiving the HPLC tubing, act to fortify the lucite housing against wear from the HPLC tubing connectors. For very short times it may be desirable to use a 2-stage mixer constructed of three parts and consisting of two sections of the mixer shown in Fig. 5. The outlet of the first mixer leads directly to the second mixer via a small hole which is, in effect, equivalent to tube DE of Fig. 2. The quenching solution is brought into the mixer in the same way as those of the first stage. Alternative mixers are HPLC "tee's", which due to their small bores, will perform quite well for applications in which mixing requirements are not too severe. These are readily available, extremely rugged and inexpensive, and can be used with organic solvents.

76

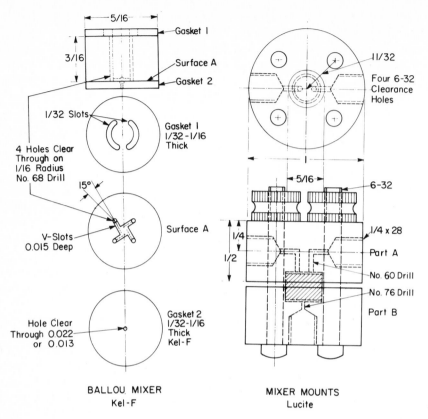

Fig. 5. Ballou mixer. The Kel-F mixer on the left is a blow-up of the shaded portion of the right-hand side. Parts A and B clamp the gaskets and mixer together. Most of the mixing takes place on surface A and in the hole on gasket 2. All measurements are in inches.

The quenched solution must be collected quantitatively. As shown in Fig. 4C, the exiting solution from the second mixer (top) is simply collected in a test tube or other vessel. A more reliable method (not shown) is to use removable needles such as Hamilton (cat. 80428). A 1/4 in. × 28 UNF male tubing fitting (Cheminert type) or needle bushing (Hamilton cat 35035) can mount the needle into the above 4-jet mixer. A similar adaptor with 10–32 threads can be used with the HPLC "tee's". This needle is then used to pierce a standard septum vial to allow efficient collection of the quenched reaction product. The needle should be of small gauge (*e.g.* 22–26-gauge) to conserve sample. Since this technique avoids spills and spattering, it should be especially valuable for working with radioactive isotopes. A particularly convenient arrangement of this idea can be realized by drilling a hole in the bottom of the water bath, plugging it with a rubber stopper, and piercing the stopper with the collection needle. Then the septum vial does not need to be placed into the bath.

Stainless steel HPLC tubing is used throughout the flow system wherever high pressure may be encountered. The bores are chosen to suit the experiment. Referring to Fig. 2, the bores of tubes between syringes A or B and mixer A should be quite small (0.01–0.02 in. diameter) so that sample is conserved. The bore of the tubing between D and E is chosen to be of a length that is convenient for handling, yet of a volume suitable for producing the desired time. In the case of "push–push", of course, the volume will be that necessary for analysis. Tubing (1/16 in. o.d.) with internal diameters listed in Table 1 is available from most HPLC distributors. Occasionally, the stainless tubing will not be inert to reactants used in quench studies. In these cases plastic tubing such as that from Cheminert, that described by Ballou and Palmer (1974), or that from Update Instruments may be useful. Alternatively, one might try glass-lined stainless tubing which is also available at HPLC distributors. This tubing is not available with the range of internal diameters that is indicated in Table 1 for the normal tubing.

3.3. Quenching design — chemical and cryogenic

Chemical quenching has been discussed above. It should be emphasized that each system being studied will have to be investigated to assure adequacy of quenching. Furthermore, the chemical quenching agent will have to be selected carefully to avoid artifacts from product conversions and other unwanted chemical events.

Electron paramagnetic resonance spectroscopy is particularly useful in studies of oxidation reactions, since information about oxidation states and environmental features of paramagnetic species can be obtained. It is difficult to apply rapid reaction techniques to EPR since (a) detection with practical sensitivity usually requires relatively long times of observation and signal averaging; (b) many species can only be observed at low temperatures; and (c) to fully characterize the observed species, the samples often need to be investigated at several incident power levels, modulation widths, and temperatures. The rapid-freezing technique developed by Bray (Bray, 1961; Bray and Peterson, 1961) brings together rapid-reaction techniques and EPR. It can also be used with Mössbauer and reflectance spectroscopy. The principles are given above (section 2.3).

To effect efficient quenching, it is necessary to spray the reactant solution into a cold liquid immiscible with water. Liquid nitrogen is unsuitable for this since when the warmer reactant solution is sprayed into it, the nitrogen would vaporize and thereby lose its efficient heat conductivity. The liquid used is usually isopentane, since although it is a liquid at room temperature, it can also be maintained at -140 to $-150°C$ without freezing. The nozzle which sprays the solution into the isopentane as a fine stream is usually constructed of plexiglass and has holes approximately 0.2 mm in diameter and 2 mm long. Usually one or two holes are used in the nozzle. It has been found that to obtain good quenching and have particles which can be packed well, a flow rate of approximately 1 ml/sec/hole is reasonable. It may be possible to use HPLC tubing (0.01 in. i.d.) as a nozzle with the provision that the tubing be insulated to prevent the liquid from freezing inside the tubing. The only

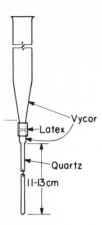

Fig. 6.   Special EPR tube for rapid freezing. The bottom portion of the tube is a standard X-band EPR tube. The upper portions are designed to hold a reservoir of the quenching isopentane. See text for details of their use.

possible problem would be that 0.01 in. diameter would produce too coarse a particle and/or would quench the reaction too slowly. The nozzle is held directly over the special EPR tube (Fig. 6) containing the cold isopentane ($-140°C$) about 0.5 cm from the surface of the coolant. Quenching is accomplished in approximately 5 msec (Ballou and Palmer, 1974).

The cooling system consists of a metal can containing isopentane, placed in a metal dewar, and cooled to about $-145°C$ with liquid nitrogen which is in the space between the dewar and the metal can. A temperature regulator maintains constant temperature and prevents the isopentane from freezing. This bath usually holds three special EPR tubes (Fig. 6) by means of a Lucite cover which also provides some thermal insulation and prevents frost from inadvertently falling into the sample tubes. A more complete description of this system is available (Ballou, 1971).

The sample tubes (Fig. 6) containing the cold quenching isopentane, are constructed of two pieces so that the bottom section can be conveniently removed and stored in liquid nitrogen until measurement takes place. The two sections are wetted with glycerol and fastened with a short section of latex tubing which becomes rigid upon cooling. After the sample has been sprayed into the isopentane, it settles for 15–30 sec into the neck of the tube. A packer constructed of stainless steel rod (approximately 30 cm long and 1.5 mm diameter) with a Teflon tip fitting smoothly into the narrow portion of the EPR tube, is used to carefully pack the frozen crystals into the bottom of the EPR tube. Packing is the most difficult and irreproducible operation of the rapid-freezing technique, but it can be learned with practice. As can be imagined, the sample may not always be packed to the same density. This should be checked for each system and user by packing either unreacted or completely reacted sample in the same way. When packing is complete, the tube is immersed into liquid nitrogen to the depth of the packed sample. Then the isopentane is poured off. An aspirator is used to remove the final traces of isopentane, the tubes are separated and the lower part containing the sample is then stored.

3.4. Other considerations in design and use

Upon completion of a chemical quench pulse, the tubes such as DE (Fig. 2) must be cleaned before re-use; otherwise the solution in DE would contribute sample of unknown reaction age. In the single-push mode, tube DE is usually removed and cleaned before re-use. In the "push–push" mode an alternative is possible. If an additional 3-way valve and a washing syringe is placed in the line just before mixer A, it is possible to flush out the relevant part of the flow system after collecting a sample without disassembly of the tubes. After flushing, air or nitrogen can be used to dry the tubes downstream from mixer A. One must be careful in any such operation to refill the tubes as necessary so that no unwanted air spaces are left in tubes and mixers. By use of Valco 2-position valves (8-port), it is possible to effect convenient connections for two reservoir syringes and a flushing system. This more complicated tubing arrangement leads to a simpler and more fool-proof mode of operation which minimizes having valves opened or closed in undesired configurations.

In using the three (or more) syringe method in conjunction with any technique, certain precautions must be taken. In reference to Fig. 2 the following relationship must be satisfied.

$$\frac{V(\text{CE})}{V(\text{DE})} = \frac{V(\text{syringe c})}{V(\text{syringe A}) + V(\text{syringe B})} \tag{3}$$

$V(X)$ is the volume of the unit X indicated in Fig. 2. This can be accomplished in at least four different ways. (a) The volumes of the tubes can be judiciously chosen to meet this requirement and the tubes can be completely cleaned each time (which is not altogether convenient). (b) The tube, CE, is filled, but syringe C is depressed just enough that the ram does not begin driving it until the tube, DE, which is initially empty, is filled. (c) A volume, $V(\text{CE})$, satisfying Eqn. 3 is withdrawn into syringe C, expelled into the reservoir for syringe C so that the three syringe plungers are even and against the ram, and then the three syringes are pushed simultaneously. (d) The first bit of ejected solution is discarded so that all of the solution which is collected is of the same composition. The latter two methods are the easiest to use; method (d) is, of course, quite wasteful of reagent. Nevertheless, it has been used with success in conjunction with an automatic solenoid valve (Lymn et al., 1971).

4. Examples of use

4.1. Calibration

Calibration of any technique determines its ultimate value. The chemical quenching method, although simple in principle, can be difficult to apply in an unambiguous fashion. The problems of finding a suitable quenching agent and checking its reliability have been discussed above. Remaining factors include the stability and quality of flow, the efficiency of mixing, and the absolute calibration of the flow volumes. The volumes can be determined to within 5% by using the syringes to measure the volume of fluid required to just fill the tubes. Efficiency of mixing is more difficult to determine and is addressed elsewhere (Chance, 1974). The mixers described herein have been tested extensively (Ballou, 1971), showing good performance under a wide variety of conditions applicable to quenching experiments. The flow stability and quality of flow will be determined by the performance of the ram and the nature of the tubing. The system described above will perform adequately for virtually all of the standard types of quenching experiments.

Perhaps the best overall calibration is to perform studies on a well-characterized test reaction. The reaction of the hydrolysis of 2,4-dinitrophenyl-acetate in sodium hydroxide solutions (Barman and Gutfreund, 1964) is very commonly used. The 2,4-dinitrophenyl-acetate, initially concentrated in isopropanol, is diluted with 5 mM HCl so that a concentration of approximately 0.6 mM (after mixing) is achieved. A solution of sodium hydroxide (0.025–1.0 M after mixing) is mixed with the 2,4-dinitrophenyl-acetate and the reaction is quenched with 1 M HCl. The liberated 2,4-dinitrophenol can be quantitated by measuring the absorbance at 320 nm. Alternatively, the quenched reaction mixture can be adjusted to pH 4.0 with concentrated potassium acetate and the absorbance determined at 360 nm. Blanks are prepared by inverting the order of addition of the quenching acid and the "reacting" base.

The reaction is quenched at various times after mixing (at least 10) and a standard semilogarithmic first order kinetic plot is made. This procedure is repeated for two or three additional sodium hydroxide concentrations. The plots should be linear, have good precision, and converge to a single point. This point should have an ordinate equal to the blank or zero time value, i.e. the reaction curve should extrapolate to zero reaction at zero time. The abscissa of the point will lie before the apparent zero time point as calculated from Eqn. 1. This negative time is frequently attributed to the quenching time. For confirmation, the reaction should also be examined under the same conditions with the optical stopped-flow method. If it is necessary to change flow conditions (rate of flow, viscosity, etc.), the calibration should be repeated under similar circumstances.

The rapid-freezing technique uses a similar calibration method (Ballou and Palmer, 1974). The reaction of equine metmyoglobin hydrate with sodium azide at pH 7.8 in 0.02 M Tris buffer (0.1 N in sodium nitrate) at 25°C is used as a test system. This reaction shows strictly linear Arrhenius behavior with an activation energy of 12.5 kcal · mole^{-1} (typical for many biological reactions) and, with azide in excess, exhibits pseudo-first-order kinetics. The formation of the azide complex results in a conversion of the high-spin metmyoglobin hydrate to a low-spin form, a change readily measured by either EPR or optical spectroscopy. Thus the kinetics can be cross-checked with the stopped-flow technique. A second-order rate constant of 2.5 × 10^3 M^{-1}sec^{-1} is expected.

4.2. Chemical quenching

4.2.1. Determination of a mechanistic path

I would like to discuss a few examples of work using chemical quenching methods in various ways. There are, of course, numerous examples published, but a detailed review is not relevant to this chapter. Hopefully, this brief discussion will be representative of the spirit of the method.

The first system is typical of the determination of simple rapid kinetic parameters involving reactants and products most readily measured by radioactive counting techniques. The reactions catalyzed by the aminoacyl-tRNA synthetases are typical of the activation of acyl groups by ATP and are shown in Eqn. 4.

$$AA + ATP + tRNA \rightleftharpoons AA\text{--}tRNA + AMP + PP \qquad (4)$$

AA = amino acid, PP = pyrophosphate. A logical mechanism for such a reaction would be: the enzyme binds the ATP and amino acid and catalyzes the formation of an enzyme-bound aminoacyl adenylate and PP. A tRNA would then bind and would displace the adenylate and form the AA-tRNA and AMP. An alternative mechanism could involve the formation of a quarternary complex of enzyme, ATP, amino acid, and tRNA, with the reaction proceeding in a concerted fashion involving no formation of a frank AA-adenylate. Chemically this presents some difficulty since no mechanism for coupling the energy of hydrolysis of ATP to the formation of the AA-tRNA is evident.

In an elegant series of experiments Fersht (1977) has shown that the aminoacyl adenylate pathway is correct, not the concerted mechanism. It had been known that an aminoacyl adenylate does form in the absence of tRNA, and that this complex could transfer the activated amino acid to tRNA. However, these "half reaction" studies may not be completely applicable to the situation involving the simultaneous presence of all three substrates. In Fersht's studies, three essential quenching experiments were performed on isoleucyl-tRNA synthetase. First, he measured the rate of formation of [^{14}C]-isoleucyl-tRNA upon mixing tRNA^{Ile} with the enzyme–[^{14}C]isoleucyl-adenylate complex. The rate was identical to k_{cat} determined from steady-state kinetics. This reaction is therefore a reasonable candidate as a required intermediate step. In his second experiment he mixed enzyme, isoleucine, tRNA, and [γ-^{32}P]ATP and followed the kinetics of ^{32}PP release. The result was that a burst of PP was released and then a steady-state release occurred at a rate equivalent to the turnover rate. This is consistent with aminoacyl adenylate formation prior to aminoacylation of tRNA as suggested above. The alternative model consistent with these kinetics is a rapid formation of aminoacyl-tRNA and concomitant release of PP, followed by a slower release of the aminoacyl-tRNA from the enzyme. This possibility could be checked by performing the same experiment as the second (above), but using [^{14}C]-isoleucine rather than [^{32}P]ATP, and analyzing for [^{14}C]Ile-tRNA. The concerted mechanism predicts a similar burst for the charged tRNA as seen for PP. Since none was found, this mechanism could be eliminated. The Fersht (1977) reference has several excellent examples of the rapid-quenching technique and is highly recommended.

4.2.2. Determination of fraction of active substrate

The hexoses exist in two major cyclic configurations: the α- and the β-anomers. Although these forms interconvert readily, the rate is substantially lower than for enzymatic reactions. Hence, the enzyme specificity for α or β forms can often be determined easily. The phosphorylated sugars anomerize considerably faster, however, and the question of specificity is more difficult to answer. Benkovic (1979) has described very clearly how one uses the rapid-quenching technique to determine anomeric specificities of the more labile phosphorylated hexoses. The principles of the method apply equally well to other such determinations. In the example of D-fructose 1,6-bisphosphatase, the enzyme and substrate are mixed together and the reaction is quenched at various times. Given the proper conditions (substrate and enzyme concentrations, etc.) a burst of product should be formed followed by a steady-state rate which is, in fact, the rate of anomerization of the sugar. The burst will be directly related to the amount of the active anomer in the substrate pool. These methods, although frequently employing radioactive measurements, are clearly applicable to any unambiguous and sensitive technique for measuring product.

4.2.3. Identification of an intermediate in a mechanism

Frequently when using stopped-flow methods, one finds kinetic evidence for intermediates occurring in the enzymatic reaction. Although optical spectra usually can be resolved, nevertheless, the chemical identity of such species may be obscure. One such case involved the enzyme, 4-hydroxybenzoate hydroxylase. This flavoprotein mono-oxygenase, which catalyzes the NADPH-dependent hydroxylation of 4-hydroxybenzoate at the 3-position, is typical of the flavin-containing phenol hydroxylases. Its extensive study by optical and rapid kinetic methods (e.g., Entsch et al., 1976) represents one of the more thorough kinetic and spectral analyses of transient intermediates in the literature. When the reduced form of the enzyme complex with 2,4-dihydroxybenzoate (one of the possible substrates for this enzyme) is reacted with oxygen, a series of three spectrally characterized intermediates is resolved. The formation of the first intermediate is dependent upon oxygen concentration and could quite reliably be attributed to production of a C-[4a]-hydroperoxyflavin, an activated hydroperoxide. The question was: at what stage of this reaction was oxygen transferred from the hydroperoxide to the substrate? Entsch et al. (1974) performed rapid-quenching studies on the same oxygenative half-reaction (*i.e.*, reaction of oxygen with reduced enzyme—2,4-dihydroxybenzoate complex) and analyzed for product formation. In these experiments, the quenched reaction mixture was chromatographed and the product was determined. (Obviously, the use of HPLC techniques would be especially valuable for this type of analysis.) Product was formed in a monophasic reaction with a half-life of 11 msec, a rate identical to that for the appearance of the second intermediate observed spectrally. This implies that the second intermediate must be a species of flavin with only the proximal oxygen atom of the hydroperoxide remaining attached, and the substrate must be oxygenated. It is possible that the complete transformation into product had not occurred by the time that the second intermediate had formed, but that the acid quench had brought about the final conversion non-enzymatically. This study did not result in a complete determination of the structure of the second intermediate, but it did limit the possibilities. I might add that this intermediate remains the major unknown in the flavoprotein phenol hydroxylase mechanism.

4.2.4. Isotope-trapping techniques

The k_{cat} of an enzyme as determined from steady-state kinetics is only a minimum estimate for the slowest step in the reaction sequence, since it assumes that this step is significantly slower than all others. To understand the mechanism properly, it is necessary to measure all of the enzymatic steps. When some direct physical probes such as spectral changes are associated with various intermediate stages of the enzyme, of substrates, or of products, the stopped-flow technique can give direct information about the various rate constants of the reaction. Studies of the flavoprotein hydroxylases are excellent examples of this approach (see above). In these cases, the spectra of the flavins serve as signatures of the state of the flavin in the reaction. When no such signature is available, an alternative is chemical quenching. This is limited by the requirement that usually the only events which can be directly observed are those which involve product release.

Rose and colleagues have developed the isotope-trapping technique (Wilkinson and Rose, 1979; Rose, 1980) which can often be used to obtain information about steps other than those releasing product. In this method a mixture of enzyme and isotopically labeled substrate are allowed to react for a time which permits a steady state to develop, but short enough that significant product is not produced. Depending upon the turnover rate of the enzyme, this pulse can last for as little as 5 msec to as long as a few seconds. Then the mixture is rapidly diluted with a large excess of unlabeled substrate. After a few turnovers of the enzyme under these conditions, the reaction is quenched with a suitable reagent. By determining the pattern of partitioning of the isotopically labeled substrate between formation of product and dissociation from the enzyme, information about the dissociation rates from various enzyme complexes can be obtained.

This method allows evaluation of the catalytic competence and the quality of the range of complexes that may be present as enzyme–substrate complexes. This method requires good mixing and quenching for good precision. An excellent exposition of this kinetically complex method and of its limitations and sources of error is found in Wilkinson and Rose (1979) and in Rose (1980).

4.2.5. Hormone-receptor-mediated transmembrane permeability changes

The isolation of biological receptors and their reconstitution into functional artificial membrane preparations have been the subjects of a great deal of attention in recent years and constitute an active area of current research. These model systems are made up of various phospholipids and can be prepared in the form of black-lipid membranes or vesicle preparations. The main biological function that results from activation of some of these receptors is to produce an increase in membrane permeability to certain substances. For example, it is generally agreed that the binding of acetylcholine to its receptors in nerve and muscle cells induces a conformational change in the protein that results in the formation of non-conducting channels through the membrane.

The relationship between the concentration of acetylcholine and the rate of acetylcholine receptor complex-dependent ion translocation is very important since it determines the transfer of information between cells in the nervous system.

One of the problems commonly encountered in the reconstitution studies in general, and in the case of acetylcholine in particular, is that inactivation of receptor on exposure to ligand occurs rapidly. In addition, because the number of receptors incorporated in the vesicles and the size of the vesicles are both small, the translocation process is complete within a few hundred milliseconds when a saturationing concentration of the stimulus is employed (Cash and Hess, 1981). Recently, a rapid-quenching technique has been applied for the study of acetylcholine receptor-mediated transmembrane influx in membrane vesicles (Hess et al., 1982). The chemical quenching agent used in these studies is the cholinergic antagonist D-tubocurarine chloride. Employing this technique it has been possible to (a) correct for the kinetic heterogeneity of the vesicle population; (b) use the inactivation of the receptor by its natural ligand to reduce influx rate at high ligand concentrations to measurable level; and (c) determine the rate coefficients of two processes that lead to successive inactivations of the receptor and occur in different time regions.

4.3. Rapid freezing

Rapid freezing has been employed quite frequently in the study of metalloproteins. Xanthine oxidase is an enzyme containing molybdenum, flavin, and two distinct non-heme iron centers. All of these centers are paramagnetic in one or more of their oxidation states. The rapid-freezing technique was originally developed in response to investigations of this enzyme (Bray, 1961; Bray and Peterson, 1961). These studies, later studies by Bray and colleagues (Gutteridge and Bray, 1980), and those by the group at Ann Arbor, Michigan (Edmondson et al., 1973; Olson et al., 1974) represent a rather wide range of application of the rapid-freezing technique to the study of this enzyme. Since the studies are too complicated to be discussed here, the reader is encouraged to consult the above papers and other references contained in them.

In the study of non-metallic substances only free radicals will exhibit any EPR signals. Therefore the method can be very specific and unambiguous in many cases.

5. Conclusions

The chemical and rapid-freezing quenching methods are clearly of broad use in the rapid kinetic arsenal. It should be noted that the general type of apparatus described above for pulsed flow is also useful for application to a wide variety of detection methods. In recent years many of the standard techniques such as rapid-scanning optical detection, calorimetric measurements, and electrode measurements have improved to the extent that good signal-to-noise characteristics are possible with time constants of considerably less than 1 sec. Although the stopped-flow method is preferred, the time resolution of these methods does not always permit the study of reactions with rate constants of more than 5 sec^{-1}. The pulsed-flow methods, however, provide a reasonable compromise. If the measurement can be made accurately in 0.1–0.2 sec, quantities of reagents consumed will not be excessive and excellent time resolution can be obtained via the pulsed flow. Rapid scanning, for example, could be done with signal averaging to improve signal to noise. Furthermore, there would be no discrepancy in the time of the reaction at one end of the spectrum compared to the other.

Differential measurements could be made where one detector is upstream from the other. Since the same flow conditions would exist for both detectors, true differential measurements are possible. The various times of reaction are achieved by changing the reaction tube (DE of Fig. 2) and maintaining constant flow conditions. The variety of useful applications is only limited by the imagination of the experimentalist.

Acknowledgements

I wish to thank Dr. Narlin Beaty for supplying photographs (Fig. 4) and for helpful discussions.

This work was supported in part by grant GM-20877 from the National Institutes of Health.

References

Ballou, D.P. (1971) Instrumentation for the Study of Rapid Biological Oxidation–Reduction Reactions for EPR and Optical Spectroscopy, Thesis, University of Michigan, University Microfilms, Ann Arbor, MI, No. 72-14796.

Ballou, D.P. and Palmer. G.A. (1974) A practical rapid quenching instrument for the study of reaction mechanisms by EPR spectroscopy, Anal. Chem., 46, 1248–1253.

Barman, T.E. and Gutfreund, H. (1964) Rapid chemical quenching methods, in: B. Chance, R.H. Eisenhardt, Q.H. Gibson and K.K. Lonberg-Holm (Eds.), Rapid Mixing and Sampling Techniques in Biochemistry, Academic Press, New York, pp. 229–248.

Benkovic, S.J. (1979) Anomeric specificity of carbohydrate-utilizing enzymes, Methods Enzymol., 63, 370–379.

Bray, R.C. (1961) Sudden freezing as a technique for the study of rapid reactions, Biochem. J., 81, 189–193.

Bray, R.C. and Peterson, R. (1961) Electron spin resonance measurements, Biochem. J., 81, 194–199.

Cash, D.J. and Hess, G.P. (1981) Quenched flow technique with plasma membrane vesicles: acetylcholine receptor mediated transmembrane ion fluxes, Anal. Biochem., 112, 39–51.

Chance, B. (1974) Rapid flow methods, in: G.G. Hammes (Ed.), Investigations of Rates and Mechanism of Reactions, Techniques of Chemistry, Vol. VI, 3rd edn., Wiley-Interscience, New York.

Edmondson, D., Ballou, D., Van Heuvelen, A., Palmer, G. and Massey, V. (1973) Rapid freeze kinetic EPR studies on xanthine oxidase, J. Biol. Chem., 248, 6135–6144.

Entsch, B., Massey, V. and Ballou, D.P. (1974) Intermediates in flavoprotein catalyzed hydroxylations, Biochem. Biophys. Res. Commun., 57, 1018–1025.

Entsch, B., Ballou, D.P. and Massey, V. (1976) Flavin-oxygen derivatives involved in hydroxylation by p-hydroxybenzoate hydroxylase, J. Biol. Chem., 251, 2550–2563.

Fersht, A. (1977) Enzyme Structure and Mechanism, Freeman, San Francisco, CA, pp. 190 ff.

Froelich, J.P., Sullivan, J.V. and Berger, R.L. (1976) A chemical quenching apparatus for studying rapid reactions, Anal. Biochem., 73, 331–341.

Gibson, Q.H. (1969) Rapid mixing: stopped flow, Methods Enzymol., 16, 187–228.

Gutfreund, H. (1969) Rapid mixing: continuous flow, Methods Enzymol., 16, 229–249.

Gutteridge, S. and Bray, R.C. (1980) Oxygen-17 splitting of the very rapid Mo(V) EPR signal from xanthine oxidase, Biochem. J., 189, 615–623.

Hansen, R.E. (1977) The Rapid Mixing Notebook, Vol. I, Notes 1–6, Update Instruments, Inc., Madison, WI.

Hartridge, H. (1923) The coincidence method for the wavelength measurement of absorption bands, Proc. Roy. Soc. (London), A102, 575–587.

Hartridge, H. and Roughton, F.J.W. (1923) A method of measuring the velocity of very rapid chemical reactions, Proc. Roy. Soc. (London), A104, 376–394.

Hess, G.P., Pasquale, E.B., Walker, J.W. and McNamee, M.G. (1982) Comparison of acetylcholine receptor-controlled cation fluxes in membrane vesicles from Torpedo californica and Electrophorus electricus: chemical kinetic measurements in the milliseconds region, Proc. Natl. Acad. Sci. (U.S.A.), 79, 963–967.

Lymn, R.W., Gibson, G.H. and Hanacek, J. (1971) A chemical stop rapid flow apparatus, Rev. Sci. Instr., 42, 356–358.

Olson, J.S., Ballou, D.P., Palmer, G. and Massey, V. (1974) The reaction of xanthine oxidase with molecular oxygen, J. Biol. Chem., 249, 4350–4362.

Rose, I.A. (1980) The isotope trapping method: desorption rates of productive E · S complexes, Methods Enzymol., 64, 47–59.

Wilkinson, K.D. and Rose, I.A. (1979) Isotope trapping studies of yeast hexokinase during steady-state catalysis, J. Biol. Chem., 254, 12567–12572.

C · ER :

Application of pulse radiolysis to biochemistry

KAZUO KOBAYASHI and KOICHIRO HAYASHI

Institute of Scientific and Industrial Research,
Osaka University, Mihogaoka, 8–1, Ibaraki, Osaka 567, Japan

Contents

1. Introduction 88
2. Principles and methods 88
    2.1. Pulsed radiation sources 88
    2.2. Detection methods 88
        2.2.1. Spectrophotoelectric recording 89
        2.2.2. Spectrographic recording 90
        2.2.3. Resonance Raman scattering 90
        2.2.4. Flow radiolysis 91
    2.3. Chemical techniques 91
3. Applications of pulse radiolysis to biochemistry 92
    3.1. One-electron reduction in oxyform of hemoproteins 93
    3.2. One-electron reduction in flavoproteins 96
        3.2.1. Flavodoxin 97
        3.2.2. D-Amino acid oxidase 98
    3.3. Kinetics of oxygen and carbon monoxide binding to partially reduced methemoglobin 99
    3.4. The electron pathway of redox proteins having multiple electron-accepting sites 101
        3.4.1. Cytochrome oxidase 101
        3.4.2. Copper oxidases 103
    3.5. Superoxide and superoxide dismutase 104
        3.5.1. The reaction O_2^- with ferrihemoproteins 104
        3.5.2. Superoxide dismutase 106
References 107

R.I. Sha'afi and S.M. Fernandez (Eds.), Fast Methods in Physical Biochemistry and Cell Biology
© *1983 Elsevier Science Publishers*

1. Introduction

The pulse radiolysis technique is based on a perturbation of a system by a short pulse (< 1 μsec) of high-energy radiation. This technique has been mainly applied to the study of fast reactions of excited molecules, ions, and free radicals induced by ionizing radiation, much of which is of interest to the physical chemist and radiation chemist. Land and Swallow were the first to apply this technique to biomolecules such as NAD (Land and Swallow, 1968), flavin (Land and Swallow, 1969), ubiquinone (Land and Swallow, 1970) and cytochrome c (Land and Swallow, 1971). This technique has now been employed for the study of the mechanism of action of the oxidation-reduction proteins with high time resolution.

In this chapter we discuss the application of pulse radiolysis to problems of biological interest.

2. Principles and methods

2.1. Pulsed radiation sources

Pulses of ionizing radiation can be provided by a variety of sources. Pulse radiolysis studies have typically employed Van de Graaff and microwave linear accelerators; detailed characteristics of these sources have been given by Matheson and Dorfman (1969). The pulse duration should be short with respect to the lifetime of the species generated, and pulse widths ranging from 1 nsec to 1μsec are readily available for the study of biomolecules. The sample should be irradiated uniformly with a radiation pulse intense enough to produce an easily detectable concentration of a transient species. The yield of the intermediate species generated by pulse radiolysis depends on peak current, not electron energy; adequate electron energy, however, is required for sufficient penetration of the sample. A 100-nsec pulse of 1 A and 1 MeV produces 6×10^{-6} moles of radicals in a volume of 10 ml of aqueous solution.

2.2. Detection methods

Physicochemical properties of the chemical products which result from pulse radiolysis can be observed by a variety of techniques which include optical absorption or emission; resonance Raman scattering (Dallinger et al., 1979); electron paramagnetic resonance (Fessenden, 1973; Verma and Fessenden, 1973) and electrical conductivity measurements. Optical absorption techniques are most often employed. Two methods of optical detection are used: in one method the absorption of the transient species at a given wavelength after pulse radiolysis is recorded as a function of time with a photodetector (spectrophotoelectric recording), and in another method the transient absorption at a given time is recorded as a function of wavelength with a spectrographic plate (spectrographic recording). The formation and decay of transient species can be studied rapidly and accurately by the method of spectrophotoelectric recording. This approach is more sensitive to small changes in absorption than that of the spectrographic technique. Spectrographic recording, however, is useful to identify transient absorption spectra, although it has the drawback of requiring large sample volumes.

2.2.1. *Spectrophotoelectric recording*

A typical experimental arrangement for spectrophotoelectric recording is shown in Fig. 1. The electron beam is absorbed in an irradiation cell. A source of visible or ultraviolet light provides an analyzing beam which traverses the cell at right angles to the direction of the electron beam. The analyzing beam is directed out of the irradiation area via a system of mirrors and is finally focused on the entrance slit of a monochromator, the exit of which is coupled to a photomultiplier tube. The photomultiplier signals are amplified and displayed in an oscilloscope as voltage vs. time. Changes in voltage are proportional to the photocurrent produced by the analyzing light, thus, the absorption at any time can be calculated from a given change of voltage. Oscilloscope traces are photographed and used to calculate reaction rates. Absorption spectra, at any desired time after the electron pulse are reconstructed from repeated oscilloscope measurements from a series of wavelengths while monitoring the dose given to the solution by each pulse.

The light source for the analyzing beam should be as stable as possible over the period of measurement. Xenon or high-pressure mercury lamps are suitable for this purpose. For the detection of small absorption changes or the measurement of nsecond-resolved spectra, it is necessary to use an efficient optical system and a high-intensity lamp. To achieve sufficiently high light intensities without saturating the photomultiplier, a millisecond pulse of light can be used. Over times of a few hundred μsec, fluctuations in lamp output due to power supply variation become problematic. DC tungsten lamps are especially useful in this regard but their use is limited to the visible and infrared region.

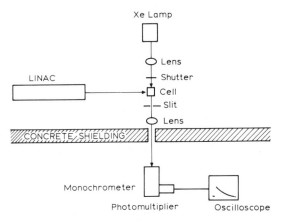

Fig. 1.   Typical pulse radiolysis detection system.

Continuous illumination with light of high intensity causes photochemical changes in some solutions. This effect can be reduced by employing selected filters and a remotely actuated shutter between the lamp and the irradiation cell, which can be synchronized to the arrival of the electron pulse.

A reservoir containing sample solution outside of the irradiation area is connected to a flow cell; the solution in the irradiated cell can thus be replaced by fresh solution before each pulse.

Ionizing radiation from the accelerator can generate noise in the detector system, thus the detecting equipment is kept at some distance from the accelerator and shielded from its radiation. In addition, Shot noise could limit the sensitivity. Keene (1964) pointed out that signal-to-Shot-noise ratio is proportional to the square root of the light flux on the photocathode and to the square root of the circuit rise time. Thus, this Shot noise can be reduced by the use of an efficient optical system and a high intensity lamp or by reducing the bandwidth. The resulting sensitivity is such that changes in absorbance of 5×10^{-5} with a rise time of 1 μsec can be accurately measured.

As the detection system, we employ an R843 photomultiplier, the output of which is monitored by a Tektronix 7104 oscilloscope with 7A29 plug-in. This system allows kinetic experiments with time resolution of up to 700 psec (Sakurai et al., 1982).

Bronskill and Hunt (1968) reported on a new method for extending time resolution capabilities in pulse radiolysis to 20 psec. In this system, Cerenkov light which is generated synchronously with the pulse radiolysis is used as the analyzing light. A stroboscopic effect is created by varying the phase difference between the Cerenkov light and the short-lived electron pulse. In these studies, electron solvation processes were observed (Bronskill et al., 1969).

2.2.2. Spectrographic recording

A trigger pulse which is generated by the electron pulse is passed through a variable delay circuit, and then used to trigger the spectroscopic flash lamp. At a fixed time after the electron pulse, this light flash passes through the cell, and then the spectrum is recorded on spectrographic plate. The time interval between the electron pulse and the spectroscopic flash should be controlled and variable. If a series of these experiments with differing delay times are performed, the time-resolved absorption spectra of the transient species can be obtained. Either photographic film or spectroscopic plates can be used for recording the transient spectra. Recently, image dissectors or optical multichannel analyzers have been employed for spectrographic recording. Automated systems have been developed for recording, storing, reading, and evaluating spectra. These methods, however, are not sensitive to small changes in absorption.

2.2.3. Resonance Raman scattering

Recently time-resolved resonance Raman scattering technique was devised by Dallinger et al. (1979). A laser pulse (7 nsec, 531.8 nm frequency-doubled Nd; YAG) was synchronized with the electron beam to strike the sample when the concentration of the intermediate species is near maximum. The Raman photons were detected using a visicon spectrograph (Woodruff and Farquharson, 1978). In this approach, it becomes necessary to employ detector gating electronics to reject the Cerenkov light.

2.2.4. *Flow radiolysis*
Another interesting approach is the flow radiolysis method (Bielski and Gebicki, 1977). This technique entails rapid mixing of the species generated by the pulse radiolysis with a second solution. The second solution, which contains the appropriate biological compounds, whole cells or microorganisms, is thus protected from the effects of radiation. Only relatively long-lived transients can be studied with this approach. The mechanism of oxidation of NADH (Bielski and Chan, 1973) and the reaction of ascorbate-free radical with a variety of solutes have been studied with this technique (Bielski et al., 1975).

2.3. *Chemical techniques*

The first step resulting from the interaction of high-energy electrons with the system is the production of ions, electrons, and excited species. Since the experiments in biological systems are usually carried out in aqueous solution, the energy is almost exclusively absorbed by the water molecules. Thus, we observe the reaction of solutes with the intermediates formed from water. Direct interactions of the solute, at concentrations of the order 10^{-5} M, with the high-energy electrons are negligible. In aqueous solution, several reactive species are produced as follows:

$$H_2O \xrightarrow{\hspace{1cm}} H\cdot, \ OH\cdot, \ e_{aq}^- \tag{1}$$

These are atomic hydrogen, hydroxyl-free radical and the hydrated electron (e_{aq}^-). The yields of water products from radiolysis are conveniently expressed in terms of the G value, defined as the number of molecules or radicals per 100 eV of energy absorbed. Generally accepted values in water irradiated at neutral pH are as follows:

$$G(\cdot OH) = 2.8, \ G(e_{aq}^-) = 2.7, \ G(\cdot H) = 0.7$$

Among these species, the reaction of an individual free radical and a solute can be isolated under appropriate conditions. For example, in order to investigate the reaction of e_{aq}^- with a solute we add 0.1 M *tert*-butylalcohol to the solution reaction 2) (Dorfman and Adams, 1973).

$$\cdot OH + CH_3-\underset{\underset{CH_3}{|}}{\overset{\overset{CH_3}{|}}{C}}-OH \longrightarrow H_2O + \cdot CH_2-\underset{\underset{CH_3}{|}}{\overset{\overset{CH_3}{|}}{C}}-OH \tag{2}$$

$$k = (4 - 6) \times 10^8 \text{ M}^{-1} \text{ sec}^{-1}$$

The OH radical reacts with the *tert*-butylalcohol within 1 μsec under the experimental conditions and yields an unreactive radical. The production of hydrogen radical is small compared with that of e_{aq}^-. The e_{aq}^- has a broad absorption with a peak near 700 nm and is the most potent reductant known ($E^0 \leqslant 2.67$ V, $e_{aq}^- + H_3O^+ \rightarrow \frac{1}{2} H_2 + H_2O$) (Matheson and Dorfman, 1969). Therefore the kinetics of e_{aq}^- as the reductant with a solute can be measured directly by optical detection methods.

In order to isolate the reaction of $\cdot OH$ with a solute, the solution is saturated with N_2O. N_2O reacts with e_{aq}^- to form the hydroxyl radical (Anbar et al., 1973).

$$e_{aq}^- + N_2O \longrightarrow N_2 + OH^- + \cdot OH \tag{3}$$

$$k = (5 - 8) \times 10^9 \text{ M}^{-1} \text{ sec}^{-1}$$

The electron is converted into the $\cdot OH$ radical. Thus in N_2O-saturated water, less than 10 nsec after irradiation, $\cdot OH$ constitutes about 90% of the radical.

In oxygenated aqueous solution, the reducing species e_{aq}^- and H$\cdot$ react with oxygen to produce the superoxide radical. This technique provides a tool for the generation of O_2^- in a very short time (Behar et al., 1970).

$$e_{aq}^- + O_2 \rightarrow O_2^- \qquad k = 2 \times 10^{10} \text{ M}^{-1} \text{ sec}^{-1} \tag{4}$$

$$H \cdot + O_2 \rightarrow O_2^- \qquad k = 2 \times 10^{10} \text{ M}^{-1} \text{ sec}^{-1} \tag{5}$$

$$HO_2 \cdot \underset{\substack{\longleftarrow \\ pK = 4.8}}{\overset{\longrightarrow}{}} O_2^- + H^+ \tag{6}$$

3. Applications of pulse radiolysis to biochemistry

By virtue of its ability to generate e_{aq}^- as a reductant, the pulse radiolysis technique is well suited to investigate a number of biological processes. For example, assume that the reduction of a redox protein occurs as shown in Eqn. 7,

$$E_{ox} \xrightarrow{k_{7a}} \text{“}E_{red}\text{”} \xrightarrow{k_{7b}} E_{red} \tag{7}$$

where E_{ox} is the oxidized form of the protein, "E_{red}" is the unstable intermediate after reduction, i.e., non-equilibrium conformation of the protein structure, and E_{red} is the final product after reduction. In this scheme it is difficult to separate these two steps. The difficulty stems from the fact that the reduction of E_{ox} by the chemical reductants is the rate-determining step in reaction 7 ($k_{7b} \ll k_{7a} [E_{ox}]$). In contrast, e_{aq}^- reduces E_{ox} rapidly within the μsec time scale. Thus, one might expect to observe the unstable intermediates "E_{red}" after reduction and the subsequent reaction steps from "E_{red}" to E_{red}. For example, slower conformational changes following the rapid reduction of the heme iron in cytochrome c (Pecht and Faraggi, 1972; Wilting et al., 1974; Land and Swallow, 1974) and met Hb (Wilting et al., 1974; Raap et al., 1977a, b; Van Leeuwen et al., 1981) have been detected, where the rate constants for the conformational changes of cytochrome c and met Hb were 1.3×10^5 and 4.6×10^4 sec^{-1} respectively. The release of heme-bound water from met Hb and met Mb after reduction has been observed. The rate constant of this reaction at pH 6 was 2.8×10^3 sec^{-1} (Wilting et al., 1974). In the presence of a high concentration of methanol (1 M), the hemochrome spectrum characteristic of the ferrous low-spin state was observed after reduction of met Hb, but not after reduction of met Mb. This hemochrome spectrum decayed to the deoxy spectrum with a rate constant of 1.6×10^4 sec^{-1} at pH 7.0 (Raap et al., 1977a, b; Van Leeuwen et al., 1981). Other interesting examples are examined in sections 3.1, 3.2 and 3.3.

This technique is also useful for investigating electron transfer reactions in a variety of biological systems. These include proteins containing multiple electron-accepting sites per molecule such as cytochrome oxidase and laccase, and complexes of two different proteins such as cytochrome c – cytochrome oxidase. In these systems it can be anticipated that e_{aq}^- reacts with the primary redox site and subsequently flows to other electron-accepting sites by inter- and intramolecular migration until equilibrium is reached, as in Eqn. 8. Typical examples are presented in section 3.4.

$$\text{\textcircled{E_{1ox}}\textcircled{E_{2ox}}} \xrightarrow{e_{aq}^-} \text{\textcircled{E_{1red}}\textcircled{E_{2ox}}} \rightleftharpoons \text{\textcircled{E_{1ox}}\textcircled{E_{2red}}} \qquad (8)$$

Another interesting application of pulse radiolysis is the reaction of the superoxide anion O_2^- with biomolecules. A typical case is discussed in section 3.5. The recent finding of the enzymatic and non-enzymatic generation of O_2^- and its scavenging enzyme, superoxide dismutase, in living organisms may help explain the molecular mechanism of oxygen toxicity, and also the oxygen effect observed in radiation biology (McCord and Fridovich, 1968; McCord et al., 1971).

3.1. One-electron reduction in oxyform of hemoproteins

Horseradish peroxidase (HRP), which contains iron(III) protoporphyrin, catalyzes the oxidation of a wide variety of hydrogen donors by H_2O_2. The widely accepted

94

mechanism of HRP reaction, shown in Eqns. 9–12 was derived principally from the work of Chance (1943), and Yamazaki and Piette (1961).

$$HRP \qquad + H_2O_2 \rightarrow compound\ I \qquad\qquad (9)$$

$$Compound\ I\ + AH_2 \rightarrow compound\ II + AH\cdot \qquad\qquad (10)$$

$$Compound\ II + AH_2 \rightarrow HRP + AH\cdot \qquad\qquad (11)$$

$$2AH\cdot \qquad \rightarrow A + AH_2 \qquad\qquad (12)$$

The AH_2 is hydrogen donor, such as ascorbate, phenol and aromatic amines. HRP is a unique enzyme in that it possesses five oxidation-reduction states, which are ferrous, ferric, compound I, compound II and compound III (oxyperoxidase).

Yamada and Yamazaki (1974) have proposed a general peroxidase mechanistic scheme. Fig. 2 shows the relationships among the five oxidation-reduction states of HRP. In this scheme, it has been thought that oxyperoxidase is at one equivalent oxidized state above compound I (Tamura and Yamazaki, 1972). The 1-electron reduction of oxyperoxidase to compound I presumably occurs according to

$$Fe^{2+} - O_2 \rightarrow compound\ I \qquad\qquad (13)$$

where it has been proposed that compound III is an oxyform structure like oxymyoglobin (Yokota and Yamazaki, 1965; Wittenberg et al., 1967), and compound I is a ferryl π cation radical (Dolphin et al., 1971; La Mar et al., 1981). Reaction 13, however, has not yet been observed directly. This difficulty stems from the

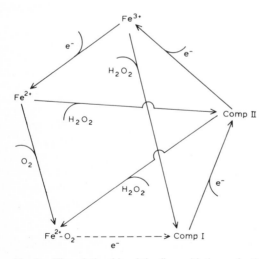

Fig. 2.   The relationship of the five oxidation-reduction states of HRP.

fact that the reaction products, compound I and compound II, further react with the reducing agents and oxyperoxidase (Tamura and Yamazaki, 1972). We therefore undertook an investigation of the reaction of e^-_{aq} with the oxyforms of hemoproteins such as oxyperoxidase and oxymyoglobin (MbO_2) by the pulse radiolysis method (Kobayashi and Hayashi, 1981). As the native oxyperoxidase is unstable, the oxyform of an artificial peroxidase, in which the 2,4-diacetyldeuterohemin replaced the protohemin IX of natural peroxidase, was employed in these experiments. The oxyform of this peroxidase is very stable and can be kept at room temperature for a few hours (Makino et al., 1976).

We found that the oxyform of diacetyldeuteroperoxidase was reduced by e^-_{aq} to form compound I of this enzyme, which was demonstrated by the kinetic difference spectra shown in Fig. 3. Similarly MbO_2 was reduced by e^-_{aq} to form the hydrogen peroxide-induced compound, so-called "ferryl" Mb, not deoxy Mb or met Mb, which in turn was established by the kinetic difference spectra shown in Fig. 4. Both reactions were found to follow second-order kinetics with rate constants of 4×10^{10} M^{-1} sec^{-1} at pH 7.4, without detection of intermediates. That is to say, e^-_{aq} apparently reduces the redox side of oxyheme in a direct reaction. These reactions, however, can be interpreted in terms of the following sequence of events:

$$e^-_{aq} + Fe^{2+} - O_2 \xrightarrow[k_{14a}]{} Fe^{2+} - O^-_2 \xrightarrow[k_{14b}]{} \begin{array}{l} \text{ferryl Mb} \\ \text{Compound I} \end{array} \tag{14}$$

An initial step is the formation of $Fe^{2+}-O^-_2$, followed by an intramolecular electron transfer from the heme iron to the ligand forming $Fe^{3+}-O^{2-}_2$ or $Fe^{4+}-O^{3-}_2$. Subsequent rapid association of H^+ and loss of water would then lead to the observed final product. The rate constant k_{14b} would be greater than 10^7/sec, thus the reac-

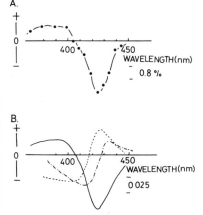

A.

B.

Fig. 3. (A) Kinetic difference spectrum of pulse radiolysis of oxydiacetyldeuteroperoxidase. The spectrum is taken at 8 μsec after pulse radiolysis. (B) Difference spectra of oxydiacetyldeuteroperoxidase minus ferric peroxidase (. . . .), compound I (——), and compound II (-· · ·-).

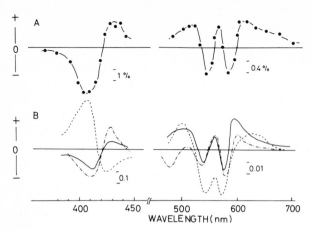

Fig. 4.   (A) Kinetic difference spectrum of pulse radiolysis of MbO$_2$. The spectrum is taken at 8 μsec after pulse radiolysis. (B) Difference spectra of MbO$_2$ minus met Mb (– – –), deoxy Mb(– · – · –), and hydrogen peroxide-induced compund (——).

tion of e^-_{aq} with oxyhemoproteins is the rate-limiting step in the overall process.

The reaction of oxyheme with e^-_{aq} is analogous to that proposed for the oxidation of substrate by cytochrome P-450 (Tyson et al., 1972; Ishimura et al., 1971), and for hydroxylation of the α-methene bridge of heme in the heme oxygenase reaction (Yoshida and Kikuchi, 1978; Yoshida et al., 1980). The identity of the 1-electron reduction state in these oxygenated heme enzymes has attracted much attention. From our results, it is suggested that the 1-electron reduction state of oxygenated cytochrome P-450 and heme oxygenase may also be the ferryl state of these enzymes.

3.2.   One-electron reduction in flavoproteins

Flavoproteins, which contain the flavin coenzymes FAD or FMN, catalyze redox reactions during which either one or two electrons from the electron donor are transferred transiently to the isoalloxazine nucleus of the flavin coenzyme and then to the electron acceptor. It is well known that a 1-electron reduction of these flavoproteins produces a stable free radical of the flavin molecule, which may be protonated (the blue semiquinone) or unprotonated (the red semiquinone) in the physiological pH range. The spectra and the structures of the two radicals are shown in Fig. 5. Most flavoproteins form only blue or red semiquinones of the radical species independently of the external pH (Massey and Palmer, 1966; Massey et al., 1969). One exception is the semiquinone of the flavin of glucose oxidase, which changes from the blue to the red semiquinone (Stankovich et al., 1978). Land and Swallow (1969), and Faraaggi and Pecht (1973) have demonstrated some of the advantages of the pulse radiolysis technique for determining the spectral and kinetic behavior of 1-electron reduction products of flavin. Employing pulse radiolysis,

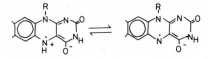

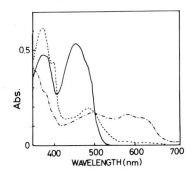

Fig. 5. Structure and optical absorption spectra of the blue semiquinone (– · – · –), the red semiquinone (– – – –) and the oxidized state (——) of FAD bound to D-amino acid oxidase.

Meisel and Neta (1975) obtained the redox potentials of riboflavin by determining both the equilibrium constants and the kinetics of approach to equilibrium of reaction 15:

$$H_2Rf\cdot + A \rightleftharpoons HRf + A^{\bar{\cdot}} + H^+ \tag{15}$$

where $H_2Rf\cdot$ is the semiquinone of riboflavin and A is the electron acceptor, such as duroquinone or 9,10-anthraquinone-2 sulfonic acid.

The spectral characteristics of the blue semiquinone (FADH·) appear in the reaction of e_{aq}^- with FAD at acidic pH. This indicates that in the reduction of FAD the production of the red semiquinone (FAD·) is followed by immediate protonation and rapid establishment of equilibrium between the blue and the red species, i.e.,

$$FAD \xrightarrow[e_{aq}^-]{} FAD\cdot \overset{H^+}{\rightleftharpoons} FADH\cdot \tag{16}$$

In contrast to protein-free FAD, the formation of the semiquinones of flavoproteins is more complex. The following section deals with the 1-electron reduction of flavoproteins such as flavodoxin and D-animo acid oxidase.

3.2.1. Flavodoxin

Flavodoxin, which contains one equivalent FMN, serves as an electron carrier between several oxidation-reduction proteins. This protein appears to function

98

physiologically in a 1-electron cycle involving the blue semiquinone. The reaction kinetics of fully oxidized flavodoxin with e_{aq}^- employing pulse radiolysis techniques were investigated by Faraggi and Klapper (1979). In this study, four spectrally distinct processes with ultimate formation of the blue semiquinone were observed. The spectral differences among these species were not as drastic as would be expected from a change in protonation of the flavin. Faraggi et al. (1975) proposed that these transitions reflect conformational changes leading to the stable blue semiquinone.

3.2.2. D-Amino acid oxidase

D-Amino acid oxidase (DAAO) catalyzes the oxidation of D-amino acid to yield pyruvate and NH_4^+. The semiquinone form of DAAO is the red species in the whole pH range, however, it is converted into the blue species upon formation of the enzyme–benzoate complex (Yagi et al., 1972). Here, benzoate is the competitive inhibitor of this enzyme. The reaction kinetics of DAAO with e_{aq}^- have been investigated both in the absence and presence of benzoate (Kobayashi et al., 1983). The e_{aq}^- did not reduce the flavin moiety in DAAO, and reacted with the amino acid residues in the protein; e.g., aromatic amino acids and cysteine (Adams, 1972). The transient spectrum obtained after 1 μsec of the pulse exhibits a peak near 400 nm (Fig. 6); this spectrum may be due to the radical ions of cysteine and histidine. It is known that this enzyme contains six sulfhydryl groups per protein (Fonda and Anderson, 1969). In contrast to DAAO, the FMN of the flavodoxin can be reduced by e_{aq}^- (Faraggi and Klapper, 1979), as mentioned above. This difference can be explained by the fact that in the flavodoxin case the isoalloxazine ring is found at the periphery of the molecule with the dimethylbenzene end accessible to solvent, as has been suggested by X-ray analysis (Burnett et al., 1974). This reduction is assigned mainly to a direct reaction proceeding via the exposed edge of the isoalloxazine ring. The isoalloxazine of DAAO is considered to be masked by the protein moiety and is not directly exposed to the solvent. On the other hand, in the presence of excess benzoate, the e_{aq}^- first reacts with benzoate to yield benzoate anion radical, which as an absorption maximum at 450 nm (Sangster, 1966). Subsequently, the benzoate anion radical reduces the DAAO-benzoate complex to form the red semiquinone with a rate constant of 2 $\times$ 10^9 M^{-1} sec^{-1} at pH 8.3. Since the value of the second-order rate constant is

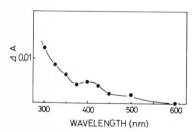

Fig. 6.   Primary transient spectra from reaction of e_{aq}^- with DAAO at pH 8.3.

near that expected for a diffusion-controlled process, this reaction cannot be attributed to the exchange between benzoate bound to the enzyme and benzoate anion radical as depicted in Eqn. 17.

$$
\begin{bmatrix} FAD \\ Bz \end{bmatrix} + Bz^{\overline{\cdot}} \rightarrow \begin{bmatrix} FAD \\ Bz^{\overline{\cdot}} \end{bmatrix} + Bz \rightarrow \begin{bmatrix} FAD\cdot \\ Bz \end{bmatrix} \tag{17}
$$

Similar rates for the reaction of the benzoate anion radical with cytochrome c and met Mb have been obtained (Simic and Taub, 1978). In this case, it was proposed that interactions between the donors and the porphyrin ring play an important role. In the case of DAAO, the mechanism for electron transfer from benzoate anion radical to FAD is not clear at present.

The remarkable finding of our work is that the red semiquinone is found transiently in the DAAO–benzoate complex, and is subsequently converted to the blue species, as evidenced by the difference spectra of Fig. 7. The formation of the blue semiquinone obeyed first-order kinetics. This rate constant increased with increasing benzoate concentration. This suggests that this process is due to complex formation between the red semiquinone and free benzoate with a rate constant of $5.6 \times 10^2 \ M^{-1}sec^{-1}$ at pH 8.3. The reaction paths may be schematized as follows:

$$
Bz + e_{aq}^{-} \rightarrow Bz^{\overline{\cdot}} \tag{18}
$$

$$
Bz^{\overline{\cdot}} + \begin{bmatrix} FAD \\ Bz \end{bmatrix} \underset{(i)}{\rightarrow} \begin{bmatrix} FAD\cdot \\ Bz \end{bmatrix}
$$

$$
(ii) \Big\downarrow Bz \tag{19}
$$

$$
\begin{bmatrix} FADH\cdot \quad H^+ \\ Bz \end{bmatrix} \underset{(iii)}{\longleftarrow} \begin{bmatrix} FAD\cdot \\ Bz \end{bmatrix}
$$

In this scheme we were unable to detect from absorption changes of the enzyme, process (ii), in which the bound benzoate dissociates after reduction.

3.3. Kinetics of oxygen and carbon monoxide binding to partially reduced methemoglobin

It is well known that oxygen binding to Hb is characterized by cooperativity. The binding of O_2 to one subunit increases the O_2 affinities to the remaining subunits. Fully ligated Hb exists in a high-affinity state and the unligated Hb exists in a low-

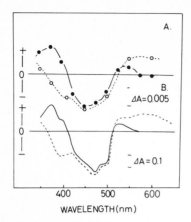

WAVELENGTH(nm)

Fig. 7. (A) Kinetic difference spectra of pulse radiolysis of DAAO. The spectra are taken at 100 μsec (–•–•–) and 20 msec (––o––o––) after pulse radiolysis. (B) Difference spectra of fully oxidized DAAO minus the red semiquinone (——) and the blue semiquinone (– – –).

affinity state. The Hb tetramer can be described as composed of subunits which can exist in either of two quaternary structures, a high-affinity (R state) or a low-affinity one (T state). During the ligation process, a transition between these two structures occurs.

Until recently, kinetics of the binding of CO or O_2 to deoxy Hb have been studied by rapid-mixing and flash photolysis techniques. A new approach to the study of CO and O_2 binding was offered by pulse radiolysis (Ilan et al., 1976, 1978; Chevion et al., 1979; Rollema et al., 1976). In this approach, met Hb is irradiated in the presence of CO or O_2. Within a μsec the e_{aq}^- first reduces the ferric heme of met Hb to yield valency-hybrid Hb molecules in which one subunit per tetramer is ferrous heme and the other subunits are ferric heme. The fraction of reduced met Hb is kept low (< 0.03) so that the concentration of Hb molecules with two reduced heme groups can be neglected. Subsequently, CO or O_2 binds to the singly reduced subunit within the tetramer as shown in Eqn. 20.

$$Hb[(Fe^{3+})_4] \xrightarrow[e_{aq}^-]{} Hb[(Fe^{3+})_3(Fe^{2+})] \begin{array}{c} \xrightarrow[O_2]{} Hb[(Fe^{3+})_3(Fe^{2+} - O_2)] \\ \xrightarrow[CO]{} Hb[(Fe^{3+})_3(Fe^{2+} - CO)] \end{array} \qquad (20)$$

The second-order rate constants for the binding of O_2 and CO obtained by the use of this method are 2.3×10^7 $M^{-1}sec^{-1}$ and 5.1×10^6 $M^{-1}sec^{-1}$ at pH 6.8 respectively. It is interesting to compare these values with those obtained for the first oxygen molecule which binds to Hb (T state) in reaction 21 and the fourth one which binds to Hb (R state) in reaction 22 (Ilgenfritz and Schuster, 1974).

$$Hb[(Fe^{2+})_4] + O_2 \rightarrow Hb[(Fe^{2+})_3(Fe^{2+} - O_2)_4] \tag{21}$$

$$k_{21} = 9 \times 10^6 \ M^{-1} \ sec^{-1}$$

$$Hb[(Fe^{2+} - O_2)_3 Fe^{2+}] + O_2 \rightarrow Hb[(Fe^{2+} - O_2)_4] \tag{22}$$

$$k_{22} = 4 \times 10^7 \ M^{-1} \ sec^{-1}$$

The rate constant for the oxygentation in process 20 is about 3 times higher than k_{21}, and rather similar to the value of k_{22}. On this basis it is concluded that one-site-reduced met Hb is in an R quaternary structure. This is also supported by the kinetic difference spectrum of deoxy Hb (T state) minus met Hb (Rollema et al., 1976). A similar intermediate spectrum for CO-Hb was obtained employing a laser photolysis method by Sawicki and Gibson (1976). The effects of pH, inositolhexaphosphate, and temperature on these reactions have also been examined (Ilan et al., 1978).

The kinetics of CO binding to hemoglobin following fast reduction of the valency hybrids ($\alpha_2^+ \beta_2^{CO}$) and $\alpha_2^{CO} \beta_2^+$ by e_{aq}^- have been investigated (Rollema et al., 1976) at different degrees of reduction. The singly reduced subunits of these valency-hybrid Hb reacted rapidly with CO at a rate characteristic of the R state ($k = 4 \times 10^6$ $M^{-1}sec^{-1}$). As the degree of reduction increased, biphasic binding kinetics were observed, reflecting the presence of fast ($k = 4 \times 10^6 \ M^{-1}sec^{-1}$) and slow reacting forms ($k = 4.1 \times 10^5 \ M^{-1}sec^{-1}$), which correspond to the R and T conformations respectively. From these results the following reaction schemes were proposed,

$$\alpha_2^+ \ \beta_2^{CO} \xrightarrow{e_{aq}^-} \alpha_2\beta_2^{CO}(R^*) \begin{cases} \longrightarrow \alpha_2\beta_2^{CO}(R) \xrightarrow[CO]{} \alpha_2^{CO}\beta_2^{CO} \\ \longrightarrow \alpha_2\beta_2^{CO}(T) \xrightarrow[CO]{} \alpha_2^{CO}\beta_2^{CO} \end{cases} \tag{23}$$

where R* represents the intermediate species immediately after reduction. From this scheme the following conclusions were derived. At pH 6.0, the intermediates $\alpha_2^{CO} \beta_2^+$ and $\alpha_2^+ \beta_2^{CO}$ bind CO with a rate characteristic of the T state, while in the case of $\alpha_2^{CO} \beta_2^+$ at pH 7.0 the R and T state are equally populated. A similar observation has been reported for cyanomet valency hybrids by Cassoly and Gibson (1972).

3.4. The electron pathway of redox proteins having multiple electron-accepting sites

3.4.1. Cytochrome oxidase

Cytochrome oxidase, which lies in the mitochondrial inner membrane catalyzes the 4-electron reduction of molecular oxygen to 2 molecules of water. It has been established that each of the four electron-accepting sites per molecule of the protein, two heme a and two copper atoms, represent unique environments. Furthermore, the two heme a components are distinguishable on the basis of their respective ligand-binding properties.

The reaction of cytochrome oxidase with e_{aq}^- has been investigated in the presence or absence of cytochrome c (Van Buuren et al., 1974; Van Gelder et al., 1979). It was found that e_{aq}^- does not readily reduce the heme of cytochrome oxidase, which suggests that this heme is not directly exposed to the solvent. Another possibility would be that e_{aq}^- is repulsed by the negative charge on cytochrome oxidase, since the entrance to the electron path in the oxidase is surrounded by negative charge, which attracts the positive charge on cytochrome c. Furthermore in the complex of cytochrome oxidase and cytochrome c, neither heme a of cytochrome oxidase nor ferric heme of cytochrome c is significantly reduced by e_{aq}^-. These facts indicate that cytochrome c which is bound to cytochrome oxidase tightly, has its reduction site completely masked.

On the other hand, with excess of cytochrome c, the free cytochrome c is first reduced by e_{aq}^-, and then it transfers an electron to heme a in the cytochrome oxidase–cytochrome c complex with rate constants k_{on} and k_{off} of $2 \times 10^7 \, M^{-1}sec^{-1}$ and $50 \, sec^{-1}$ respectively at pH 7.0. It is interesting that the values of these rate constants are similar to those obtained by the rapid-mixing methods between ferrous cytochrome c and cytochrome oxidase (Gibson et al., 1965). These findings demonstrate that the rate constant is independent of whether cytochrome c is bound to cytochrome oxidase or not. In other words, cytochrome oxidase possesses another binding site for cytochrome c; Van Gelder et al. (1979) concluded that this site is the primary electron-accepting site on this enzyme. Recently, studies of steady-state kinetics of cytochrome oxidase have revealed two binding sites for cytochrome c on cytochrome oxidase, which are a high-affinity site ($K_m \sim 50 \, mM$) and a low-affinity site ($K_m \sim 1 \, \mu M$) (Ferguson-Miller et al., 1976; Errede et al., 1976).

In the reaction of the reduced cytochrome c with cytochrome oxidase, two equivalents of cytochrome c are oxidized per equivalent of heme a reduced. This fact indicates that another electron acceptor in cytochrome oxidase, probably one of the copper atoms, takes up an electron concomitant with the reduction of heme a, as in Eqn. 24.

$$2 \text{ cytochrome } c \text{ (Fe}^{2+}) + \text{heme a (Fe}^{3+})Cu^{2+}$$

$$\to 2 \text{ cytochrome } c \text{ (Fe}^{3+}) + \text{heme a (Fe}^{2+})Cu^{1+}$$

$$(24)$$

It has also been found that the EPR signal assigned to Cu^{2+} is diminished rapidly in reduction experiments (Beinert et al., 1976). Whether heme and copper are reduced simultaneously or consecutively cannot be decided from these experiments.

After initial reduction, a slow absorption change was observed with a rate constant of the order of $1 \, sec^{-1}$. Van Buuren et al. (1974) proposed that this slow change was due to transfer of electrons from the second electron acceptor to the heme until an equilibrium is reached.

3.4.2. Copper oxidases

In copper oxidases, Cu^{2+} ion oxidizes the substrate by accepting an electron, and the Cu^{1+} thus formed is then reoxidized by molecular oxygen to a water molecule. For example, laccase, a copper oxidase which contains four copper ions per molecule of protein, catalyzes the electron transfer from p-diphenols and related reductants to molecular oxygen (Malmström et al., 1968). Ceruloplasmin, which contains eight copper ions per molecule of protein (Deinum and Vanngard, 1973), catalyzes the electron transfer from ferrous ion or a variety of other electron donors to molecular oxygen. The copper atoms in these proteins are bound to three distinct sites with the following characteristics: type 1, "blue" copper, absorbing around 610 nm, very small hyperfine splitting in its EPR signal and marked positive Cu(II)/Cu(I) redox potential ($E_o' = 0.78$ V); type 2, similar to that generally found for simple Cu^{2+} complexes and ability to interact with a number of anions such as N_3^- and F^-; type 3, the EPR non-detectable copper ions.

Pecht and Faraggi (1971, 1972) applied the pulse radiolysis technique to the study of these blue copper oxidases. The e_{aq}^- was found to reduce type 1 Cu^{2+} in laccase and ceruloplasmin indirectly by an intramolecular electron transfer from the primary electron adducts. In contrast to type 1 Cu^{2+}, no absorption changes could be observed at the 330 nm band, which is associated with the two electron-accepting sites of laccase (Malkin et al., 1969). These results show that the type 1 Cu^{2+} of laccase appears to accept electrons faster than the other sites, and are compatible with the anaerobic reduction of the copper oxidase by the substrate.

In the presence of F^- the reduction of type 1 Cu^{2+} appeared unaffected, but a transient absorption from 400 to 300 nm appeared and then decayed in a second-order process (Pecht and Faraggi, 1971). It is known that the coordination environment of type 2 Cu^{2+} changes upon the binding of F^- concomitantly with the decrease of redox potential for type 3 Cu^{2+} from 782 mV to 570 mV (Reinhammar, 1972). However, it is not clear that this absorption change is due to the reduction of type 2 or type 3 Cu^{2+}.

The reoxidation of the reduced blue oxidase by molecular oxygen has been studied by the same technique (Goldberg and Pecht, 1978). In this study, two distinct processes of reduction and reoxidation of type 1 copper were observed; the reoxidation was strongly inhibited by F^-. The reduction was attributed to the reaction of type 1 Cu^{2+} with e_{aq}^- or O_2^-, and the reoxidation to the reaction of type 1 Cu^{1+} with molecular oxygen. This oxidation process, however, was independent of O_2 concentration ($k = 8.3 \pm 0.4 \, \text{sec}^{-1}$), and occurred extremely slowly compared to the process that results from the rapid mixing of the reduced enzyme with O_2 ($k - 9.6 \times 10^3 \, \text{M}^{-1}\text{sec}^{-1}$) (Malmström et al., 1968). These differences can be explained as follows: in the pulse radiolysis method, the type 1 copper is reduced and the other ones are oxidized, whereas in the second case the electron-accepting sites of the copper oxidase are fully reduced. Goldberg and Pecht (1978) proposed the following scheme for the reoxidation process:

$$\begin{array}{ccccc}
\text{Type 1 Cu}^{2+} & \text{Type 1 Cu}^{1+} & k_{25a} & \text{Type 1 Cu}^{2+} & k_{25b} & \text{Type 1 Cu}^{2+} \\
\text{X(ox)} & \xrightarrow{\quad} \text{X(ox)} & \rightleftharpoons & \text{X(red)} & \rightleftharpoons & \text{X(red)} \cdot O_2 \quad (25) \\
& e_{aq}^- & k_{25\text{-}a} & & k_{25\text{-}b} &
\end{array}$$

X = type 2 or type 3. In this scheme, e_{aq}^- reacts with only type 1 Cu^{2+}, not type 2 or type 3. Subsequently, O_2 reacts with a solvent-accessible redox site (type 2 site), which is in a state of electron exchange equilibrium with the type 1 site. The rate-determining step in this reaction scheme is the intramolecular electron transfer from the type 1 site to the site of O_2 interaction ($k_{ox} = k_{25a} << k_{25b} [O_2]$9. This interpretation is in agreement with the steady-state experiments (Malmström et al., 1968; Carrico et al., 1971).

3.5. Superoxide and superoxide dismutase

3.5.1. The reaction O_2^- with ferrihemoproteins

The reactions between superoxide anion, O_2^-, and ferrihemoproteins have received much attention from the viewpoint of oxygen toxicity (McCord and Fridovich, 1968; McCord et al., 1971). It is of particular interest to determine whether O_2^- reacts with the ferric state (Fe^{3+}) to form the oxyform (Fe^{2+}–O_2) directly; i.e.,

$$Fe^{3+} + O_2^- \rightarrow Fe^{2+} - O_2 \tag{26}$$

The presence of such reactions has already been suggested for HRP (Yamazaki and Piette, 1963) and more recently for indoleamine deoxygenase(Hirata et al., 1977). This question is also important in regard to oxygen-carrying hemoproteins such as Hb and Mb, since O_2^- is generated from their oxyforms during autooxidation (Misra and Fridovich, 1972a, b; Wallace et al., 1974). Thus the possible existence of reaction 26 gives these hemoproteins the additional role of providing protection against O_2^- in situ.

Hayashi et al. (1978) have studied the time course of the absorption change of ferri-HRP and met Mb by irradiation in low-oxygen (10^{-5} M) and oxygen-saturated conditions. In low-oxygen conditions two distinct steps are observed in the absorption change, an initial rapid one followed by a slow change. The difference spectra obtained at 20 and 90 μsec after irradiation are shown in Fig. 8A. The spectrum in Fig. 8B is the difference between the spectra obtained 10 and 200 μsec after the pulse and shows the intermediate responsible for the slow absorption change. The 420-nm absorption can be identified as the oxyform (Fe^{2+}–O_2) of HRP. This is verified by the difference spectrum between the oxyform minus ferric state shown in Fig. 8C. Thus at low-oxygen level, the reduction of ferri-HRP and the formation of the oxyform is observed. The rate constant of reaction 26 is estimated to be 10^8 $M^{-1}sec^{-1}$, from which the following alternate possibilities can be excluded.

$$Fe^{3+} + O_2^- \rightarrow Fe^{2+} + O_2, Fe^{2+} + O_2 \rightarrow Fe^{2+} - O_2 \tag{27}$$

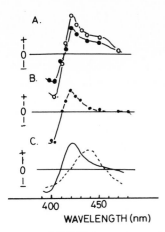

A.

B.

C.

400   450
WAVELENGTH (nm)

Fig. 8.   (A) Difference spectra obtained at 20 μsec (–●–●–) and 90 μsec (–o–o–) after the pulse. (B) Difference spectra between 10 and 200 μsec after the pulse. (C) Difference spectra of the ferrous minus ferric state (– – – –) and oxyform minus ferric state (——).

$$Fe^{3+} + e^-_{aq} \rightarrow Fe^{2+}, \qquad Fe^{2+} + O_2 \rightarrow Fe^{2+} - O_2 \tag{28}$$

In both cases of reactions 27 and 28, the rate-determining step is the combination reaction of ferro HRP with oxygen ($Fe^{2+} + O_2 \rightarrow Fe^{2+}$–$O_2$; $k = 4 \times 10^4\ M^{-1}sec^{-1}$ (Tamura and Yamazaki, 1972). Sawada and Yamazaki (1973) have obtained a rate constant of $3 \times 10^7\ M^{-1}sec^{-1}$ for reaction 26 using an O_2^--generating system, which is in good agreement with the results of Hayashi et al. (1978).

Under oxygen-saturated conditions, ferri-HRP is converted to the hydrogen peroxide compound, compound I. The time course of this absorption change is relatively slow in the millisecond time range. Similar results were obtained by Bilski (1972). The appearance of compound I shows the formation of H_2O_2 in the solution, which can be generated by the disproportionation reaction of O_2^-, as shown by reactions 29 and 30.

$$2O_2^- \xrightarrow{\ 2H^+\ } H_2O_2 + O_2 \tag{29}$$

$$H_2O_2 + Fe^{3+} \longrightarrow compound\ I \tag{30}$$

Here, the rate-determining step in the formation of compound I is the disproportionation reaction of O_2^-. This is supported by the fact that, under similar conditions, the rate of decay of O_2^-, in the absence of ferri-HRP, observed at 280 nm, is similar to that of the formation of compound I.

With met Mb, however, the formation of the oxy compound could not be observed by the pulse radiolysis study. The appearance of the "ferryl compound" suggests that O_2^- is very unreactive with met Mb, and that its disappearance is the result of interactions with one another.

3.5.2. Superoxide dismutase

A number of biological reactions in aerobic organisms have proposed to involve the generation of superoxide (Bray et al., 1964: Massay et al., 1969; Misra and Fridovich, 1972a, b; McCord and Fridovich, 1968). The O_2^- is potentially hazardous to living matter. McCord and Fridovich (1968) identified, isolated and named superoxide dismutase (SOD) whose function is the enzymatic dismutation of the superoxide anion. This enzyme is widely distributed in all oxygen-metabolizing cells; it contains two copper and two zinc ions in eukaryotes and manganese or iron ions in prokaryotes (Yost and Fridovich, 1974; Steinman and Hill, 1973). The net catalytic reaction of this enzyme is:

$$2O_2^- + 2H^+ \xrightarrow[\text{SOD}]{} O_2 + H_2O_2 \tag{31}$$

The disappearance of O_2^- followed at 250 nm, obeys pseudo-first-order reaction kinetics because of the high O_2^- concentration relative to enzyme. The second-order rate constant is estimated to be $2 \times 10^9 \text{ M}^{-1}\text{sec}^{-1}$ (Rotilio et al., 1972). The catalytic reaction 31, however, consists of two reaction steps:

$$Cu^{2+} + O_2^- \longrightarrow Cu^{1+} + O_2 \tag{32}$$

$$Cu^{1+} + O_2^- \xrightarrow{2H^+} Cu^{2+} + H_2O_2 \tag{33}$$

Klug-Roth et al. (1973) have given values for the second-order rate constants for reactions 32 and 33 as $(1.2 \pm 0.2) \times 10^9 \text{ M}^{-1}\text{sec}^{-1}$ and $(2.2 \pm 0.5) \times 10^9 \text{ M}^{-1}\text{sec}^{-1}$ respectively. These were obtained by following absorption changes et 650 nm where the copper of SOD has maximum absorption. Employing EPR spectroscopy in conjunction with pulse generation of O_2^- and rapid-freezing techniques (Fielden et al., 1974; Bray et al., 1973, 1974), it was found that the EPR signal of the Cu^{2+} of SOD decreased in the reaction of SOD with O_2^- (Rotilio et al., 1972). Fielden et al. (1974) concluded from the absorbance at 680 nm or the EPR signal of Cu^{2+}, that after obtaining a steady state by repeated pulses, only half of the copper atoms participate in turnover. The reaction on one of two initially identical copper ions, Cu^{2+} or Cu^{1+}, renders the other transiently non-reactive toward O_2^- according to the following scheme:

$$\boxed{\begin{array}{c} R\text{–}Cu^{2+} \\ \hline R\text{–}Cu^{2+} \end{array}} \xrightleftharpoons[]{O_2^-} \boxed{\begin{array}{c} R\text{–}Cu^{1+} \\ \hline N\text{–}Cu^{2+} \end{array}} \xrightleftharpoons[]{} \boxed{\begin{array}{c} N\text{–}Cu^{1+} \\ \hline R\text{–}Cu^{2+} \end{array}} \xrightleftharpoons[]{O_2^-} \boxed{\begin{array}{c} R\text{-}Cu^{1+} \\ \hline R\text{–}Cu^{1+} \end{array}} \tag{34}$$

Each box represents one subunit; R and N represent copper, reactive or non-reactive, respectively, toward O_2^-.

A similar mechanism was offered (Lavelle et al., 1977) for the enzymatic reaction of an iron-containing SOD, which is composed of two identical subunits. Turnover experiments demonstrate that only the fraction of ferric iron that is reduced by H_2O_2 is involved in the catalysis, which is alternately oxidized and reduced by O_2^-.

For the Mn-SOD isolated from *E. coli* or *Bacillus stearothermophilus* a mechanism similar to Fe-SOD or Cu-SOD does not appear to be applicable (Pick et al., 1974; Michael et al., 1977). When the initial concentration of O_2^- is less than 10 times the total concentration of the enzyme, a first-order reaction was observed with a rate constant of $1.5 \pm 0.15 \times 10^8 M^{-1}sec^{-1}$. When the concentration of O_2^- is larger than 15 times the concentration of SOD, a biphasic process was observed. Excess O_2^- was removed by a less efficient reaction. From these results, Pick et al. (1974) postulated the existence of four second-order oxidation-reduction reactions, and three oxidation states of manganese. At present, it is not possible to identify these oxidation states.

References

Adams, G.E. (1972) Radiation chemical mechanism in radiation biology, in: M. Bruton and J.L. Magee (Eds.), Advances in Radiation Chemistry, Wiley-Interscience, New York, pp. 125–153.

Anbar, M., Bambenek, M. and Ross, A.B. (1973) Selected specific rates of reactions of transient form of water in aqueous solution, Vol. I, Hydrated Electron, NSRDS-NBS, 43, Washington, DC.

Behar, D., Czapski, G., Rabani, J., Dorfman, L.M. and Schwarz, H.A. (1970) The acid dissociation constant and decay kinetics of the perhydroxyl radical, J. Phys. Chem., 74, 3209–3213.

Beinert, H., Hansen, R.E. and Hartzell, C.R. (1976) Kinetics studies on cytochrome *c* oxidase by combined EPR and reflectance spectroscopy after rapid freezing, Biochim. Biophys. Acta, 43, 339–355.

Bielski, B.H.J. and Chan, P.C. (1973) Enzyme-catalyzed free radical reactions with nicotinamide-adenine nucleotides, 1. Lactate dehydrogenase-catalyzed chain oxidation of bound NADH by superoxide radical, Arch. Biochem. Biophys., 159, 873–879.

Bielski, B.H.J. and Gebicki, J.M. (1977) Application of radiation chemistry to biology, in: W.A. Pryor (Ed.), Free Radicals in Biology, Vol. III, Academic Press, New York, pp. 2–48.

Bielski, B.H.J., Comstock, D.A., Harber, A. and Philip, C. (1974) Study of peroxidase mechanisms by pulse radiolysis, II. Reaction of horseradish peroxidase compound I with O_2^-, Biochim. Biophys. Acta, 350, 113–120.

Bielski, B.H.J., Richter, H.W. and Chan, P.C. (1975) Some properties of the ascorbate free radical, Ann. N.Y. Acad. Sci., 258, 231–237.

Bray, R.C., Palmer, G. and Beinert, H. (1964) Direct studies on the electron transfer sequence in xanthine oxidase by electron paramagnetic resonance spectroscopy, J. Biol. Chem., 239, 2267–2676.

Bray, R.C., Lowe, D.J., Capeililere-Blandin, C. and Fielden, E.M. (1973) Trapping of short-lived intermediates in enzymic by rapid freezing: combination of electron paramagnetic resonance with pulse radiolysis, Biochem. Soc. Trans., 1, 1067–1072.

Bray, R.C., Cockle, S.A. and Calabrese, L. (1974) Reduction and inactivation of superoxide dismutase by hydrogen peroxide, Biochem. J., 139, 49–60.

Bronskill, M.J. and Hunt, J.W. (1968) A pulse-radiolysis system for the observation of short-lived transients, J. Phys. Chem., 72, 3762–3766.

Bronskill, M.J., Wolf, R.K. and Hunt, J.W. (1969) Subnanosecond observation of the solvated electron, J. Phys. Chem., 73, 1175–1176.

Burnett, R.M., Darling, G.D., Kendall, D.S., LeQuesne, M.E., Mayhew, S.G., Smith, W.W. and Ludwid, M.L. (1974) The structure of the oxidized form of Clostridial flavodoxin at 1.9 Å resolution, Description of the flavin mononucleotide binding site, J. Biol. Chem., 249, 4383–4392.

Carrico, R.J., Malmström, B.G. and Vanngard, T. (1971) A study of the reduction and oxidation of human ceruloplasmin, Evidence that a diamagnetic chromophore in the enzyme participates in the oxidase mechanism, Eur. J.Biochem., 22, 127–133.

Cassoly, R. and Gibson, Q.H. (1972) The kinetics of ligand binding to hemoglobin valency hybrids and the effect of anions, J. Biol. Chem., 247, 7332–7341.

Chance, B. (1943) The kinetics of the enzyme–substrate compound of peroxidase, J. Biol., Chem., 151, 553–577.

Chevion, M., Ilan, Y.A., Samuni, A., Navok, T. and Czapski, G. (1979) Quaternary structure of methemoglobin, Pulse radiolysis study of the binding of oxygen to the valency hybrid, J. Biol. Chem., 254, 6370–6374.

Dallinger, R.F., Guanci Jr., J.J., Woodruff, W.H. and Rodgers, M.A.J. (1979) Vibrational spectroscopy of the electronically excited state, Pulse radiolysis/time resolved resonance Raman study of triplet β-cartene, J. Am. Chem. Soc., 101, 1355–1357.

Deinum, J. and Vanngard, T. (1973) The stoichiometry of the paramagnetic copper and the oxidation-reduction potentials of the type 1 copper in human ceruloplasmin, Biochim. Biophys. Acata, 310, 321–330.

Dolphin, D., Forman, A., Borg, D.C., Fajer, J. and Felton, R.H. (1971) Compound I of catalase and horseradish peroxidase: π-cation radicals, Proc. Natl. Acad. Sci. (U.S.A.), 68, 614–618.

Dorfman, L.M. and Adams, G.E. (1973) Reactivity of hydroxyl radicals in aqueous solution, NSRDS-BNS, 46, Washington, DC.

Errede, B., Haight Jr., G.P. and Kamen, M.D. (1976) Oxidation of ferrocytochrome c by mitochondrial cytochrome c oxidase, Proc. Natl. Acad. Sci. (U.S.A.), 73, 113–117.

Faraggi, M. and Klapper, M.H. (1979) One electron reduction of flavodoxin, A fast kinetic study, J. Biol. Chem., 254, 8139–8142.

Faraggi, M. and Pecht, I. (1973) The electron pathway to Cu (II) in ceruloplasmin, J. Biol. Chem., 248, 3146–3149.

Faraggi, M., Hemmerich, P. and Pecht, I. (1975) O_2-affinities of flavin radical species as studied by pulse radiolysis, FEBS Lett., 51, 47–51.

Ferguson-Miller, S., Brautigan, D.L. and Margoliash, E. (1976) Correlation of the kinetics of electron transfer activity of various eukaryotic cytochrome c with binding to mitochondrial cytochrome c oxidase, J. Biol. Chem., 251, 1104–1115.

Fessenden, R.W. (1973) Time resolved ESR spectroscopy, I. A kinetic treatment of signal enhancements, J. Chem. Phys., 58, 2489–2500.

Fielden, E.M., Roberts, P.B., Bray, R.C., Lowe, D.J., Mautner, G.N., Rotilio, G. and Calabrese, L. (1974) The mechanism of action of superoxide dismutase from pulse radiolysis and electron paramagnetic resonance, Evidence that only half the active sites function in catalysis, Biochem. J., 139, 49–60.

Fonda, M.L. and Anderson, B.M. (1969) D-Amino acid oxidase, IV. Inactivation by maleimides, J. Biol. Chem., 244, 666–674.

Gibson, Q.H., Greenwood, C., Wharton, D.C. and Palmer, G. (1965) The reaction of cytochrome oxidase with cytochrome c, J. Biol. Chem., 240, 888–894.

Goldberg, M. and Pecht, I. (1978) The reaction of "blue" copper oxidase with O_2 — A pulse radiolysis study, Fast Biochemical Reactions in Solutions, Membranes and Cells., Airline Virginia, p. 179.

Hayashi, K., Lindenau, D. and Tamura, M. (1978) A pulse radiolysis study on active oxygen, in: A. Silver, M. Erecinska and H.I. Bicher (Eds.), Advances in Experimental Medicine and Biology, Vol. 94, Plenum, New York, pp.353–359.

Hirata, F., Ohnishi, T. and Hayaishi, O. (1977) Indoleamine 2,3-dioxygenase, Characterization and properties of enzyme · O_2^- complex, J. Biol. Chem., 252, 4637–4642.

Ilan, Y.A., Rabani, J. and Czapski, G. (1976) One electron reduction of metmyoglobin and methemoglobin and methemoglobin and the reduced molecule with oxygen, Biochim. Biophys. Acta, 446, 277–286.

Ilan, Y.A., Samuni, A., Chevion, M. and Czapski, G. (1978) Quaternary states of methemoglobin and its valence-hybrid, A pulse radiolysis study, J. Biol. Chem., 253, 82–86.

Ilgenfritz, G. and Schuster, T.M. (1974) Kinetics of oxygen binding to human hemoglobin, Temperature jump relaxation studies, J. Biol. Chem., 248, 2959–2973.

Ishimura, Y., Ullrich, V. and Peterson, J.A. (1971) Oxygenated cytochrome P-450 and its possible role in enzymic hydroxylation, Biochem. Biophys. Res. Commun., 42, 140–146.

Keene, J.P. (1964) Pulse radiolysis apparatus, J. Sci. Inst., 493–496.

Klug-Roth, D., Fridovich, I. and Rabani, J. (1973) Pulse radiolytic investigation of superoxide catalyzed disproportionation, Mechanism for bovine superoxide dismutase, J. Amer. Chem. Soc., 95, 2786–2790.

Kobayashi, K. and Hayashi, K. (1981) One-electron reduction in oxyform of hemoproteins, J. Biol. Chem., 256, 12350–12354.

Kobayashi, K., Hirota, K., Ohara, H., Hayashi, K., Miura, R. and Yamano, T. (1983) One electron reduction of D-amino acid oxidase, The conversion from red semiquinone to blue semiquinone, to be published.

La Mar, G.N., de Ropp, J.S., Smith, K.M. and Langry, K.C. (1981) Proton nuclear magnetic resonance investigation of the electronic structure of compound I of horseradish peroxidase, J. Biol. Chem., 256, 237–243.

Land, E.J. and Swallow, A.J. (1968) One-electron reactions in biochemical systems as studied by pulse radiolysis, I. Nicotinamide-adenine dinucleotide and related compounds, Biochim. Biophys. Acta, 162, 327–337.

Land, E.J. and Swallow, A.J. (1969) One-electron reactions in biochemical systems as studied by pulse radiolysis, II. Riboflavin, Biochemistry, 8, 2117–2125.

Land, E.J. and Swallow, A.J. (1970) One-electron reactions in biochemical systems as studied by pulse radiolysis, III. Ubiquinone, J. Biol. Chem., 245, 1890–1894.

Land, E.J. and Swallow, A.J. (1971) One-electron reactions in biochemical systems as studied by pulse radiolysis, V. cytochrome c, Arch. Biochem. Biophys., 145, 365–372.

Land, E.J. and Swallow, A.J. (1974) One-electron reactions in biochemical systems as studied by pulse radiolysis, VI. Stages in the reduction of ferricytochrome c. Biochim. Biophys. Acta, 368, 86–96.

Lavelle, F., McAdame, M.E., Fielden, E.M. and Roberts, P.B. (1977) A pulse radiolysis study of the catalytic mechanism of the iron-containing superoxide dismutase from *Photobacterium leiognathi*, Biochem., J., 161, 3–11.

Makino, R., Yamada, H. and Yamazaki, I. (1976) Effects of 2,4-substituents of deuteroheme upon the stability of the oxyform and compound I of horseradish peroxidase, Arch. Biochem. Biophys., 173, 66–70.

Malkin, R., Malmström, B.G. and Vanngard, T. (1969) The reversible removal of one specific copper (II) from fungal laccase, Eur. J. Biochem., 7, 253–259.

Malmström, B.G., Reinhammer, B. and Vanngard, T. (1968) Two forms of copper (II) in fungal laccase, Biochim. Biophys. Acta, 156, 66–76.

Massey, V., and Palmer, G. (1966) On the existence of spectrally distinct classes of flavoprotein semiquinones, A new method for the quantitative production of flavoprotein semiquinones, Biochemistry, 5, 3181–3188.

Massey, V., Mullerm F., Feldberg, R., Schuman, N., Sullivan, P.A., Howell, L.G., Mayhew, S.G., Matthews, R.G. and Foust, G.P. (1969) The reactivity of flavoproteins with sulfite, Possible relevance to the problem of oxygen reactivity, J. Biol. Chem., 244, 3999–4006.

Matheson, M.S. aand Dorfman, L.M. (1969) Pulse Radiolysis, M.I.T. Press, Cambridge, MA, pp. 6–19.

McCord, J.M. and Fridovich, I. (1968) The reduction of cytochrome c by milk xanthine oxidase, J. Biol. Chem., 243, 5753—5760.

McCord, J.M., Keele, B.B. and Fridovich, I. (1971) An enzyme-based theory of obligate anaerobiosis: The physiological function of superoxide dismutase, Proc. Natl. Acad. Sci. (U.S.A.), 68, 1024–1027.

Meisel, D. and Neta, P. (1975) One-electron reduction potential of riboflavine studied by pulse radiolysis, J. Phys. Chem., 79, 2459–2461.

Michael, E., Fox, R.A., Lavelle, F. and Fielden, E.M. (1977) A pulse radiolysis study of the manganese-containing superoxide dismutase from *Bacillus stearothermophilus*, Biochem. J., 165, 71–79.

Misra, H.P. and Fridovich, I. (1972a) The generation of superoxide radical during the autoxidation of hemoglobin, J. Biol. Chem., 247, 6960–6962.

Misra, H.P. and Fridovich, I. (1972b) The univalent reduction of oxygen by reduced flavins and quinones, J. Biol. Chem., 247, 188–192.

Pecht, I. and Faraggi, M. (1971) Reduction of copper (II) in fungal laccase by hydrated electrons, Nature (London), 233, 116–118.

Pecht, I. and Faraggi, M. (1972) Electron transfer to ferricytochrome *c*: reduction with hydrated electrons and conformational transitions involved, Proc. Natl. Acad. Sci. (U.S.A.), 69, 902–906.

Pick, M., Rabani, J., Yost, F. and Fidrovich, I. (1974) The catalytic mechanism of the manganese-containing superoxide dismutase of *Escherichia coli* studied by pulse radiolysis, J. Am. Chem. Soc., 96, 7329–7333.

Raap, A., Van Leeuwwen, J.W., Rollema, H.S. and De Bruin, S.H. (1977a) Pulse radiolysis studies on the spin-state transitions in aquomethemoglobin after reduction of a single heme group, FEBS Lett., 81, 111–114.

Raap, A., Van Leeuwen, J.W., Van Eck-Schouten, T., Rollema, H.S. and De Bruin, S.H. (1977b) Heterogeneity in the kinetics of oxygen binding to paretially reduced human methemoglobin, Eur. J. Biochem., 81, 619–626.

Reinhammar, B.R.M. (1972) Oxidation-reduction potentials of the electron acceptors in laccases and stellacyamin, Biochim. Biophys. Acta, 275, 245–259.

Rollema, H.S., Scholberg, H.P.F., De Bruin, S.H. and Raap, A. (1976) The kinetics of carbon monoxide binding to partially reduced methemoglobin, Biochem. Biophys. Res. Commun., 71, 997–1003.

Rotilio, G., Bray, R.C. and Fielden, E.M. (1972) A pulse radiolysis study of superoxide dismutase, Biochim. Biophys. Acta, 268, 605–609.

Sakurai, H., Kawanishi, M., Hayashi, K., Okada, T., Tsumori, K., Takeda, S., Kimura, N. Yamamoto, T., Hori, T., Ohkuma, J. and Sawai, T. (1982) Recent development in the Osaka University picosecond single bunch electron linear accelerator, Mem. Inst. Sci. Ind. Res. Osaka, Univ., 39, 21–40.

Sangster, D.F. (1966) Absorption spectra and transient species found in the pulse radiolysis of alkaline aqueuos benzoate solutions, J. Phys. Chem., 70, 1712–1717.

Sawada, Y. and Yamazaki, I. (1973) One electron transfer reactions in biochemical systems, VII. Kinetic study of superoxide dismutase, Biochim. Biophys. Acta, 327, 257–265.

Simic, M.G. and Taub, I.A. (1978) Fast electron transfer process in cytochrome *c* and related metalloproteins, Fast Biochemical Reactions in Solutions, Membrane and Cells, Airline Virginia, pp. 100–106.

Stankovich, M.J., Schopfer, L.M. and Massay, V. (1978) Determination of glucose oxidase oxidation-reduction potentials and the oxygen reactivity of fully reduced and semiquinone form, J. Biol. Chem., 253, 4971–4979.

Steinman, H.M. and Hill, R.L. (1973) Sequence homologies among bacterial and mitochondrial superoxide dismutase, Proc. Natl. Acad. Sci. (U.S.A.), 70, 3725–3729.

Tamura, M. and Yamazaki, I. (1972) Reaction of the oxyform of horseradish peroxidase, J. Biochem., 71, 311–319.

Tyson, C.A., Lipscomb, J.D. and Gunsalus, I.C. (1972) The roles of putidaredoxin and P 450_{cam} in methylene hydroxylation, J. Biol. Chem., 297, 5777–5784.

Van Buuren, K.J.H., Van Gelder, B.F., Wilting, J. and Braams, R. (1974) Biochemical and biophysical studies on cytochrome *c* oxidase, XIV. The reaction with cytochrome *c* as studied by pulse radiolysis, J. Biol. Chem., 333, 421–429.

Van Gelder, B.F., Veerman, E., Wilms, J. and Dekker, H.L. (1979) The electron-accepting site of cytochrome oxidase, in: B. Chance, T.E. King, K. Okunuki and Orii (Eds.), Cytochrome Oxidase, Elsevier, Amsterdam, pp. 305–313.

Van Leeuwen, J.W., Butler, J. and Swallow, A.J. (1981) A non-equilibrium state of deoxyhemoglobin, Temperature-dependence and oxygen binding, Biochim. Biophys. Acta, 667, 185–196.

Verma, N.C. and Fessenden, R.W. (1973) Time resolved ESR spectroscopy, II. The behavior of H atom signals, J. Phys. Chem., 58, 2501–2506.

Wallace, W.J., Maxwell, J.S. and Caughey, W.S. (1974) The mechanism of hemoglobin autoxidation, Evidence for proton assisted nucleophilic displacement of superoxide by anions, Biochem. Biophys. Res. Commun., 57, 1104–1110.

Wilting, J., Raap, A., Braams, R., De Bruin, S.H., Rollema, H.S. and Janssen, H.M. (1974) Conformational changes and ligand dissociation kinetic following rapid reduction of human aquomethemoglobin and horse aquometmyoglobin by hydrated electrons, J. Biol. Chem., 249, 6325–6330.

Wittenberg, J.B., Noble, R.W., Wittenberg, B.A., Antonini, E., Brunori, M. and Wyman, J. (1967) Studies on the equilibria aand kinetics of the reaction of peroxidase with ligands, II. The reaction of ferroperoxidase with oxygen, J. Biol. Chem., 242, 626–634.

Woodruff, H. and Farquharson, S. (1978) Time-resolved resonance Raman spectroscopy of hemoglobin derivatives: Heme structure changes in 7 nanoseconds, Science, 201, 831–833.

Yagi, K., Takai, A. and Ohishi, N. (1972) Conversion of the red semiquinone of D-amino acid oxidase to the blue semiquinone by complex formation, Biochim. Biophys. Acta, 289, 37–43.

Yamada, H. aand Yamazaki, I. (1974) Proton balance in conversions between five oxidation-reduction states of horseradish peroxidase reaction, Biochim. Biophys. Acta, 50, 62–69.

Yokota, K. and Yamazaki, I. (1965) The activity of the horseradish peroxidase compound III, Biochem. Biophys. Res. Commun., 18, 48–53.

Yoshida, T. and Kikuchi, G. (1978) Feature of the reaction of heme degradation catalyzed by the reconstituted microsomal heme oxygenase system, J. Biol. Chem., 253, 4230–4236.

Yoshida, T., Noguchi, M. and Kikuchi, G. (1980) Oxygenated form of heme–heme oxygenase complex and requirement for second electron to initate heme degradation from oxygenated complex, J. Biol. Chem., 255, 4418–4420.

Yost, F.J. and Fridovich, I. (1974) Superoxide radical and phagocytosis, Arch. Biochem. Biophys., 161, 395–401.

CHAPTER 6

Flash photolysis studies in heterogeneous systems

MAMORU TAMURA, TOSHIHISA ISHIKAWA,
SUSUMU TSUBOTA and ISAO YAMAZAKI

*Biophysics Division, Research Institute of Applied Electricity,
Hokkaido University, Sapporo 060, Japan*

Contents

1. Introduction 113
2. Flash photolysis of ferrihemoprotein complexes 115
3. Reactions with membrane-bound enzymes 118
4. Reactions in gels 121
5. Reactions of crystalline myoglobin 122
6. Reactions in frozen state 125
7. Flash photolysis in cells and cellular organelles 126
8. Flash photolysis in organs and tissues 128
9. Special applications to cell physiology 134
10. Summary 134
References 135

1. Introduction

The principle of the flash photolysis method is to employ light to displace an equilibrium or to disturb a steady state in a time which is short compared to the time required for the subsequent dark relaxation. The dark reaction is typically monitored by measuring an absorbance change as a function of time. This method is capable of picosecond time resolution when a mode-locked pulsed laser is employed as the flash source.

Flash photolysis has proven to be a powerful tool to analyze the mechanisms of many light-related biological phenomena, such a photosynthesis and visual processes. These topics are reviewed in Chapter 10 of this volume, and experimental details of this method are presented in Chapter 2. Here we will deal mainly with kinetic studies of ligand binding to hemoproteins. Most experiments have been performed with ferrous hemoproteins and a typical example can be represented by the following reactions:

R.I. Sha'afi and S.M. Fernandez (Eds.), Fast Methods in Physical Biochemistry and Cell Biology
© 1983 Elsevier Science Publishers

$$Fe_p^{2+} \cdot L \xrightarrow{h\nu} Fe_p^{2+*} + L \tag{1}$$

$$Fe_p^{2+*} \longrightarrow Fe_p^{2+} \tag{2}$$

$$Fe_p^{2+} + L \longrightarrow Fe_p^{2+} \cdot L \tag{3}$$

where Fe_p^{2+}, L and Fe_p^{2+*} denote respectively a ferrous unligated form, photodissociable ligand, and an unstable ferrous unligated species which is formed immediately after departure of the ligand. Reaction 2, a protein conformational change, occurs rapidly at room temperature; its relaxation time being shorter than $1\,\mu sec$ (Alpert et al., 1974). Therefore, except in special cases, such as in the reaction of hemoglobin (Sawicki and Gibson, 1976), stopped-flow and flash photolysis methods give similar reaction kinetics in a time range from 0.1 msec to sec. These kinetics, for example, would apply only to reaction 3. The experiments discussed in this chapter will be limited to reactions occurring within this time range.

Since in flash photolysis the reaction is initiated by light, it becomes possible to apply this method to reactions which cannot be initiated by mixing the reactants; reactions in frozen states and heterogeneous systems belong to such cases. Reactions in biology occur principally in heterogeneous systems. For instance, reactions of immobilized enzymes such as those that are membrane-bound or in crystalline states, as well as enzymes incorporated into artificial vesicles such as liposomes, all occur in heterogeneous systems.

Organs and tissues consist of cells, and a cell consists of various organelles such as nucleus, mitochondria and endoplasmic reticulum. Enzymes and substrates are not uniformly distributed in the cell, but are located in cell compartments which offer unique physicochemical environments different from those which arise under experimental conditions in vitro. The functions of organs and tissues depend on this cellular architecture. Thus, non-invasive techniques become essential for kinetic studies in vivo. Flash photolysis is well suited for this purpose.

This chapter will deal with topics on flash photolysis in heterogeneous systems, and in particular with the photodissociation of nitric oxide (NO)–ferrihemoprotein complexes. Several studies which have been reviewed in detail elsewhere (Alpert et al., 1974; Sawicki and Gibson, 1976; Ke, 1972; Gibson, 1978; De Vault, 1978) have applied flash photolysis to the investigation of ferrous hemoproteins in homogeneous systems. We will discuss here the application of this technique to ferrihemoproteins, although these reactions are carried out in homogeneous systems.

The techniques and apparatus for flash photolysis have been reviewed by many investigators (Ke, 1972; Gibson, 1978; De Vault, 1978) and are also described in greater detail in Chapter 2 of this volume.

2. Flash photolysis of ferrihemoprotein complexes

Photodissociation has been believed for a long time to be characteristic of the ferrous ligated state. The application of flash photolysis to the kinetic study of hemoproteins, however, received some impetus after the discovery of photodissociation of NO complexes from ferric hemoproteins (Tamura et al., 1978a). Fig. 1 shows the photodissociation of NO–horseradish peroxidase (HRP) from the ferric enzyme by photolysis. The absorption at 405 nm characteristic of ferriHRP jumps and then decreases with a half-time of 1.7 msec. A similar time-course of absorbance change is also seen at 425 nm, near the absorption maximum of NO–ferriHRP. The ferroHRP does not appear as can be judged by the absence of any absorbance change at 435 nm. No absorbance change is observed at 412 nm, an isosbestic point between ferriHRP and its NO complex. The difference spectrum obtained 1 msec after the flash is shown in Fig. 2A. This spectrum resembles the difference spectrum of NO–ferriHRP minus ferriHRP (solid line, Fig. 2B).

Fig. 3 shows the pH dependence of the rate constants for the recombination of NO with ferriHRP and metmyoglobin measured by the flash photolysis and the stopped-flow methods. With metmyoglobin, the rate constants obtained by the two methods are identical in the whole pH range tested. The pK value of 8.5 obtained from the curve corresponds to that of the acid–alkaline transition of metmyoglobin. With ferriHRP, the values obtained by the two methods are identical (1.9×10^5 $M^{-1}sec^{-1}$) in the pH range between 5 and 9, but not above pH 9. The value determined by flash photolysis remains constant until pH 11.8, while that obtained by stopped flow decreases above pH 10 as the pH increases, giving a pK value of 11 which corresponds to the pK for the acid–alkaline transition of ferriHRP.

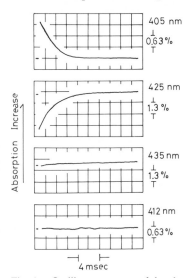

Fig. 1. Oscilloscope traces of the absorption changes after photolysis of NO–ferriHRP. HRP, $10 \mu M$; phosphate buffer (pH 7.4), 0.2 M; 25°C.

116

Unlike the ferrous unligated state where the 6th coordination position is vacant, the ferric unligated state is believed to be a hexacoordinated structure, where H_2O and OH^- coordinate at the 6th position below neutral pH and at alkaline pH, respectively. The reaction scheme is proposed in Fig. 4. First, we shall discuss the results of NO–metmyoglobin. The water molecule is taken up into the pentacoordinated form within 1 msec after departure of NO by photolysis. If one assumes that the half-time of coordination of a water molecule is less than 100 μsec, then the rate constant for the combination of a water molecule with the pentacoordinated metmyoglobin is greater than $10^2\,M^{-1}sec^{-1}$. Since in this case the stopped-flow and the flash photolysis methods give identical results, the formation of the alkaline form

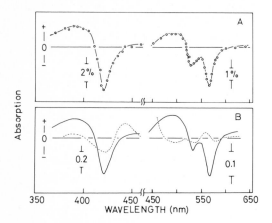

Fig. 2. (A) Kinetic difference spectrum of flash photolysis of NO–ferriHRP. The spectrum is taken 1 msec after the flash. Absorption increases in the upward direction. HRP concentrations used were 10 and 30 μM for the Soret and visible spectra, respectively. Phosphate buffer (pH 7.4), 0.2 M. (B) Difference spectra of NO–ferriHRP minus ferriHRP (solid line) and NO–ferroHRP minus ferroHRP (dashed line). The ordinate is scaled in absorbance. Phosphate buffer (pH 7.4), 0.2 M; HRP, 14 μM.

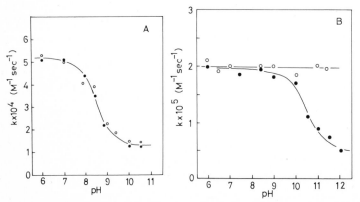

Fig. 3. The pH dependence of the rate constants for recombination of NO with metmyoglobin (A) and ferriHRP (B). Temperature, 25°C. o-o, flash photolysis method; ●-●, stopped-flow method.

probably occurs much faster than the combination of NO with metmyoglobin. Fig. 4. shows that there are alternative pathways in the formation of the alkaline form by photolysis: (a) a water molecule coordinates (i) and then a proton dissociates (iii), or (b) OH^- coordinates directly (i'). If the reaction occurs via path (i') the rate constant for OH^- binding should be larger than 10^6 $M^{-1}sec^{-1}$. If it occurs via path (iii), its rate constant is greater than 10^6 sec^{-1}.

The finding that the rate of the recombination of NO with ferriHRP after photolysis is independent of pH can be explained by assuming that the formation of the alkaline form of ferriHRP from the photodissociated product is much slower than the recombination of NO with the photodissociated product. This indicates that the conformation of this enzyme after departure of NO by photolysis differs from that of its alkaline form. In the stopped-flow measurement performed by mixing the alkaline form of the enzyme with NO, the process of OH^- dissociation from the alkaline form becomes rate-limiting and its rate constant can be determined. It is noteworthy that the flash photolysis method for ferrous ligated complexes, such as CO–myoglobin and CO–ferroHRP always gives kinetic parameters which are identical to those obtained by the flow method whenever the half-life is longer than a millisecond. Marked differences between myoglobin and HRP appear in their ferric forms. The half-life for the formation of the alkaline forms is much shorter in metmyoglobin than in ferriHRP. It is also found that a water molecule coordinates at the 6th position in metmyoglobin while it does not in ferriHRP. Successful trapping of the pentacoordinated forms was accomplished by the photolysis of NO–ferrihemoprotein complexes at 4.2°K (Kobayashi et. al., 1980).

In addition to NO-ferrihemoproteins, O_2 complexes of Co-substituted hemoproteins (Ikeda-Saito et al., 1977) and NO complexes of Mn-substituted hemoproteins (Hoffman and Gibson, 1978) are also photodissociable. These two are suited for the kinetic analysis of ligand binding by flash photolysis methods.

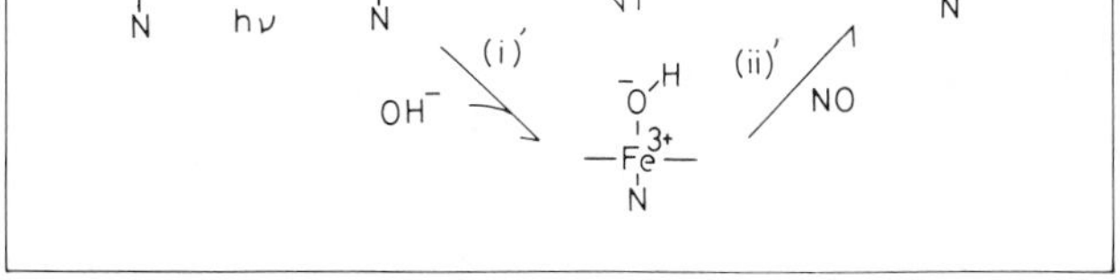

Fig. 4. Schematic presentation of the recombination of NO and ferrihemoproteins.

3. Reactions with membrane-bound enzymes

The most common type of heterogeneous enzyme system in vitro consists of immobilized enzymes. Reactions in immobilized enzyme systems have been studied by monitoring the product formation in the stationary state. Based on these studies, it has been pointed out that the diffusion of substrates is usually the rate-determining step in the overall enzyme reaction, which masks the intrinsic nature of the enzymes in their immobilized state. Here, we will present kinetic analyses of reactions of a basic HRP isoenzyme adsorbed onto a carboxymethyl cellulose membrane (Ishikawa et al., 1980). Fig. 5 shows the formation of compound I from the reaction of H_2O_2 with the bound enzyme. The membrane was soaked in a buffer solution containing H_2O_2. The reaction is:

$$Fe_p^{3+} + H_2O_2 \rightarrow \text{compound I}$$

where Fe_p^{3+} is ferriHRP. The reaction between ferriHRP and H_2O_2 is fast ($k \sim 10^7$ $M^{-1}sec^{-1}$) in a homogeneous solution, whereas that on the membrane is markedly slow. The initial rate of formation of compound I is independent of the amount of adsorbed enzyme and the reaction of the enzyme on the membrane is no longer first order even in the presence of excess H_2O_2 in the bulk solution. The apparent rate constant for this reaction is less than 1% of that in solution. This indicates that a diffusion process is rate-limiting in the reaction of enzyme adsorbed on the membrane with ligands present in a bulk solution. In other words, every molecule reaching the membrane surface is trapped by the enzyme in the stationary state of the reaction. Thus, under such conditions, we can not determine the rate constant for interaction between H_2O_2 and enzyme adsorbed on the membrane and therefore cannot say whether the immobilization causes alterations in the enzyme reactivity or not. The flash photolysis method is useful in such situations. Differences in the

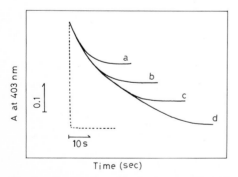

Fig. 5. The reaction of hydrogen peroxide with peroxidase adsorbed onto a carboxymethyl cellulose membrane suspended in a solution of 20 mM sodium acetate, pH 6.0. Effect of the amount of adsorbed enzyme on the rate of compound I formation. The initial adsorbance of the membrane was 0.23 (a), 0.40 (b), 0.56 (c) and 0.73 (d) at 403 nm. The compound I formation was measured at 403 nm. The medium contained 150 μM H_2O_2. Dashed line shows a reaction of 7 μM enzyme dissolved in solution.

ligand-binding reactions originated by ordinary mixing and by flash photolysis are schematically shown in Fig. 6. When a ligand is added to a bulk solution containing a membrane-bound enzyme, the diffusion of ligand to the membrane, v_{diff} limits the reaction rate (Fig. 6A). Photolysis, on the other hand, generates a concentrated layer of ligand near the membrane (Fig. 6B). In this case recombination is no longer limited by diffusion. Fig. 7A shows semilogarithmic plots for the recombination of CO with membrane-adsorbed enzyme after photodissociation. The semilogarithmic plots deviate from straight lines as the concentration of CO in the bulk solution decreases. The slopes of the initial and the stationary phases are plotted versus the concentration of CO in the bulk solution in Fig. 7B. These results show that the initial slope increases linearly with concentration of CO in the bulk solution. By extrapolation of the linear graph, it is found that flash photolysis in-

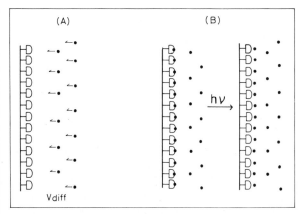

Fig. 6.   Schematic representation of a ligand-binding reaction for an enzyme adsorbed on a membrane. (A) Stopped flow. (B) Flash photolysis. In A, the diffusion of ligand (●) to the membrane surface limits the over-all reaction rate. Flash photolysis creates a concentrated ligand layer near the membrane surface (B).

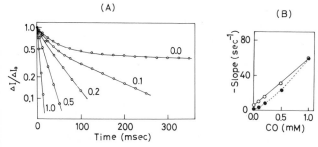

Fig. 7.   Recombination of CO and ferroHRP adsorbed to a carboxymethyl cellulose membrane after photolysis at varying concentrations of CO in the bulk solution. (A) Semilogarithmic plots at pH 8.0. The concentration (mM) of CO in the medium is indicated in the figure. (B) The slope of the initial (○) and the stationary (●) phases is plotted against the concentration of CO.

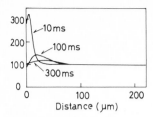

Fig. 8. Simulation of diffusion layer in the reaction of peroxidase adsorbed to a membrane. The concentration profile at indicated times was calculated according to Eqn. 4. These results correspond to the reaction (CO concentration = 0.1 mM) in Fig. 7A.

creases the effective concentration of CO near the membrane by about $100\,\mu M$ in this case. The non-linear kinetics shown in Fig. 7 can be explained by a decrease in the effective CO concentration during the reaction. Some CO apparently diffuses out to the bulk solution. This stems from the fact that the enzyme is closely packed on the flat membrane surface, where the number of enzyme layers is estimated to be about 20 on each side of the membrane. This "local concentration of enzyme" on the membrane reaches the order of mM, therefore photolysis of only 10% of CO–enzyme results in a $100\,\mu M$ increase in the local concentration of CO. From the initial rate of CO recombination and the effective concentration of CO, a second-order rate constant can be calculated and is found to be close to that measured in homogeneous solution. It can be concluded that the kinetic properties of the enzyme remain essentially unchanged after adsorption to the membrane, although the enzyme molecules cannot move freely.

In Fig. 8, the CO concentration is plotted against the distance perpendicular to the membrane surface at the times indicated. The simulation is done by solving the following one-dimensional second-order differential equation.

$$\frac{\delta c}{\delta t} = \sigma(c) + D \cdot \frac{\delta^2 c}{\sigma X^2} \tag{4}$$

where c stands for CO concentration; t, time after reaction starts; D, diffusion constant; and X, distance from the enzyme locus. The term $\sigma(c)$ denotes the rate of chemical reactions, CO recombination in this case. As shown in Fig. 8, at 10 msec after flash, the effective CO concentration is still very high in the area within $20\,\mu m$ from the surface, but it decreases rapidly with time. This is due to the diffusion of CO into the bulk solution and the recombination with the enzyme. It must be emphasized that a rapid local change in the effective concentration is characteristic of reactions in heterogeneous systems.

4. Reactions in gels

The membrane system is treated as one-dimensional with respect to the distance from the membrane surface to bulk solution as seen in Fig. 8. The "cage effect" might be observed under conditions where the enzyme and substrate are packed in a restricted space. One easy method of immobilizing enzymes is to confine them in a gel. This method was successfully applied to the immobilization of myoglobin by the use of radiation-induced polymerization of 2-hydroxyethyl-methacrylate. Fig. 9 compares time-courses of the recombination of CO with myoglobin in solution and in the gel at an identical concentration of CO (Miki et al., 1982). A section of gel larger than $2 \times 2 \times 2 \, mm^3$ was used in this experiment. More than 1 h is required for completion of the formation of CO-myoglobin complexes when the gel is soaked into the CO-saturated solution; the rate of recombination of CO with myoglobin being faster initially in the gel than in solution. The reaction in the gel, furthermore, does not obey first-order kinetics. The initial rate of CO recombination does not change significantly for CO in bulk solution concentrations between 5 and $20 \, \mu M$, whereas the rate in the slow phase depends on the CO concentration in the bulk solution. An analysis similar to the one employed in the enzyme-membrane system may be applied in this case. Since the size of the gel is relatively large compared with the diffusion layer of CO, photolysis-generated CO might not diffuse into the bulk solution. From the comparison of the gel space with the myoglobin concentration entrapped in the gel, myoglobin does not appear to disperse uniformally in each compartment of the gel. Thus, the result shown in Fig. 9 can be explained as follows: upon photolysis CO increases in compartments where myoglobin is present and the CO recombines with myoglobin in the initial phase. A portion of the CO also diffuses out into the adjacent compartments where myoglobin is absent and the slow phase of CO recombination appears. The CO concentration in

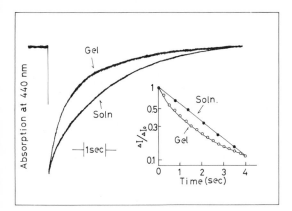

Fig. 9. CO recombination of myoglobin trapped in the gel. Absorption changes at 440 nm after flash photolysis in the gel and in solution are superimposed by normalization in the figure. The CO concentration was about $10 \, \mu M$ both in gel and in solution. The inset shows the semilogarithmic plots.

122

these compartments is already in equilibrium with that in bulk solution. Since the effective concentration of CO generated by photolysis is more than 10-fold higher than that in the bulk solution, it is reasonable that the initial rate is practically independent of the bulk CO concentration. The rate constant for the recombination of myoglobin with CO in this gel, therefore is concluded to be identical to that measured in solution. This seems to be an important conclusion derived directly from the method of flash photolysis.

5. Reactions of crystalline myoglobin

As typically seen in red blood cells where hemoglobin is concentrated to about 12 mM, it often occurs in vivo that the "local concentrations" of enzymes and substrates are much higher than the concentrations employed in in vitro experiments. For instance, catalase is present as a crystalline form in the peroxisome. Under such conditions kinetic features of enzymes may differ from those seen in vitro, since the viscosity of the medium and the solubility of the substrates differ greatly between the two cases. Thus, kinetic parameters must be determined under such in vivo conditions and compared with those determined in vitro.

The myoglobin structure has been determined by X-ray crystallography. There has been a great deal of discussion about the difference in structure and reactivity of the protein when in its crystalline state and in solution. Since kinetic parameters are very sensitive to a change of protein structure and environment, various attempts to detect alterations in the reactivity induced by crystallization have been made by many investigators. Using the stopped-flow method, Chance et al. (1966) concluded that the reactivity of metmyoglobin was lowered by crystallization. They employed a suspension of metmyoglobin crystals and measured the formation of ligand–metmyoglobin complexes with a dual wavelength spectrophotometer.

Here, we shall show results of a kinetic analysis for the recombination of CO with myoglobin in a crystal suspension which provides information on the reactivity of crystalline myoglobin.

In contrast to the case of a homogeneous solution and the transparent cellophane membrane described previously, the crystal suspension is turbid. Technical changes are required to increase the collection efficiency of the light scattered from the opaque suspension. To this end the detectors are closely coupled to the sample, but a special design for the apparatus is necessary to minimize interference due to scattering of the light flash.

Two typical kinetic traces for the recombination of CO with myoglobin in solution and in the crystalline state are compared in Fig. 10A. Fig. 10B shows optical traces of the recombination process in suspensions of crystals at varying CO concentrations in the bulk solution. The traces consist of an initial fast phase and a subsequent slow stationary phase. The semilogarithmic plots show that the reaction does not obey first-order kinetics (Fig. 11). In Fig. 12, the rates of CO recombination at two reaction stages are plotted against the CO concentration in bulk solu-

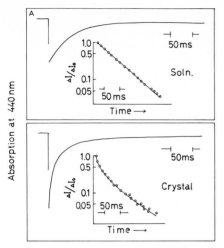

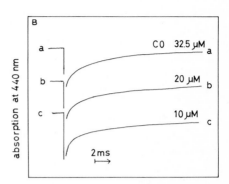

Fig. 10. CO recombination of myoglobin after photolysis in a crystal suspension. (A) Comparison of recombination of CO with myoglobin in a solution and in a crystal suspension. The myoglobin solution contained about 95% ammonium sulfate, pH 7.0. Myoglobin crystals were suspended in a saturated ammonium sulfate solution. The CO concentration was $28\,\mu M$ and the myoglobin concentration about $2\,\mu M$. Notice here that the rate constant for CO recombination is $5 \times 10^6\,M^{-1}sec^{-1}$ in a saturated ammonium sulfate solution, which is 10-fold faster than that in the usual buffer solution. The signal after photolysis was stored digitally and transferred to a disc. The traces shown were obtained by computer averaging and smoothing of more than 10 individual curves. The insets show the semilogarithmic plots. (B) CO recombination of myoglobin in crystal suspensions at varying concentrations of CO in the bulk solution.

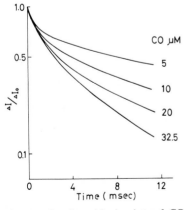

Fig. 11. Semilogarithmic plots of CO recombination. The curves are replotted from the data in Fig. 10B.

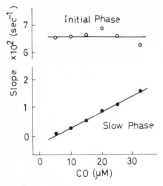

Fig. 12. Effect of the bulk solution concentration of CO on the rate of recombination of CO with myoglobin in crystal. The slope is obtained from the reaction curves in the initial and the slow phases, shown in Fig. 11.

tion. This graph shows that the initial rate does not vary with CO concentration, whereas the rate becomes proportional to the CO concentration as the recombination approaches completion. The initial fast reaction obeys second-order kinetics up to 10 msec after the flash and the rate depends on the intensity of the light flash. In contrast, the slow phase is first order and the rate is independent of the light flash intensity. These results are explained by the following considerations: the myoglobin concentration in the crystal is assumed to be about 10 mM, even 1% photolysis of CO–myoglobin yields 100 μM of CO and myoglobin simultaneously in the crystal. The concentration in the free space of the crystal may be low since its water content is approximately 50%. Thus, the concentration of CO generated by photolysis in the crystal is much higher than that in the bulk solution. Fig. 12 shows that the effective concentration of CO in the crystal is 130 μM. This may well account for the second-order kinetics of CO recombination and for the lack of dependence of the reaction rate on CO concentration in bulk solution, in the initial phase of the reaction. Since the amounts of CO and deoxymyoglobin in the crystal both depend on the degree of photolysis of CO–myoglobin, it is reasonable that the light intensity affects the initial rate. As the reaction proceeds, a portion of the CO released in the crystal diffuses into the bulk solution and the system approaches equilibrium. At this stage the rate of recombination of CO with myoglobin begins to be affected by the CO concentration in bulk solution. The principles governing the over-all recombination kinetics of the HRP-membrane system and of crystalline myoglobin appear to be similar in both cases, although the increase in the effective concentration is more dramatic in the crystalline system than in the membrane system. Rate constants for the recombination of crystalline myoglobin with CO can be calculated from both the initial rate and the rate in the slow phase. The value of 5.5×10^5 M^{-1}sec^{-1} thus obtained is very close to that found in a solution containing 95% ammonium sulfate. The reactivity of crystalline myoglobin, therefore, is similar to that in solution. It appears, then, that the lack of rotational and lateral diffusion of the myoglobin molecule does not affect the rate of CO recombination.

In separate stopped-flow experiments, where crystals and CO are mixed, the formation of CO–myoglobin occurs much more slowly than in solution or when initiated by flash photolysis. The diffusion of CO into the crystal is the rate-limiting step in this case.

6. Reactions in frozen state

Photolysis might be the only method available to initiate chemical reactions in the frozen state. Flash photolysis experiments at cryogenic temperatures were performed by Chance et al. (1965), who found that the recombination of ligands with several hemoproteins occurred at near-liquid nitrogen temperature. Low-temperature kinetic analysis have been performed with CO complexes of hemoglobin and myoglobin (Iizuka et al. 1974), and cytochrome oxidase (Sharrock and Yonetani, 1977), all of which yield multiphasic ligand-binding curves. At very low temperatures, reaction 2 is slow and Fe_p^{2+*} can be trapped. The conformational difference between Fe_p^{2+} and Fe_p^{2+*} has been discussed by several workers. Frauenfelder (1978) analyzed the recombination of CO with myoglobin, hemoglobin and heme over wide temperature and time ranges. They found that there are multiple energy barriers determining the recombination rate, which are not observable at room temperature (Austin et al., 1975). More recently, Hasinoff (1981a) has employed a photolysis technique for the study of CO and O_2 myoglobin recombination in a viscous 80% glycerol solution at $-90°C$, in which mixing the reactants seems to be practically impossible. Assignment of energy barriers for the diffusion of ligand to the protein surface differs between these authors (for details see Frauenfelder (1978) and Hasinoff (1981a)).

Chance (1978) introduced the low-temperature technique of flash photolysis to initiate ligand exchange reactions. CO–hemoprotein complexes, such as mitochondrial cytochrome oxidase–CO complex, are mixed with an oxygen-containing solution, and flash photolysis is performed. The CO complexes dissociate into CO and unligated hemoprotein which then reacts with oxygen. The principle is similar to that employed by Gibson and Greenwood (1963) and is based on the pioneer work of Hartridge and Roughton (1923), who discovered that when an oxygenated CO-hemoglobin solution was exposed to light from a carbon arc lamp, oxygen replaced CO.

In this approach, a small volume of highly concentrated cytochrome oxidase solution is placed in the bottom of an optical cell in the presence of CO and is covered with an oxygen-saturated ethylene glycol solution at $-23°C$. The two layers are mixed with a stirring rod, and the sample is immediately frozen and brought to the desired temperature; photolysis is then performed by a laser or xenon lamp. Fig. 13 shows typical recordings of the formation of oxy and peroxy compounds during the reaction of cytochrome oxidase with oxygen at $-105°C$ (Chance, 1978). The formation and disappearance of oxygenated cytochrome oxidase (compound A) is moni-

126

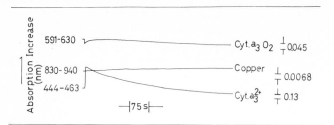

Fig. 13. Recordings of the reaction between cytochrome oxidase and oxygen in pigeon breast mitochondria, at $-105°C$.

tored by absorption at 591–630 nm. Oxidation-reduction of the oxidase is followed at 444–463 nm, and that of copper at 830–940 nm. Various intermediates are trapped and identified in the reaction of oxygen and cytochrome oxidase under these experimental conditions. These intermediates have not been detected at room temperature.

7. Flash photolysis in cells and cellular organelles

Flash photolysis has been applied to the kinetic study of mitochondria and particularly of cytochrome oxidase. The kinetics of the reaction between ferrous cytochrome oxidase and CO or O_2 were first investigated by flow methods. Since cytochrome oxidase is embedded in the inner mitochondrial membrane, the question arises whether the diffusion of O_2 or CO into the membrane controls the reaction between the oxidase and these ligands. The intrinsic concentration of O_2 or CO in the membrane, which is in equilibrium with that in the bulk solution cannot be accurately estimated. Values for the rate constant of recombination of the oxidase and CO in mitochondria are nearly identical whether measured by stopped flow or flash photolysis (Chance and Erecinska, 1971). This suggests that the increase in the effective concentration of CO by photolysis is negligible compared to the CO concentration in equilibrium with the bulk solution. The rate constants for the reaction of the oxidase with oxygen in mitochondria and in a purified preparation are 3 $\times 10^7$ and 8×10^7 $M^{-1}sec^{-1}$ respectively (Chance and Erecinska, 1971). Thus, the kinetic features of cytochrome oxidase remain unchanged when it is solubilized from intact mitochondria. Furthermore, the activities of CO and O_2 in mitochondria are similar to those in aqueous bulk solution.

It may be expected that the mobility of proteins embedded in the mitochondrial membrane is somewhat restricted and the question arises as to how such constraints may affect their kinetic features. For example, an extreme case which has already been discussed is the crystalline state. If the protein of interest is fluorescent, the techniques of dynamic fluorescence depolarization (see Chapter 9) could provide information on the mobility of the molecule. Hemoproteins, however, are generally not fluorescent. Junge and De Vault (1975) applied the technique of photoselection to CO–cytochrome oxidase complex in intact mitochondria and in a sol-ubilized state. Flash photolysis experiments were carried out with linearly polarized light employing a dye-laser as a flash source. Changes in absorbance were then fol-lowed using light polarized parallel or perpendicular relative to the flash polariza-tion. The dichroic ratio $A_\parallel/A_\perp$ thus obtained is 1.27 for CO-cytochrome oxidase in intact mitochondria. This value is close to that of 1.3 obtained in the frozen state at 77°K. Solubilized CO–cytochrome oxidase gives $A_\parallel/A_\perp = 1$ at room tempera-ture, which shows that the rotational relaxation of the oxidase occurs rapidly com-pared to the 300 nsec resolution time of the apparatus. Thus, the solubilized oxidase moves freely in solution. These workers concluded that cytochrome oxidase is capable of restricted rotation in the mitochondrial membrane around an axis perpendicular to the heme plane, although lateral movement is not allowed. This technique has wide applicability to various systems such as red blood cells where the mobility of hemoglobin molecules is of primary importance. Cytochrome P-450 embedded in microsomal and mitochondrial membranes appears to be a good candidate for study by this method.

The diffusion of oxygen and CO from the external environment across the cell membrane is rate-limiting in the combination of these ligands with hemoglobin in red blood cells. Therefore, the rate constant for reactions between hemoglobin and oxygen or CO cannot be measured by the flow method in intact cells. A photolysis experiment with single red blood cells was performed by Antonini et al. (1978), who employed a microscopic detection system. The photolysis was accomplished by opening an electromagnetic shutter placed near a mercury arc lamp which yielded a resolution time of 0.01 sec. The recombination of CO with hemoglobin in the red blood cell obeyed zero-order kinetics and the reaction rate was about one-tenth of that measured in solution. They concluded that an unstirred diffusion layer around the red blood cell slowed down the recombination rate. At present, it remains un-clear whether the rate constants for the reactions of hemoglobin with CO and O_2 are identical in red blood cells and in a purified state. Recently, Hasinoff (1981b) performed the laser photolysis of CO–hemoglobin in a red cell suspension, and concluded that the rate constant was unchanged between red cell and purified state.

Recombination processes after photolysis of CO–myoglobin complexes in car-diac cell homogenates and in cardiac cell suspensions, both in Krebs–Ringer – HEPES solution, are shown in Fig. 14A. The CO concentration was the same in both cases. Semilogarithmic plots for the CO recombination are shown in Fig. 14B. In contrast to the situation in the homogenate, in the cell suspension the reaction is not first order. The over-all pattern of CO recombination in the cell suspension re-sembles that observed in a crystal suspension of myoglobin; the characteristic fea-tures are as follows: the reaction obeys second-order kinetics until more than 90% of the reaction has been completed (Fig. 14B) and the rate is not affected by the CO concentration in bulk solution. Since the content of myoglobin in cardiac cell is 200 nmoles/g of wet weight, its concentration would be approximately 200 μM as-suming it is distributed uniformly. If the space occupied by mitochondria and con-tractile proteins in the cell is taken into account, the local concentration of myoglo-

128

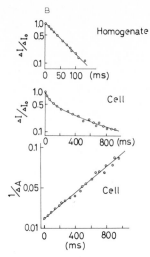

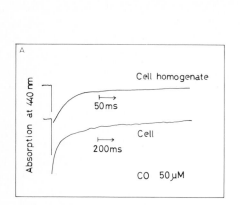

Fig. 14. Recombination of CO and myoglobin after photolysis in an isolated cell suspension and in its homogenate. (A) The CO concentration in the bulk medium was $50\,\mu M$. The myoglobin concentration in the homogenate was about $5\,\mu M$. The medium was Krebs–Ringer–HEPES buffer, pH 7.4. (B) Semilogarithmic plots for top and middle traces. The bottom trace shows that the reaction in the cell suspension was second order.

bin may be about 0.5 mM. The fact that the reaction obeys second-order kinetics until more than 90% of the reaction has been completed shows that the CO generated by photolysis is trapped in the cell and recombines with myoglobin. It appears that CO inside the cell does not escape to the bulk solution through the cell membrane until about 500 msec after the flash. This suggests that the cell membrane may act as a barrier for the diffusion of CO. Thus, the difference between the reaction patterns in crystals and cell suspensions is that CO escapes into the bulk solution faster in crystals than in cardiac cells. The rate constant for the reaction between CO and myoglobin does not differ significantly in the cell and in its homogenate.

8. Flash photolysis in organs and tissues

Unlike an isolated cell suspension, a tissue is a much more complex environment for the analysis of reactions within it. In tissues, cells are surrounded by networks of capillaries, through which oxygen and various substrates are exchanged. In cardiac tissue, for example, capillaries are about $10\,\mu m$ apart and the average diffusion path length for a ligand may be about $5\,\mu m$. Thus, the diffusion of oxygen from capillary to the interior of a cell is rate-limiting for oxygen transport. Another difficulty in tissue studies is the optical masking introduced by the circulation. The technique of hemoglobin-free perfusion for mammalian organs obviates this problem and makes

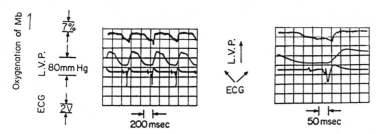

Fig. 15. Oxygenation–deoxygenation of myoglobin in relation to changes in left ventricular pressure during a single contraction–relaxation cycle. Heart was perfused with Krebs–Ringer bicarbonate buffer equilibrated with 95% O_2 + 5% CO_2, at 27°C.

it possible to apply photometric methods to kinetic studies in tissue. Here, we present kinetic studies of CO and O_2 binding to myoglobin and of electron transport in mitochondria employing the hemoglobin-free perfused rat heart. First, we compare the rate of oxygen binding to myoglobin in the beating heart measured by "stopped flow" and flash photolysis. Fig. 15 shows oxygenation–deoxygenation cycles of myoglobin in relation to changes in left ventricular pressure and the electrocardiogram during a single contraction–relaxation cycle (Tamura et al., 1978b). It is clearly seen that there is a periodic cycle of oxygenation–deoxygenation of myoglobin associated with each heart beat. This can be interpreted as follows: oxygen is supplied from capillary to cardiac cell discontinuously at each beat in early systole; thus, the mechanism may be regarded as a "repeated stopped-flow" system for the reaction of oxygen with myoglobin. From the slope of a trace of myoglobin oxygenation in Fig. 15, the rate of oxygen binding to myoglobin in the heart can be calculated. Since about 10% of the total myoglobin reacts with oxygen with a half-time of 10 msec, the rate constant of recombination of myoglobin with oxygen comes out to be about 7×10^6 $M^{-1}sec^{-1}$, assuming myoglobin and oxygen concentrations of 200 and 100 μM, respectively.

This value is significantly lower than that of 3×10^7 $M^{-1}sec^{-1}$ measured in a purified state at 25°C. The local concentration of myoglobin in the cell is probably higher than 200 μM; and the rate of oxygenation of myoglobin occurring in early systole may reflect the diffusion of oxygen from capillary to the cell interior. This would suggest that the concentration of oxygen in the cell during oxygenation is about one-fifth of that in capillary.

Fig. 16 shows a schematic diagram of a flash photolysis experiment to measure the binding of CO and O_2 to myoglobin in the perfused rat heart. The flash source is a xenon flash lamp or a dye-laser. Absorption changes related to the oxygenation state of myoglobin are measured by reflectance spectroscopy with a specially designed dual-wavelength photometer which has a time resolution of about 10 μsec.

130

Fig. 17A shows a kinetic trace of the binding of oxygen to myoglobin in the heart after photolysis of CO–myoglobin in the presence of oxygen. Infusion of 1% CO into the perfusate equilibrated with 95% O_2 + 5% CO_2 converts about 5% of the total myoglobin from oxymyoglobin to CO–myoglobin. The flash dissociates the CO–myoglobin and the myoglobin thus formed reacts with oxygen dissolved in the

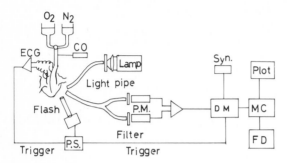

Fig. 16. Schematic diagram of a flash photolysis experiment of perfused rat heart. The absorption changes of myoglobin are measured by dual-wavelength spectrophotometry. The left ventricular wall is illuminated and the reflected light is collected by the other light pipe, which divides into two branches. Interference filters centered at 580 and 620 nm are placed in front of each photomultiplier. Outputs from the two photomultipliers serve as inputs to the difference amplifier, and the absorption difference between 580 and 620 nm is recorded. If necessary, the data can be stored in a disc prior to processing. P.M., photomultiplier; P.S., power supply; DM, digital memory; MC, microcomputer; Syn., synchroscope; FD, floppy disc; Plot, plotter.

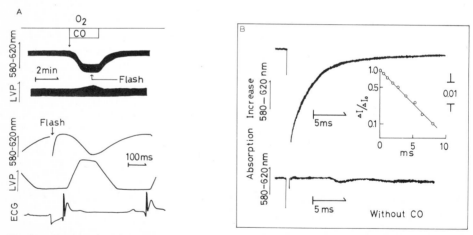

Fig. 17. Combination of oxygen with myoglobin in the heart after flash photolysis. (A) Heart was perfused with oxygen-saturated perfusate, and CO was infused to partially convert oxymyoglobin to the CO form, as shown in the 580–620 nm trace (top). The O_2 concentration was about 100 μM. The oscilloscope trace (bottom) shows the flash photolysis being performed during a single contraction–relaxation cycle. (B) Time-course of the absorption change after photolysis is shown in an expanded time scale (top). A semilogarithmic plot is shown in the inset. No change was observed without CO (bottom).

cytosol. The flash photolysis is performed during the resting state of the cardiac cycle (top trace). Upon photolysis an absorption trace at 580–620 nm jumps and then recovers to the original level with a half-time of about 2 msec. In the absence of CO, no absorption change is observed after the flash. Since this absorbance change is too fast to be explained by the recombination of myoglobin with CO, it is attributed to the formation of oxymyoglobin. This reaction of oxygen and myoglobin obeys first order kinetics with a rate constant of $3 \times 10^7 \ M^{-1}sec^{-1}$. This value is consistent with that obtained in a solution of purified myoglobin. Thus, the kinetic features of myoglobin are similar in cardiac tissue and in the purified myoglobin. Thus, we conclude that the apparent slow rate obtained from the experiment shown in Fig. 15 is due to the fact that the diffusion of oxygen from capillary to the cell is rate-limiting.

The reaction between CO and myoglobin in the heart was also studied by flash photolysis (Fig. 18). In this case, the heart was subjected to anoxic perfusate and the CO concentration in the inflow perfusate was increased in a stepwise fashion. Photolysis was performed at the times indicated by the labels on the top trace. Trace D is obtained after CO is removed from the perfusate. The incompleteness of recovery in this case is due to washout of CO from cardiac cell to capillary. Although the CO concentration in C is more than 10-fold higher than in A, the rate of CO recombination does not differ significantly. Semilogarithmic plots of the recombination do not obey first-order kinetics as shown in Fig. 19. The initial rate appears to be almost independent of CO concentration, whereas in the slow phase first-order kinetics obtain and the rate depends on the CO concentration in the per-

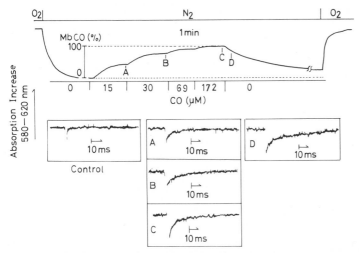

Fig. 18.   Recombination of CO with myoglobin in the heart after flash photolysis. The top trace shows formation of CO–myoglobin in the heart. The heart was subjected to an anoxic condition by perfusion with nitrogen-saturated perfusate. CO was infused stepwise at 25°C. Flash photolysis was performed in the resting state at various CO concentrations as shown by A, B, C and D. Each reaction after photolysis is shown below. Control represents the result in the absence of CO.

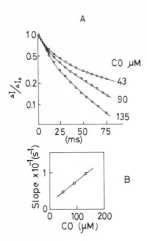

Fig. 19. (A) Semilogarithmic plots of the data in Fig. 18. (B) The dependence of slopes in the slow stationary phases on the concentration of CO in the perfusate.

fusate. The over-all pattern of the semilogarithmic plots resembles that observed in the reactions of isolated cardiac cells and myoglobin crystals. Photolysis, therefore, increases the effective concentration of CO in the cell and increases the pseudo-first-order rate constant for recombination in the initial phase. According to the trace of Fig. 18A, the half-time for the recombination is approximately 10 msec in the presence of $15 \mu M$ CO. The rate constant calculated from this CO concentration is about $5 \times 10^6 \, M^{-1}sec^{-1}$, which is 10-fold larger than that measured in a purified state. On the other hand, the rate constant can also be calculated from the slow stationary phase to be $1.8 \times 10^5 \, M^{-1}sec^{-1}$, which is close to that measured in a purified state. This discrepancy cannot be explained by the average increase in the effective CO concentration, which is $50 \mu M$ at most. The difference in rate constants might be related to the localization of myoglobin in the cell, in which case the discrepancy could be explained by postulating the appearance of a layer of locally concentrated CO after photolysis.

The rate of mitochondrial electron transport in the heart has been measured by Chance et al. (1972), who followed fluorescence changes of pyridine nucleotides and flavoproteins in mitochondria by surface fluorometry. The principle employed is similar to that of flash photolysis, namely, the activation of a $CO-O_2$ exchange reaction with cytochrome oxidase. Perfused heart is subjected to a CO-saturated perfusate which converts the cytochrome oxidase and the myoglobin to their CO form. Oxygen is then introduced into the heart and the cytochrome oxidase is freed from CO by flash photolysis. The oxygen oxidizes the oxidase, which in turn triggers reactions in the mitochondrial electron transport system, such as the oxidation of pyridine nucleotides and flavoproteins. As previously shown in Fig. 17, CO–myoglobin is converted to oxymyoglobin simultaneously with the above reaction, which masks the absorption change due to the cytochrome oxidase reaction. Mea-

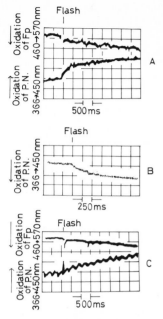

Fig. 20. Oxidation of pyridine nucleotides and flavoproteins in the heart after photolysis of CO–cytochrome oxidase. Fluorescence was excited at 366 and observed at 450 nm for the pyridine nucleotides. The corresponding wavelengths were 450 and 560 nm for the flavoproteins. Rat heart was perfused with Krebs–Ringer bicarbonate buffer. (A) Fluorescence change after photolysis. (B) Expanded time scale of the jump in the fluorescence of pyridine nucleotides. The data were obtained by averaging 10 traces. (C) In the absence of CO.

surements of pyridine nucleotide fluorescence on the other hand, can reveal changes in the oxidation-reduction of the mitochondrial component independently of the oxygenation of myoglobin in the heart. As seen in Fig. 20A, the fluorescence trace of pyridine nucleotides jumps after the flash. Fig. 20B shows a fast record obtained by averaging ten individual traces in order to improve the signal-to-noise ratio (Tamura and Chance, unpublished observation). The half-time of the oxidation of pyridine nucleotides by flash photolysis is approximately 100 msec; without CO, no fluorescence jump is seen upon photolysis (Fig. 20C). Since myoglobin oxygenation occurs much faster than the fluorescence change (cf. Fig. 17), the slow change seen in Fig. 20 is due to the electron flow in mitochondria triggered by the rapid oxidation of cytochrome oxidase by molecular oxygen. Isolated mitochondria give a half-time of about 100 msec for the oxidation of pyridine nucleotides by oxygen. This value agrees with that obtained in the perfused heart. Thus, we have come to the conclusion that kinetic parameters of purified mitochondrial electron transport systems do not differ from those in vivo.

134

9. Special applications to cell physiology

The use of flash photolysis in conjunction with other than optical detection systems considerably enhances the versatility of this technique. For example, reaction mechanism involved in hormonal responses and electrophysiological process can be investigated. Lester and Chang (1977) applied this type of approach to measure the

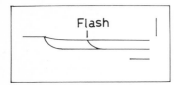

Fig. 21. Flash-induced relaxations associated with jumps of membrane voltage and of agonist concentration at Electrophorus electroplaques. At the start, medium contained inactive *cis* form of bis-Q. A flash light increased the *trans*-bis-Q (active form) concentration to about 240 nM. Following the concentration jump, the conductance increased exponentially. Calibration marks: 24 msec and 24 mA/cm². Temperature, 8°C.

mean acetylcholine open-channel lifetime in single cells of Electrophorus by exploiting the photochemical transition of the probe, bis-Q-(3,3′-bis(2-trimethylammonium-methyl)azobenzene), which is a powerful agonist of the acetylcholine receptor in its *trans* isomer form. After prolonged exposure to 320 nm light, the probe exists predominantly in its *cis* inactive form. Concentration jumps can then be elicited by a light flash at 420 nm, which causes an isomerization from the *cis* to the *trans*, active form. As seen in Fig. 21 (Barrantes, 1979), at constant voltage, the agonist-induced current increases along an exponential time-course with a time constant that reflects the relaxation of the conductance mechanism. This in turn is related to the mean open-channel lifetime. This photo-induced "concentration jump" method is ideally suited to measure rates of elementary processes in receptor–ligand interactions and awaits further development in cell physiology.

10. Summary

Flash photolysis techniques are indispensable tools for kinetic studies of biological reactions occurring in heterogeneous systems. The data obtained by applying these techniques to various reactions in crystals, on membrane-bound enzymes, and in cells and organs show that the rate constants for reactions of proteins with ligands are not significantly changed even if their motions are restricted in such heterogeneous systems. Intrinsic rate constants cannot be measured by flow methods when the acceptor sites are locally concentrated.

References

Alpert, B., Banerjee, R. and Lindqvist, L. (1974) The kinetics of conformational changes in hemoglobin, studied by laser photolysis, Proc. Natl. Acad. Sci. (U.S.A.), 71, 558–562.

Antonini, E., Brunori, M., Gialdina, B., Benedetti, P.A., Bianchini, G. and Grassi, S. (1978) Kinetics of the reaction with CO of human erythrocytes, Observations by single cell spectroscopy, FEBS Lett., 86, 209–212.

Austin, R.H., Beeson, K.W., Eisenstein, L., Frauenfelder, H. and Gunsalus, I.C. (1975) Dynamics of ligand binding to myoglobin, Biochemistry, 14, 5355–5373.

Barrantes, F.J. (1979) Endogeneous chemical receptors, Some physical aspects, Annu. Rev. Biophys. Bioeng., 8, 287–321.

Chance, B. (1978) Cytochrome kinetics at low temperatures, Trapping and ligand exchange, Methods Enzymol., 54-E, 102–111.

Chance, B. and Erecinska, M. (1971) Flow flash kinetics of the cytochrome a_3–oxygen reaction in coupled and uncoupled mitochondria using the liquid dye laser, Arch. Biochem. Biophys., 143, 675–687.

Chance, B., Schoener, B. and Yonetani, T. (1965) The low-temperature photodissociation of cytochrome $a_3^{2+} \cdot$ CO, in: T.S. King, H.S. Mason and M. Morrisson (Eds.), Oxidase and Related Redox Systems, Vol. 2, John Wiley, New York, pp. 609–621.

Chance, B., Ravilly, H. and Rumen, N. (1966) Reaction kinetics of a crystalline hemoprotein: an effect of crystal structure on reactivity of ferrimyoglobin, J. Mol. Biol., 17, 525–534.

Chance, B., Sarkovitz, I.A. and Kovach, A.G.B. (1972) Kinetics of mitochondrial flavoproteins and pyridine nucleotide in perfused heart, Am. J. Physiol., 223, 207–218.

De Vault, D. (1978) Nanosecond absorbance spectrophotometry, Methods Enzymol., 54-E, 32–46.

Frauenfelder, H. (1978) Principles of ligand binding to heme proteins, Methods Enzymol., 54-E, 506–532.

Gibson, Q.H. (1978) Flash photolysis techniques, Methods Enzymol., Vol. 54-E, 93–101.

Gibson, Q.H. and Greenwood, C. (1963) Reactions of cytochrome oxidase with oxygen and carbon monoxide, Biochem J., 86, 541–555.

Hartridge, H. and Roughton, F.J.W. (1923) The velocity with which carbon monoxide displaces oxygen from combination with hemoglobin, Proc. Roy. Soc. London, Ser. B, 94, 336–367.

Hasinoff, B.H. (1981a) Flash photolysis reactions of myoglobin and hemoglobin with carbon monoxide and oxygen at low temperature, Evidence for a transient diffusion-controlled reaction in supercooled solvents, J. Phys. Chem., 85, 526-531.

Hasinoff, B.H. (1981b) Kinetics of carbon monoxide and oxygen binding to hemoglobin in human red blood cell suspensions, studied by laser flash photolysis, Biophys. Chem., 13, 173–181.

Hoffman, B.M. and Gibson, Q.H. (1978) On the photosensitivity of liganded hemoproteins and their metal-substituted analogues, Proc. Natl. Acad. Sci. (U.S.A.), 75, 21–25.

Iizuka, T., Yamamoto, H., Kotani, M. and Yonetani, T. (1974) Low temperature photodissociation of hemoproteins: Carbon monoxide complex of myoglobin and hemoglobin, Biochim. Biophys. Acta, 371, 126–139.

Ikeda-Saito, M., Iuzuka, T., Yamamoto, H., Kane, F.J. and Yonetani, T. (1977) Studies on cobalt myoglobins and hemoglobins, Interaction of sperm whale myoglobin and glycera hemoglobin with molecular oxygen, J. Biol. Chem., 252, 4882–4887.

Ishikawa, T., Tamura, M. and Yamazaki, I. (1980) A kinetic study on the diffusion-coupled reaction of a basic horseradish peroxidase adsorbed on the carboxymetylcellulose membrane, J. Biol. Chem., 255, 10764–10770.

Junge, W. and De Vault, D. (1975) Symmetry, orientation and rotational mobility in the a_3 heme of cytochrome oxidase in the inner membrane of mitochondria, Biochim. Biophys. Acta, 408, 200–214.

Ke, B. (1972) Flash kinetic spectrophotometry, Methods Enzymol. 26B, 25–53.

Kobayashi, K., Tamura, M., Hayashi, K., Hori, H. and Morimoto, H. (1980) Electron paramagnetic resonance and optical absorption spectrum of the pentacoordinated ferrihemoproteins, J. Biol. Chem., 255, 2239–2242.

Lester, H.A. and Chang, H.W. (1977) Response of acetylcholine receptors to rapid photochemically induced increase in agonist concentration, Nature (London), 266, 373–374.

Miki, M., Kobayashi, K., Tamura, M. and Hayashi, K. (1982) Biochim. Biophys. Acta, in press.

Sawicki, C.A. and Gibson, Q.H. (1976) Quaternary conformational changes in human hemoglobin studied by laser photolysis of carboxyhemoglobin, J. Biol. Chem., 251, 1533–1542.

Sharrock, M. and Yonetani, T. (1977) Low temperature flash photolysis studies of cytochrome oxidase and its environment, Biochim. Biophys. Acta, 462, 718–730.

Tamura, M., Kobayashi, K. and Hayashi, K. (1978a) Flash photolysis studies on nitric oxide ferrihemoprotein complexes, FEBS Lett., 88, 124–126.

Tamura, M., Oshino, N., Chance, B. and Silver, I. (1978b) Optical measurements of intracellular oxygen concentration of rat heart in vitro, Arch. Biochem. Biophys., 191, 8–22.

Time-resolved X-ray scattering from solutions using synchrotron radiation

JOAN BORDAS[1] and ECKHARD MANDELKOW[2]

[1]European Molecular Biology Laboratory, Hamburg Outstation,
Notkestrasse 85, D-2000 Hamburg,
and [2]Max-Planck-Institute for Medical Research,
Jahnstrasse 29, D-6900 Heidelberg, F.R.G.

Contents

1. Properties of synchrotron radiation 138
   1.1. SR has a continuous wavelength spectrum 138
   1.2. SR has excellent directional properties 139
   1.3. SR is polarized 139
   1.4. SR has a well defined time structure 139
   1.5. SR is an intense source 140
   1.6. SR has a small source size 140
2. Instrumentation 140
   2.1. The technical problem 140
   2.2. X-Ray optics 140
   2.3. Position-sensitive detector 145
   2.4. Data-acquisition system 146
   2.5. Temperature-jump device 147
   2.6. Data reduction 149
3. Solution scattering 150
   3.1. General aspects 150
   3.2. Instrumental parameters 153
   3.3. Contrast and resolution 153
   3.4. Radiation damage 155
4. X-Ray studies of microtubule assembly 156
   4.1. Introduction to microtubules 156
   4.2. Protein preparation 159
   4.3. Assembly and disassembly induced by temperature jump 161
   4.4. Difference intensity patterns 162

Abbreviations: GTP, guanosine triphosphate; MAPs, microtubule-associated proteins; SR, synchrotron radiation.

R.I. Sha'afi and S.M. Fernandez (Eds.), Fast Methods in Physical Biochemistry and Cell Biology
© *1983 Elsevier Science Publishers*

138

    4.5. Intensity-versus-temperature plots: hysteresis 164
    4.6. Correlation analysis 165
    4.7. Radius of gyration of tubulin 166
    4.8. Implications for models of microtubule assembly 167
5. Conclusion 169
Acknowledgements 169
References 170

1. Properties of synchrotron radiation

Whenever a charged particle changes its speed it emits electromagnetic radiation. This phenomenon can be explained from Maxwell's equations of electromagnetism, and manifestations of it are abundant: radio antennas, microwave ovens, lasers, X-ray generators, etc. The effect is also responsible for the synchrotron radiation (SR) emission by particle accelerators. These devices keep particles circulating in a closed orbit by conferring to them a centripetal acceleration by means of magnetic fields. Energy is lost in the form of synchrotron radiation and has to be restored by electric fields. This nuisance to the high-energy physicist is a bonus to the spectroscopist or the X-ray diffractionist. The smaller the rest mass of a circulating particle the more efficient is the conversion of its energy into SR. Consequently only electron or positron accelerators are useful in practice as SR sources.

The properties of SR are discussed in several reviews (for example, Godwin, 1969; Kunz, 1979), and information on future applications of SR sources can be found in the proposal by the European Science Foundation (ESF) for a European Synchrotron Radiation Facility (ESRF) (1979). The most relevant characteristics of SR are summarized below.

1.1. SR has a continuous wavelength spectrum
From simple considerations one might expect that an observer would see emission at a frequency corresponding to the circulating one as the particle goes around orbit. However, since the particle approaches him at relativistic speeds as it bends around the tangent point, the Doppler blue shift induces a continuum of frequencies which extend into the very short wavelength region. One witnesses the equivalent phenomenon when a car blowing its horn approaches us: then the sound shifts towards high pitches, i.e. short wavelengths. How far into the hard-radiation region the emitted light extends depends mainly on the particle energy and the bending radius of its orbit at the point of observation.

The spectral profile of SR is strictly calculable (Fig. 1a). An important parameter for an SR source is the critical wavelength, λ_c, which is related to the hardness of the radiation. It is given by

$$\lambda_c = 5.6 \ R/E^3 \tag{1}$$

where E is the energy of the particle in GeV, and R is the orbital radius in meters.

One can see from above that the higher the energy of the accelerator and the smaller the orbital radius, the shorter will be the critical wavelength.

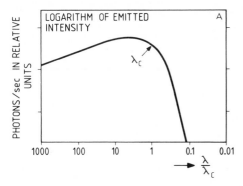

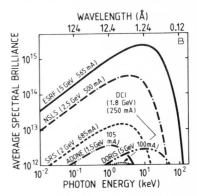

Fig. 1. (a) General emission spectrum for an electron or positron accelerated in a curved orbit. The maximum of intensity is between 1 and 10 times the critical wavelength. (b) Comparison of spectral brilliance for several existing synchrotron radiation facilities: ADONE (Frascati, Italy), DCI (Orsay, France), DORIS (Hamburg, F.R.G.), SRS (Daresbury, Great Britain); under commission: NSLS (Brookhaven, U.S.A.); or projected: ESRF (Europe). Both figures from ESF Proposal (1979).

1.2. SR has excellent directional properties

The radiation coming from circulating electrons is focused forwards towards the observer. In a storage ring the radiation emitted at $\lambda = \lambda_c$ emerges at a mean angle

$$\theta = m_0 c^2 / E \tag{2}$$

where m_0 is the rest mass of the particle (0.511 MeV for an electron or a positron), E is its energy, and c is the speed of light.

1.3. SR is polarized

An observer looking at the moving particle along a tangent to its orbit sees it as a classical dipole emitting radiation. As a result the radiation is 100% polarized in the plane of the orbit. Away from the orbital plane the radiation is elliptically polarized to a degree determined by the viewing angle.

1.4. SR has a well defined time structure

The particles stored in the storage ring circulate in compact groups (bunches) in synchrony with a radiofrequency electric field which restores to them the energy they have lost by SR emission. The bunch length ranges typically between 5 psec to 1 nsec and the time between them can range from a few nanoseconds to a few microseconds.

1.5. SR is an intense source

The intensity radiated is proportional to the number of particles stored, i.e. the circulating current. The number of photons per second emitted at λ_c, with a wavelength bandpass of 0.1% and integrated over the vertical aperture is given by:

$$N(\lambda) = 1.6 \cdot 10^{10} \cdot E \text{ per mrad horizontal} \tag{3}$$

1.6. SR has a small source size

The source size depends on the extent of the particle bunches. Electron-focusing techniques indicate that it should be possible to contain the emitting bunches within a cross-section of dimensions smaller than 0.1 mm × 0.1 mm. This value is not reached in present-day SR sources, and their performance depends on the design of the accelerator. For DORIS (DESY, Hamburg) where we have conducted our experiments, the following applies. Energies range up to 5 GeV, although normal running conditions for SR work have so far been around 3.6 GeV. Circulating currents have reached 100 mA. At 5 GeV and 100 mA one could obtain 8×10^{12} photons/sec, mrad horizontal aperture, 0.1% bandpass. The degree of polarization of the light emitted by DORIS is about 80%. This emission is vertically contained within 2–4 mrad for the critical wavelength. For DORIS, λ_c varies between 0.25 nm at 3 GeV and 0.054 nm at 5 GeV. The size of the source in DORIS is around 2.1 mm × 4.5 mm at our particular tangent point. These values can vary a great deal depending on running conditions of the machine.

From Liouville's theorem one can prove that the brightness of a source, defined as photons/unit area, unit solid angle, unit time, is a constant. Independently of whatever manipulations one cares to perform (lenses, collimators, monochromators, etc.) the brightness cannot be increased. The reverse is not true and one will always observe some deterioration due to the imperfections of the optical elements. These concepts lead to brightness as a measure of the merit of a source. A convenient definition of the figure of merit (FOM) of a white source is its spectral brilliance defined as brightness, in units of photons/sec/mm²/mrad² in 0.1% wavelength bandpass. Taking the particular case of DORIS at 5 GeV and 100 mA circulating current, we find that the figure of merit is 2.9×10^{12} for 10 KeV photons ($\lambda = 0.124$ nm). This value is several orders of magnitude higher than the most powerful X-ray generators, therefore opening up the possibility of obtaining X-ray-scattering patterns in the time-resolved mode.

In order to put the present status of SR sources into context we show in Fig. 1b the spectral brilliance for a variety of existing sources (ADONE, DCI, DORIS, SRS), sources under commission (NSLS) or projected (ESRF).

2. Instrumentation

2.1. The technical problem

The maximum intensity of solution scattering occurs at very small angles (see section 5). The smaller the angle of measurement the better an appreciation one can obtain about the overall size of the object. Thus the camera must provide a beam which is collimated enough and with a cross-section small enough to separate the primary beam from the one scattered at very small angles. Failure to achieve adequate conditions in these points results in poor angular resolution and/or the need to use large volumes.

It is also desirable to make measurements at larger scattering angles, which yield information about changes in the internal structure of the object in solution. These requirements determine detector length and resolution, as well as the ideal wavelength bandpass.

The intensity scattered by protein solutions is weak, ranging between 10^{-4} and 10^{-8} times the incident flux. This defines the need for an intense primary beam. The requirements for high flux arise also from the fact that the specimen is in a solvent and inside a container whose absorption reduces the incoming intensity and contributes to the unwanted background.

Finally many interesting specimens are available in small quantities and one wants to contain them in a small as possible volume. This demands a small cross-section of the beam at the specimen position.

The above boundary conditions require the production of a quasimonochromatic X-ray beam with a high figure of merit (i.e. the higher the photons/sec//unit area/ unit solid angle the better the experimental accuracy). The considerations derived from Liouville's principle indicate that the figure of merit of the source is the ultimate limitation.

Since SR has a continuous spectrum it is important to consider the wavelength bandpass when a spectral continuum is available. Increasing the bandpass results in a higher intensity but also in a deterioration of the resolution. This effect is illustrated by an example in Fig. 2. The information lost can be recovered in part by deconvolution methods, but this requires higher counting statistics which offset the gain in primary intensity. (As discussed by Luzzatti and Tardieu (1980), the statistical error in the data limits the number of structural parameters which can be deduced from a solution-scattering pattern.) Ideally one would like to work with a variable bandpass monochromator system in order to gain higher intensity when the needs for resolution do not preclude it. Unfortunately the technology to achieve this with X-ray monochromators, without undue deterioration of the figure of merit of the beam, has not yet been developed.

2.2. X-Ray optics

In scattering experiments the information on some periodicity in the structure is related to the scattering angle 2θ and the wavelength λ by Bragg's law:

$$n\lambda = 2\,d\,\sin\theta \tag{4}$$

We will refer to the quantity $S = 2\sin\theta/\lambda$ as the Bragg spacing. A structure determination consists of measuring the scattered intensities at all possible values of S. This can be achieved in two ways. Either one irradiates the specimen with the continuum of wavelengths provided by SR and analyzes the energies of photons scat-

142

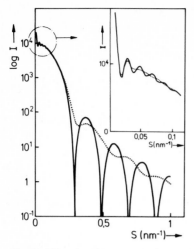

Fig. 2. Wavelength spread and spatial resolution. The theoretical scattering is shown for a model structure consisting of two spheres of diameter 5 nm separated by 50 nm. The mean wavelength of X-rays is 0.15 nm, with a wavelength spread $\Delta\lambda/\lambda$ of 0 (solid line) or 30% (dotted line). The main curve is a logarithmic plot of the scattering intensity up to $S = 1.0$ nm^{-1}. The outer part is dominated by the scattering of the spheres which is independent of their relative arrangement. Increasing the wavelength spread averages the intensities out, and peaks beyond the first subsidiary are shifted away from their correct positions. The loss of information could be retrieved by deconvolution methods only if intensities are measured with higher statistical accuracy. Thus the gain in primary intensity obtained by allowing a wide bandpass is countered by the loss of structural information and/or time resolution. The inset shows a linear plot of the small angle region up to $S = 0.1$ nm^{-1}. It is dominated by the interference between the two spheres. As before, beyond the first subsidiary the maxima become averaged out and shifted by the wavelength spread. Thus increasing the bandpass degrades the information both on the overall structure and on the subunit shape. The deterioration of relative contrast becomes worse with increasing S, and only the zero-angle scatter remains unaffected.

tered into a fixed angle 2θ. This is the energy-dispersive method (Bordas et al., 1976; Bordas and Randall, 1978). Alternatively one uses X-ray optical elements to produce a monochromatic beam and measures the angular dependence of scattered radiation. The camera X13 is based on the latter principle, combining totally reflecting X-ray mirrors to focus the beam in the vertical plane and a crystal monochromator to focus in the horizontal plane. This approach was developed to study muscle and viruses with rotating anode X-ray sources (Holmes and Barrington-Leigh, 1974; see Huxley and Brown, 1967; Finch and Holmes, 1967). The adaptation of the double focusing principle to a SR source has undergone a series of refinements (Rosenbaum et al., 1971; Barrington-Leigh and Rosenbaum, 1974; Haselgrove et al., 1977; Webb et al., 1977). The present instrument X13 was constructed on the basis of this previous experience adapted to the geography of the EMBL hall at DORIS (Fig. 3) (Hendrix et al., 1979).

In Fig. 4a we show the reflectivity of quartz mirrors as a function of wavelength for several angles of incidence. At 0.15-nm wavelength the angle of incidence in the

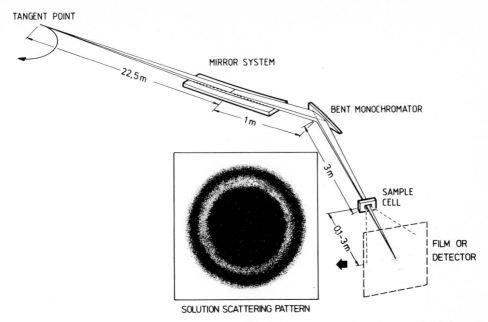

Fig. 3. Layout of instrument X13 illustrating its construction principles. The radiation emitted along a tangent to the orbit is vertically focused by a set of totally reflecting X-ray mirrors and horizontally by a bent crystal monochromator. The crystal is set to reflect quasi-monochromatic radiation of 0.15 nm wavelength and a bandpass of 0.1%. The higher orders (λ/n) are rejected by the mirror (Fig. 4a). The detector is placed at the focal plane while the specimen intercepts the beam 3 m behind the monochromator. The solution-scattering pattern is rotationally symmetric so that a detector placed along a radial line records all available information. From Bordas et al. (1980).

mirror should be of the order of 3 mrad for total reflection. This also eliminates the unwanted monochromator harmonics at shorter wavelengths. In order to increase the vertical angle of acceptance the instrument X13 contains 8 segmented mirrors distributed tangentially to the surface of a cylindrical ellipsoid. This is the ideal shape for focusing a point source (which SR effectively is for our purposes) into a line focus (Fig. 4b).

The crystal monochromator intercepts the white SR reflected off the mirror surfaces, selects from the continuum a certain quasimonochromatic wavelength and diverts it into angles given by Bragg's law. For our aims a Germanium (111) monochromator, with its surface cut parallel to the 111 planes or in some cases at an angle relative to the crystallographic planes, appears to be the most suitable choice. The amount of asymmetry in the cut is largely dictated by the size of the available specimens (see Hendrix et al., 1979).

The focusing of the monochromatic beam can be simply achieved by cutting the monochromator in a triangular shape, holding its base firmly and pushing the tip of the triangle. This results in a bending of the monochromator into a perfect circle

144

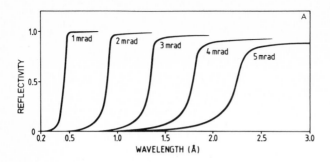

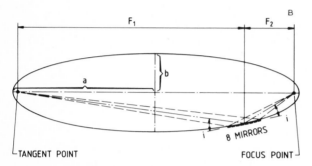

Fig. 4. Total reflection from mirrors. (a) Reflectivity of quartz as a function of wavelength for several angles of incidence. At 3 mrad the mirror eliminates the spectrum below 0.13 nm, and the monochromator located downstream selects only the wavelengths of 0.15 nm. From Hendrix et al. (1979). b shows the positions of the 8 mirror segments. The ideal shape for line focusing of a point source is a cylindrical ellipsoid. This ideal shape is approximated by positioning the segmented mirrors along tangents to the ellipsoidal surface defined by the source at one of the foci and the beam focus at the other, with the constraint that the angle of incidence (i) on the mirrors is around 3 mrad. From Hendrix et al. (1979).

which at the distances involved is a good approximation to the ideal surface shape.

In practice one has achieved a quasimonochromatic beam with a band pass of $\Delta\lambda/\lambda = 0.1\%$, focused over an area of approximately 1 mm^2 or less and with a vertical and horizontal convergence greater than or equal to 3 mrad and 1.25 mrad resp. In useful running conditions this beam contains between 10^{10} and 10^{12} photons/sec depending on the energy and the current in the accelerator. This means that one can work from 80 nm down to atomic resolutions. If one desires to work at even smaller angles one is forced to further collimate the beam at the expense of losing primary intensity.

A better geometry for SR work consists of inverting the order of the optical elements, i.e. monomochromator followed by mirror system. This achieves cleaner background, better angular resolution, and longer lifetimes for the delicate mirror surfaces. Such a device has recently been commissioned at one of the new EMBL beam lines (X33, Koch and Bordas, 1983).

2.3. *Position-sensitive detector*

In order to make optimal use of the incoming intensity the scattering patterns must be recorded by a position-sensitive detector. Ideally one would like a two-dimensional detector which counts every photon (100% efficiency) with negligible coincidence loss. The spatial resolution of the detector should be at least as good as the size of the focal spot, with an area as large as the limit of resolution provided by the specimen. The detector system should be able to count as fast as the arrival of scattered quanta. To our knowledge, such a device has not yet been built. The technical difficulties mainly arise from the high speed required in the digitization of the positional information and from the size of storage needed for the large amount of data. Other problems arise from the need to contain a gas at high pressure using a window which is thin enough to transmit the X-rays (for a review of the state of the art, see Bordas et al., 1982).

For our applications we have relied on linear position determination. Since solution scattering patterns have circular symmetry (Fig. 3) a detector measuring along a radial line will collect all available information. However, the statistical accuracy will be considerably reduced over what one would obtain covering the whole scattering plane.

The detector (Fig. 5) is based on the delay-line method (Gabriel, 1977). When an X-ray photon traverses the gas of the detector chamber it loses energy due to ionization of gas molecules. In a Xe–CO_2 mixture this is 1 electron-hole pair for about 27 eV of energy loss. By means of a thin anode wire at high voltage (+4 kV) one produces a region of high electric field in the gas. In the proximity of the wire the charges are amplified in an avalanche-like fashion. On the cathode side of the detector one makes these charges propagate along a delay line. The differences in time of arrival at the respective ends of the delay line correspond to the positions at which the photons arrived.

For an efficient collection of photons one requires a high stopping power in the gas, i.e. high pressure and/or depth. Since an increase of detector depth may result in loss of resolution and parallax problems, it is better to improve stopping power by increasing pressure. The detector we have used is equipped with a 400 nsec total delay which means that no more than 2.5×10^5 positional determinations/sec can be performed with less than 20% coincidence or dead time loss.

The detector-associated electronics have to digitize the differences in time of arrival at the ends of the delay line. Classical methods have relied on time-to-pulse-height conversion followed by pulse-height analysis. A considerable improvement in speed was recently achieved by direct time digitization (Bordas et al., 1980). The system is capable of counting about 3×10^5 events/sec and enables us to obtain the statistical accuracy needed for the interpretation of transient states in the sample. A promising recent development in this field has been the design of a very fast analog-to-digital converter (ADC) which is capable of producing 256 picture elements on a 50-nsec delay line with a total dead time of less than 80 nsec (J. Hendrix, personal communication).

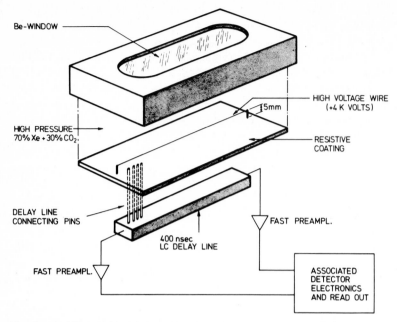

Fig. 5. Position-sensitive detector. This diagram illustrates the principles behind the linear position-sensitive detector. A mixture of Xenon (70%) and CO_2 (30%) is contained inside the chamber at 4 atm. The chamber has a beryllium window (thickness 0.3 mm) which allows the transmission of X-ray quanta. A thin steel wire (diameter 20 μm) is held at high voltage (+4 kV). Because of the small diameter of the wire there is a very high electric field in its vicinity. When a photon strikes the gas molecules at a certain point they become ionized. The charge is further multiplied by the high field in an avalanche-like fashion. A resistive coating on the backing plate of the chamber is tapped by pins that transmit the created charges into a delay line. Its purpose is to delay the propagation of charges to speeds that become electronically measurable. We have used an LC delay line with a total propagation time of 400 nsec. The charges emerging at either end of the delay line are amplified by fast preamplifiers and their differences in time of arrival are then determined by the associated detector electronics and readout (Fig. 6). The implementation of this detecting method is due to A. Gabriel and described in Gabriel (1977).

2.4. Data-acquisition system

In a typical experiment we want to record the scattering pattern (256 channels) as a function of time (256 time frames of programmable length) and the physical or chemical state of the specimen. Thus one has to record the time dependence of two types of information: position (scattering angle) and physical–chemical parameters. In the case of microtubule assembly the parameters are temperature and the time at which assembly is initiated, either by temperature jump or rapid mixing.

The block diagram of the organization of the data-acquisition system is given in Fig. 6 (details in the caption and in Bordas et al., 1980). One of the major difficulties in this type of experiment is the sheer volume of information one has to collect for a proper characterization of the sample. With 256 detector channels and 16 calibration channels one experiment contains up to $256 \times (256 + 16) = 69\,632$ data items (16 bit each), and several dozen experiments can be performed during one 8-h shift of beam time. The present system can cope with this data at a rate of about 1 MHz, and time framing can be chosen between 2 μsec and 90 min.

In general the limitation on the rate of data collection arises from the part of the system with the slowest time constant (as a rule of thumb, percent dead time = 2 × counting rate × slowest time constant). From the point of view of the chamber the rate is limited by space charge buildup (around 10^4 events/mm^2/sec). For our detector arrangement this translates into a maximum of the order of 10^6 events/sec. A 400-nsec delay line will only allow 2.5 × 10^5 events/sec at 20% dead time. The time digitizer has a comparable maximum rate, 3 × 10^5, for the same dead time. The data acquisition system is 4 times faster (approx. 10^6 events/sec). Under optimal storage ring conditions we have been able to carry measurements at rates corresponding to a total scattered intensity of about 2.5 × 10^7 photons/sec, corresponding to 10^{11}–10^{12} incident photons. This data rate would have to be collected by the ideal two-dimensional detector. However, because of the limitations in detector and associated electronics we have to content ourselves with a maximum rate of the order of 10^5.

For this reason one is forced to use a narrow slit in front of the detector which reduces the flux by a factor of more than 10^2. For solution-scattering patterns, and in the absence of the ideal 2D detector, an effective line scan through the pattern is necessary also in order to avoid deconvolution problems (Luzzatti, 1960). One can see that major improvements in the various aspects of detector technology are required to make use of the intensity available even in the present SR sources.

2.5. Temperature-jump device

In principle the chemical reactions to be studied by time-resolved X-ray scattering could be induced by any of the classical perturbation methods, such as temperature jump, rapid mixing, or pressure jump, etc. (see Eigen and DeMaeyer, 1974). One severe constraint is that the glass or sapphire windows used for optical detection methods are opaque for X-rays. Thus one has to replace them by thin sheets of mica, Lindemann glass, or beryllium. This may lead to transient changes in effective cell thickness which can obscure the signals arising from the chemical reaction. The problem is noticeable in mixing devices and makes pressure-jump experiments very difficult.

For most experiments on microtubule assembly we have used temperature jumps or scans. The specimens are contained in a thermostatted cell with mica windows which are sufficiently transparent to X-rays (30–50 μm). The cell is constructed in a hollow copper block, through which one drives either very cold ($-20°C$), cold ($0°C$), warm ($37°C$) or hot ($70°C$) water or water/glycerol mixtures. With a combination of fast valves and electronic servoloops one can change and stabilize the temperatures with time constants of about 1–2 sec. Temperature jumps were typically performed between 0 and $37°C$ (see Renner et al., 1983).

148

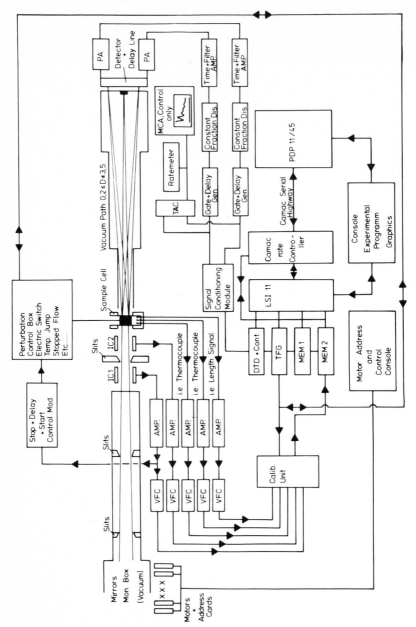

Fig. 6. Block diagram for the control and data acquisition system for time-resolved X-ray scattering studies. The control of the X-ray optical elements (top left-hand side) is achieved by a series of motors which have to move mechanical parts in the non-accessible areas (vacuum and high radiation levels). This includes the motors for adjusting the positions of mirrors, monochromator, guard slits, etc. Each motor has an address card which can be activated from a remote control console by the experimentalist. Detector and data acquisition system: There are two types of information, positional information of an

2.6. Data reduction

The data collected during an experiment are stored on magnetic disks attached to a PDP 11/45 computer. They are processed using a general data-handling program, INSCOM (Koch and Bendall, 1981). This involves several steps: correction of input data, determination of kinetic parameters, extraction of structural information and comparison with models. During the initial stage the following operations are performed.

(a) All spectra are normalized with respect to the incident intensity which decays slowly due to loss of positrons in the storage ring.

(b) The spectra are divided by the detector efficiency measured separately by exposing the detector to a homogeneous source of radiation (^{57}Fe at a distance of 50 cm).

(c) The spectra are corrected for background due to scattering from solvent and optical elements, measured with a buffer-filled cell.

---

event in the detector, and calibration information of the state of the experiment, such as temperature of the specimen, time elapsed since the initiation of the reaction and primary beam intensity at the specimen. There are two ionization chambers (IC1, IC2), one before and one after the guard slits immediately upstream from the sample. They provide a current from which the incoming intensity is obtained. They also measure the stability of the beam position which is of critical importance in the subsequent data analysis. The signals are amplified (AMP.), and the voltage output of AMP. is converted into a frequency by a voltage-to-frequency converter (VFC). The signals from the VFC are digitized by counters in the calibration channel unit module (Calib. Unit). The digitized number is given a time information by the time frame generator (TFG) so that for any time frame the recorded calibration information contains the time information. The data are stored in the local memory MEM2. The same frame time is given to the positional information which comes down from the position-sensitive detector (top right). This data is stored in another fast local memory MEM1. The position of the scattered photons is digitized by using the delay-line method (Fig. 5). The signals emerging at either end of the delay line are preamplified with a fast preamplifier (PA) and conditioned by a time and filter amplifier, constant fraction discriminator, and gate and delay generator. In order to measure the differences in time of arrival of the pulses at either end of the delay line one has to provide additional electronics logic, such as rejection of any pulses that arrive before the processing time for the previous one is over. This is performed by the signal-conditioning module. Finally the positional information is digitized by a time digitizer module and controller (DTD + Cont.). The time frame generator confers to it the time information. For convenience the signals can be fed into a time-to-amplitude converter (TAC) after they pass through the gate and delay generators. This output goes into a rate meter and a multichannel analyzer (MCA). This option is used only for control and diagnosis. The data acquisition is programmed at a central PDP 11/45 which controls the experiments via the CAMAC serial highway. The programs are deposited through the serial highway and a CAMAC rate controller into a local large scale integrated microprocessor (LSI 11) which takes over the control of the experiment. The measurement cycle is initiated by a signal from the perturbation control system (top middle). Its function depends on the experiment, e.g. temperature jump, stopped flow, etc. The initiation is either manual or electronic. From that moment on, scattering patterns are recorded in sequential time frames, with frame lengths and wait times as programmed previously. When the experiment is over the data contained in MEM1 and MEM2 are transferred to the magnetic disks in the PDP 11/45 via the CAMAC highway. During a measurement the experiment and graphics console can communicate either with the LSI11 or retrieve data from the computer disk for graphics display by simply setting a switch. This feature is important when the current experiment has to be compared with previous ones in order to obtain immediate results so that subsequent experiments can be planned. Experiment control is achieved by CATY programming (CATY is a BASIC-like language allowing easy access to CAMAC hardware, see Golding, 1982).

(d) Normalization with respect to the protein concentration is done when the concentration dependence of some reaction is under study.

After this stage we have the corrected intensity

$$I_{corr}(ch,fr) = \frac{1}{c} \cdot \frac{1}{I_{dr}(ch)} \cdot \left[\frac{I_p(ch,fr)}{I_{0,p}(fr)} - \frac{I_{bg}(ch,fr)}{I_{0,bg}(fr)} \right] \tag{5}$$

where c = concentration, I_{dr} = detector response, I_p = scattering from protein sample, I_{bg} = background, I_0 = incident beam intensity, and the dependence on scattering angle and time is indicated by ch = channel and fr = time frame. The corrected data are then used to derive the structural and kinetic parameters of the reaction. Statistical noise can be reduced by curve fitting using cubic spline functions (Powell, 1967).

3. Solution scattering

3.1. General aspects
There are several reviews on the subject of solution-scattering theory (Guinier and Fournet, 1955; Kratky, 1963; Finch and Holmes, 1967; Stuhrmann and Miller, 1978; Luzzatti and Tardieu, 1980; and others) and the use of SR in small-angle scattering from biological objects (Stuhrmann, 1980; Rosenbaum and Holmes, 1980; Koch et al., 1983). We recall only a few points relevant for time-resolved studies.

The intensity along a line through the center of the scattering pattern from a solution is described by

$$I(S) = \text{const} \cdot I_0 \cdot c \cdot i(S) \tag{6}$$

where $S = 2 \sin \theta / \lambda$ is the Bragg spacing, $I(S)$ is the scattered intensity at S, I_0 is the primary intensity, c is the concentration of the particles in the solution, and $i(S)$ is a function of the shape of the object.

For a monodisperse solution of particles with random orientations and large interparticle separations, $i(S)$ is given by Debye's formula (see Guinier and Fournet, 1955):

$$i(S) = 1/f_{tot} \cdot \sum_i \sum_j f_i f_j \sin (2 \pi S r_{ij})/(2\pi S r_{ij}) \tag{7}$$

The sums extend over all atoms in the particle. Their scattering power is given by f_i, r_{ij} is the distance between scatterers i and j, and $f_{tot} = \sum_i \sum_j f_i f_j$ normalizes the intensity to unity at the origin. The formula illustrates that the scattering intensity is built up of components representing the interference between pairs of scatterers.

For a polymerizing system the time dependence is introduced as follows (Mandelkow et al., 1980; Bordas et al., 1983). Let c_0 be the concentration of all subunits irrespective of their state of aggregation. Since the solution may contain a mixture of aggregates we define x_k = the fraction of subunits incorporated into aggregates of type k, $i_k(S)$ = the corresponding shape function and p_k the degree of polymerization, i.e. the number of subunits per aggregate of type k. The weight concentration of aggregate k is $c_0 x_k$, and its number concentration is proportional to $c_0 x_k / p_k$. Then

$$I(S,t) = \text{const} \cdot I_0 c_0 \sum_k x_k(t) p_k(t) i_k(S) \tag{8}$$

In this equation we express the fact that during assembly the subunits change from one state (e.g. monomeric) to another one, differing in extent of reaction (x_k), degree of polymerization (p_k), and shape (i_k). In practice there is a maximum (S_{max}) and minimum (S_{min}) value of the scattering that defines the range of measurements. S_{min} is determined by the angular resolution of the camera (i.e. the edge of the beam stop), while S_{max} depends on the scattering power of the specimen and/or the length of the detector.

It is instructive to consider some properties of the above equation (dropping the constants before the sum). (a) The shape function $i_k(S)$ is normalized at the origin so that the intensity at very small angles approaches

$$I(S,t) = \sum_k x_k(t) p_k(t) \tag{9}$$

Thus the central scatter is proportional to the weight average degree of polymerization and therefore is a good indicator of polymerization or depolymerization. It corresponds closely to the scattering of visible or UV light (Berne, 1974; for a comparison see Finch and Holmes, 1967). Since the true zero-angle scatter cannot be observed because of the beam stop, changes in particle dimensions can only be observed in general below the limit of $1/S_{min}$ ($\cong$ 80 nm).

(b) Elongation of filaments F_n by endwise addition of subunits S is a bimolecular reaction:

$$S + F_n \rightarrow F_{n+1} \tag{10}$$

The scattering of this system is

$$I(S,t) = x_{sub}(t) i_{sub}(S) + x_{fil}(t) p_{fil}(t) i_{fil}(S) \tag{11}$$

Once the filaments are larger than the cutoff $1/S_{min}$ the maxima of the product $p_{fil} \cdot i_{fil}(S)$ are nearly independent of length (Fig. 7). Thus the changes in scattered intensity are simply due to the incorporation of subunits into the filaments or the depletion of the pool of free subunits.

(c) At large angles the scattering intensity is independent of the state of aggregation (see Fig. 2, outer part). The influence of overall size vanishes roughly at angles

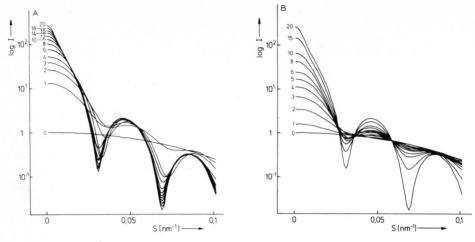

Fig. 7.   Scattering from cylinders of different lengths. (a) Logarithmic plot of scattering intensities vs. spacing $S = 2 \sin \theta/\lambda$. All models are constructed from spheres of radius 4 nm. Curve 0 is the scattering from monomeric subunits, normalized to 1 at the origin $S = 0$. The other models are composed of stacked rings of subunits. Each ring has a mean radius of 13 nm and contains 13 subunits. The axial separation of rings is 8 nm. At low resolution the models approximate the shape of microtubules of different lengths. The shape functions $i_k(S)$ have been multiplied by their degree of polymerization $p_k = 13n$, n being the number of stacks. Thus all traces refer to the same total concentration of subunits (all $x_k = 1$ so that $I(S) = p_k i_k(S)$). The numbers on the left indicate the stacking number n, 1–20. The central scatter at $S = 0$ rises proportionally with the degree of polymerization. The subsidiary maxima asymptotically approach values which are independent of length and reflect only the particle diameter. For the first subsidiary peak the final value is reached at about $n = 5$, corresponding to a "microtubule" of length 40 nm. Beyond this length the subsidiary peaks cease to be sensitive to particle lengths. The intervening minima continue to approach the asymptotic value of 0, but in a practical experiment this is not measurable because of noise. (b) This set of traces represents the elongation of cylinders by endwise addition of subunits. Trace 0 is from depolymerized subunits, normalized to 1 at $S = 0$. Trace 1 is obtained by assuming that 5% of the subunits form single rings while 95% remain dispersed ($x_{sub} = 0.05$, $x_{fil} = 0.95$). In trace 2 an additional 5% of subunits are added on to the single rings, resulting in 10% 2-stacked rings and 90% free subunits. Finally, in trace 20 all subunits are assembled into particles of 20 stacks of rings each. The main difference with respect to a is that the maxima increase smoothly with the degree of polymerization. Since the first subsidiary is insensitive to length beyond $n = 5$ (see a) its increase is due to the decrease in the pool of free subunits so that their rate of assembly may be measured.

where $1/S$ is smaller then the size of the subunit (and thus the intersubunit spacing within an aggregate).

(d) Most aggregates contain characteristic intraparticle distances, e.g. the diameter of a ring or a cylinder. The pattern therefore contains a series of maxima and minima whose angular positions are determined by the particle cross-section. Any change of shape in a reaction sequence will result in a shift of the subsidiary maxima, combined with a change in height. Together with the central scatter they constitute the most sensitive indicator of assembly. By using Debye's formula to calculate $i_k(S)$, weighted by the degree of polymerization p_k and the extent of reaction x_k one can calculate the scattering from model solutions containing any mixture of aggregates (Fig. 7).

3.2. Instrumental parameters

As mentioned in section 2 the instrumental configuration depends strongly on the type of experiment. For the problem of microtubule assembly the following criteria were applied.

(a) Scattering intensities were required at the lowest possible angles in order to be sensitive to overall polymerization. Since parasitic scatter from optical elements increases sharply near the center it was necessary to use a large specimen-to-detector distance ($D = 300$ cm), with $1/S_{min} = 80$ nm.

(b) The highest spatial resolution is given by the maximum scattering angle. Since the useful length of the detector is $L = 8$ cm it follows that $1/S_{max} = \lambda D/L = 5.6$ nm. This resolution is sufficient to record changes in particle shape, but not for intersubunit spacings (see Fig. 8) which would require either a longer detector or a shorter camera length.

(c) The detector is subdivided into 256 channels. This amounts to sampling the scattering pattern in steps of $\Delta S = (8$ cm$/256)(\lambda \times 300$ cm$) = 1/(1400$ nm$)$. Considering that particle cross-sections are below 40 nm the sampling steps need not exceed $1/(80$ nm$)$ according to Shannon's (1949) theorem. In practice some oversampling is of advantage if one is interested in the shape of the maxima rather than just their integrated intensity, and this is important when comparing the data with model curves. The signal/noise ratio can be improved by averaging over several channels at the data-processing stage.

In time-resolved measurements factors other than spatial resolution may be of equal importance. One is the amount of computer processing required to analyze the data. For example, at the above setting the beam focus size is small compared to the scattering pattern. Thus it is not necessary to deconvolute the data for the shape of the beam so that the scattering traces are directly comparable to the theoretical calculations (this point has been verified experimentally with proteins of known radius of gyration).

3.3. Contrast and resolution

In time-resolved X-ray experiments one tries to obtain both structural and kinetic information of some specified spatial and temporal resolution. The visibility of a given structure and/or transition will depend on the magnitude of the intensity change in the scattering pattern — this is referred to as contrast. On the other hand, the types of structures or transitions which can be observed depend on the angular range covered by the detector — this defines the resolution. The two quantities are dependent on each other, on instrumental parameters, and on the nature of the specimen.

In order to observe a signal from the protein solution we require that it should exceed the noise by a factor of, say, five. The noise is given by the square root of the number of events counted during a time frame in one channel, $\sqrt{(I \cdot \Delta t)}$. Contrast is defined as the ratio of some intensity change relative to a reference intensity, $\Delta I/I$. The absolute contrast $C_a = (I_{sample}-I_{background})/I_{background}$ determines whether we see a signal arising from the protein solution at all. For time-resolved studies we are more interested in changes of intensities with respect to some initial level so that the relative contrast $C_r = (I_1-I_2)/I_1$ is more appropriate. Intensity changes are measurable when the frame length is at least

$$\Delta t_{min} = 5^2/(IC^2) \tag{12}$$

where I refers to the intensity of the sample, and C is the contrast.

Table 1 shows that time resolutions obtainable for several scattering angles. The time required for statistically significant results increases rapidly with scattering angle. Thus there is a roughly reciprocal relationship between temporal and spatial resolutions. This is due to the fact that the contrast of fine detail is usually much lower than the contrast between the shape of the particle and its surroundings.

TABLE 1
Scattering intensities at selected angles

Spacing S (nm^{-1})	0.021	0.037	0.078	0.117
Counts/sec, channel				
Cold solution	728	336	69	29
Warm solution	851	179	41	19
Buffer	311	54	12	6
Relative contrast				
$C_r = (I_c-I_w)/I_c$	0.17	0.47	0.41	0.34
Δt_{min} (sec)	1.20	0.34	2.20	7.25
$1/\Delta t_{min}$ (sec^{-1})	0.83	2.94	0.45	0.14
Absolute contrast				
$C_a = (I_c-I_{bg})/I_{bg}$	1.34	5.22	4.75	3.83
Δt_{min} (sec)	0.045	0.017	0.092	0.28
$1/\Delta t_{min}$ (sec^{-1})	22.4	58.9	10.8	3.53

Run TO3OPO.703, July 3, 1981, positron energy 3.3 GeV, current 70 mA, camera length 3 m, protein concentration 23 mg/ml, depth of sample cell 1.5 mm, thickness of both mica windows 40 μm. The scattering of the sample is more than 5 times higher than that of the buffer, even though the protein contributes only 2% in mass. The minimum frame length is calculated from $\Delta t_{min} = 25/(I \cdot C^2)$, where I = reference intensity (cold solution for relative contrast and buffer for absolute contrast), and C is the contrast. The time resolution $r_t = 1/\Delta t_{min}$ decreases as the spacial resolution $S = 1/d$ increases, illustrating their reciprocal behavior. At $S = 0.037$ nm^{-1} the signal is particularly large because here a peak of the cold solution pattern is replaced by a minimum of the warm solution. Note also that even though the signals at 0.037 and 0.078 nm^{-1} are similar their time resolutions differ by a factor of 6 because the absolute intensity is lower at 0.078 nm^{-1}. The area of measurement in the detector plane was only 1 mm $\times$ 80 mm. If an efficient two-dimensional detector were available the time resolution could be improved proportionally to the increase in area.

A formula similar to the one above is often used to describe the relationship between radiation dose and spatial resolution in electron microscopy (e.g. Isaacson, 1977). In our case both temporal and spatial resolutions are ultimately limited by the damage which results from the dose required to visualize features of low relative contrast (see below).

3.4. Radiation damage

A major concern in applying synchrotron radiation to biological objects is that of radiation damage. In the case of microtubules we may take the loss of polymerizability as an indicator. We find that radiation damage becomes noticeable after about 15 min of exposure at optimum beam conditions and thus may be neglected in most experiments.

The dose can be roughly quantified by the following consideration. An absorbed photon of wavelength 0.15 nm deposits in the specimen an energy

$$E = h\nu = 1240/\lambda \ [\text{nm}] = 8267 \ \text{eV} \tag{13}$$

Using the definition of

$$1 \ \text{rad} = 10^{-5} \ \text{J/g} = (1.6 \cdot 10^{-14})^{-1} \ \text{eV/g} \tag{14}$$

1 photon/g is equivalent to $1.32 \cdot 10^{-10}$ rad. The number of incident and absorbed photons are related by

$$N_{abs} = N_{inc} \ (1 - \exp(-\mu(\lambda)t)) \tag{15}$$

where $\mu(\lambda)$ is the linear absorption coefficient and t is the thickness of the sample. The energy density is $E/V\varrho)$ where $\varrho \ (\cong 1 \ \text{g/ml})$ is the mass density and V the irradiated volume. Combining these quantities we have the dose in rad,

$$D = N_{inc} \ (1 - \exp(-\mu)\lambda)t)) \cdot (1240/\lambda \ [\text{nm}]) \cdot 1.6 \cdot 10^{-14}/(\varrho V \ [\text{g}]) \tag{16}$$

We have a beam cross-section of 1 cm $\times$ 0.3 cm and a specimen thickness $t = 0.15$ cm so that the volume is $V = 0.045$ ml. At 0.15 nm the absorption coefficient is $\mu = 10 \ \text{cm}^{-1}$ so that $\exp(-10 \times 0.15) = 22\%$ of the incident photons are transmitted and 78% absorbed. Thus the dose is

$$D = N_{inc} \cdot 2.3 \cdot 10^{-9} \ \text{rad} \tag{17}$$

An incident flux of about 4×10^{11} photons/sec represents a dose rate of 10^3 rad/sec, or 60 000 rad/min.

This number can be compared to published reports on radiation damage (for a comparison of various X-ray wavelengths and electrons, see Sayre et al., 1977). Doses around 10^9 rad destroy the structure of organic molecules, and 10^6 rad kills living cells. Levels of 30 000–50 000 rad (delivered at rates around 2000 rad/min) suffice to impair the in vitro polymerizability of microtubules after irradiation with medical X-ray sources (Coss et al., 1981; Zaremba and Irwin, 1981). These doses would be obtained within 10^6 sec (= 12 days), 10^3 sec (= 17 min), and 50 sec of exposure time on instrument X13, respectively. It appears that although under SR conditions the dose rate is about 30 times higher than with medical sources the assembly competence is retained for similar periods due to a corresponding increase in radiation tolerance. This is explained in part by the mechanism of radiation damage. While the dose (= energy per mass) is easily quantified, the damage is the end result of a chain of chemical reactions mediated by the ion pairs created by the photoelectric effect (measured in units of Roentgen, 1 R = 2.58×10^{-7} Asec/g = 1.61×10^{12} ion pairs/g). For water and biological tissue 1 R corresponds to 93–98 erg/g (close to 1 rad), based on about 37 eV per ion pair (cf. Niemann, 1977). The major part of the radiation damage is caused by the free radicals generated from the radiolysis of water. Their action is time dependent which explains the importance of the dose rate. Their efficiency is enhanced by impurities such as oxygen and diminished by radical scavengers. For this reason we have routinely included 1 mM dithiothreitol (DTT) which acts both as a radical scavenger and protects sulfhydril groups.

The considerations above illustrate the difficulty of relating a given dose to the resulting radiation damage. It seems that each type of specimen has to be tested for its individual dose response. However, our experience with microtubules has confirmed earlier observations with biological objects (muscle, collagen) that their damage is greatly reduced when they are irradiated with high dose rates obtained from SR (Bordas and Randall, 1978).

4. X-Ray studies of microtubule assembly

4.1. Introduction to microtubules

Several networks of protein filaments which are responsible for maintaining cell shape, movement, transport processes, to name a few, are found inside eukaryotic cells. Microtubules form one class of these cytoskeletal elements. They are most conspicuously seen as the spindle fibers responsible for chromosome separation during mitosis, they constitute the motor of cilia, and they are involved in transport of material along nerve cells. Microtubules display several features in vivo which one would like to understand in molecular terms (for reviews see Kirschner, 1978; Roberts and Hyams, 1979; DeBrabander and DeMey, 1980).

They are composed of roughly globular subunits, tubulin. The chemical building block is a heterodimer of α- and β-tubulin of molecular weight 100 000.

In many instances one finds a dynamic equilibrium between microtubules and a pool of subunits so that the fibers can be rapidly assembled for a specific purpose — e.g. mitosis — and are depolymerized thereafter.

Attached to the outer surface of microtubules are a variety of additional proteins which are thought to mediate the function of microtubules. They are collectively termed microtubule-associated proteins (MAPs).

A common source of tubulin for in vitro studies is mammalian brain (Weisenberg, 1972). Microtubules are depolymerized by cold temperature or calcium; they can be reassembled in warm temperatures (37°C) in the presence of magnesium and guanosine triphosphate (GTP) (Fig. 8). In the presence of MAPs the critical concentration is low, about 0.2 mg/ml. MAPs facilitate the nucleation of microtubule assembly, but purified tubulin is also capable of forming microtubules at higher concentrations. The mixture of tubulin and MAPs is referred to as microtubule protein (Fig. 8a). Cold microtubule protein contains a mixture of 6S-tubulin, the heterodimers, and ring-shaped aggregates composed of tubulin and MAPs (Fig. 8b). This close association combined with the observation that MAPs facilitate assembly has led many authors to propose assembly models in which rings feature as assembly intermediates or nucleation centers (reviewed by Kirschner, 1978). As a result there has been a debate about whether microtubules may be regarded as a self-assembly system in the strict sense, and how the principles of nucleated assembly apply.

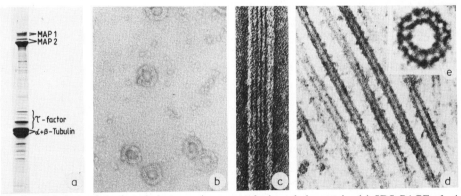

Fig. 8. Gel electrophoresis and electron microscopy of microtubule protein. (a) SDS–PAGE of microtubule protein, prepared as described (E.-M. Mandelkow et al., 1980) following the method of Borisy et al. (1975). α- and β-tubulin (mol. wt. 50 000) are resolved into two bands. The main microtubule-associated proteins are Tau (Weingarten et al., 1975) and the high molecular weight MAP1 and MAP2 (Sloboda et al., 1976). (b) Electron micrograph of cold solution of microtubule protein, negatively stained with 1% uranyl acetate. About 40% of the protein is aggregated into rings of diameter 35–40 nm. They also contain the microtubule-associated proteins. Rings have played a key role in many models of microtubule assembly since tubulin assembly is most efficient when starting from a ring-containing cold solution. Magnification × 116 000. (c) Microtubule repolymerized in vitro at 37°C, negatively stained, showing longitudinal protofilaments spaced about 5 nm. Each protofilament consists of an alternating sequence of α- and β-tubulin molecules. The chemical subunit is the α–β heterodimer. Magnification × 261 000. (d) Microtubule pellet, embedded, thin sectioned, and negatively stained. The protrusions seen around the cylindrical microtubule core (diameter about 25 nm) are microtubule-associated proteins. Magnification × 81 200. (e) Cross-section through a microtubule showing 13 protofilaments in projection. Magnification × 406 000. All micrographs provided by Dr. E.-M. Mandelkow and reproduced from Czihak et al. (1981).

158

Following the early description of microtubules by Ledbetter and Porter (1963) their structure and subunit arrangement has been studied by electron microscopy and image reconstruction (Amos and Klug, 1974) and X-ray fiber diffraction (Cohen et al., 1971; Mandelkow et al., 1977). They are hollow cylinders of about 25-nm mean diameter. The wall consists of 13 longitudinal protofilaments, each of which is formed by a linear array of subunits of length 8 nm, width 5 nm, and thickness 6 nm (Fig. 8c, e). MAPs protrude from the outside surface (Fig. 8d).

Fig. 9 shows the X-ray fiber pattern and the model deduced from it. Fig. 10 is a diagram of the principal reflections out to the first layer line (4-nm spacing). The solution-scattering pattern contains the same reflections, but smeared out into rings because of disorientation (Fig. 11). The merging of rings originating from different layer lines is one reason why solution-scattering patterns are not directly interpretable. Fig. 10 illustrates however that out to 5-nm resolution all reflections arise from the equator and are therefore unique. They are the origin peak (central scatter), three subsidiary maxima of the zero-order Bessel term (J_0) indicating the particle diameter (intermediate angular region), and the J_{13} term arising from the 5-nm spacing of protofilaments ("high"-angle region for our present discussion). The J_0 maxima have been used by Fedorov et al. (1977) to determine the radial density distribution of microtubules from solution scattering. Most of the results to be described are based on recording the time dependence of the intensity in the low and intermediate angular range.

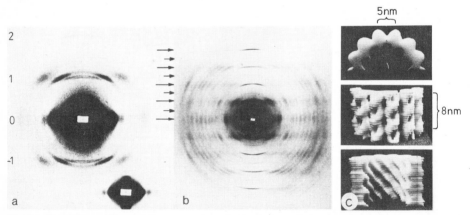

Fig. 9. X-ray fiber diffraction from an oriented gel of microtubules (obtained in collaboration with Drs. J. Thomas and C. Cohen, Brandeis University). Main layer lines occur at orders of 4 nm, indicating the axial repeat of tubulin molecules (α or β). The left photograph shows layer lines $l = 0, 1, 2$, the center up to $l = 4$ (1 nm resolution, long arrows). Secondary layer lines at orders of 8 nm are seen midway between the main layer lines. They correspond to the axial repeat of the α–β heterodimer (center, short arrows). Inset: Short exposure showing the low-angle equatorial maxima. Compare Fig. 10. (c) Model of microtubule structure derived from X-ray fiber diffraction. The diameter is about 25 nm, the lateral separation of protofilaments is about 5 nm, and α–β heterodimers are spaced 8 nm along the protofilaments. From Mandelkow et al. (1977).

4.2. Protein preparation

Microtubule protein was prepared from pig brain in Heidelberg as described (Mandelkow et al., 1980), using two temperature-dependent cycles of assembly and disassembly (modified after Borisy et al., 1975). The preparation was characterized by a number of techniques, including gel chromatography, analytical ultracentrifugation, electron microscopy, X-ray fiber diffraction and light scattering. Protein pellets were stored in liquid nitrogen and transported to Hamburg. Before an experiment the protein was unfrozen, cycled once more and redissolved in the appropriate buffer. At this stage variations in assembly conditions were intro-

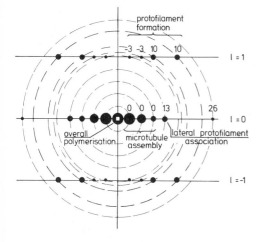

Fig. 10.   Diagram illustrating the relationship between the oriented fiber pattern and the solution-scattering pattern of microtubules. The main reflections of the fiber pattern are indicated by dots whose size is roughly proportional to their intensity. They are: the central scatter (partly obscured by the beam stop) and three subsidiary peaks, all belonging to the zero-order Bessel function J_0; the equatorial peaks at 5.3 nm (J_{13}) and 2.5 nm (J_{26}) and several spots on the 4-nm layer line (J_{-3}, J_{10}). The solution-scattering pattern of microtubules is obtained qualitatively by distributing the intensity of each spot around a circle (dashed). This means that circles from different layer lines are superimposed on each other and illustrates the loss of resolution resulting from disorientation. Only the low-angle reflections belonging to the J_0 and J_{13} terms remain unambiguous. The fiber-diffraction pattern can be interpreted on the basis of helical diffraction theory (Klug et al., 1958), and deviations from ideal orientation are compensated by a disorientation correction (Holmes and Barrington-Leigh, 1974). In a solution-scattering pattern the orientations are completely random, but the radial density distribution of the particle may be obtained if the intensities are weighted by their spacing S (Fedorov and Aleshin, 1967). (We note in passing that there are fluctuations around the mean in a randomly oriented sample which contain structural information and are accessible by SR. See Kam et al., 1981.) In an assembly experiment the central scatter measures the overall degree of polymerization; the three J_0 subsidiaries are nearly independent of microtubule length (see Fig. 7a) and therefore measure incorporation of subunits. The J_{13} term indicates lateral association of protofilaments, and the 4-nm layer line spots measure the longitudinal growth of protofilaments. Thus the reflections fall into three categories, indicating average particle size, particle shape, and subunit arrangement. From Mandelkow et al. (1980).

160

duced: protein concentration, pH, ionic strength, divalent salts, nucleotide analogs, etc. In a typical experiment the protein solution was filled into the pre-cooled (0–4°C) sample cell and polymerized in the beam, either by heating to 37°C in a (rapid) temperature jump or in a (slow) temperature scan. Alternatively the warm protein was mixed with assembly factors such as Mg^{2+}-GTP in the stopped-flow device.

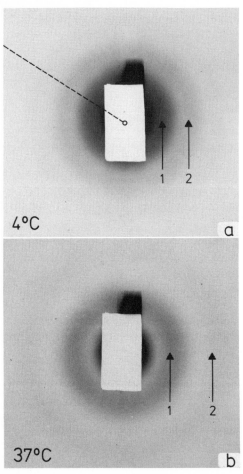

Fig. 11. X-Ray patterns of solutions of microtubule protein. (a) Initial cold solution containing rings of mean diameter 37 nm. Outside the backstop (white rectangle) there are two circular subsidiary reflections of the J_0 term indicated by arrows. In a time-resolved experiment the film is replaced by a linear detector whose position is shown by the dashed line. The circle in the center indicates the position of the direct beam. (b) Final warm solution containing microtubules of mean diameter 26 nm. The two arrowed subsidiaries of the J_0 term are at larger distances from the center than their counterparts in a, reflecting the decrease in mean diameter. Instrument X13 at EMBL Hamburg, exposure time 10 min, specimen-to-film distance 3 m. From Mandelkow et al. (1980).

4.3. Assembly and disassembly induced by temperature jump

During a temperature jump the scattering patterns were collected every 2–15 sec, depending on the desired time resolution. Fig. 12 is a projection plot of two cycles of assembly and disassembly. Microtubules are formed in the warm, as seen from the increase in central scatter and the appearance of characteristic maxima. They are again depolymerized by returning to the cold temperature. During each cycle 1 mole of GTP is consumed for every mole of tubulin. The GDP produced is an inhibitor of assembly so that the second polymerization is slower than the first (the effect is not due to radiation damage).

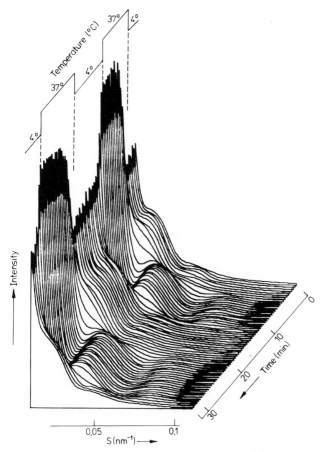

Fig. 12. Projection plot of scattering traces obtained during two cycles of assembly and disassembly. Reassembly buffer 0.1 M PIPES, 1 mM each of GTP, dithiothreitol, EGTA, and $MgSO_4$, pH 6.94. Data were recorded in 15-sec intervals, but only every second frame is shown here. Temperature is indicated on top (4°C or 37°C). The main features are the rise in central scatter following the temperature jump to 37°C, indicating the increase in overall polymerization, and the change in position and height of the subsidiary maxima which arise from tubulin rings at 4°C and from microtubules at 37°C. The curves have been smoothed using spline functions. From Mandelkow et al. (1980).

162

The maxima seen in the initial (cold) pattern arise from tubulin rings. Their degree of polymerization is lower than that of microtubules, corresponding to the lower central scatter. However, their diameter is about 25% larger than microtubules so that the peak positions are closer to the center.

Fig. 13 shows a superposition of 4 scattering traces. a is the initial cold solution, d is the final warm solution, b and c are two intermediate patterns. We note that curve b lies below a. Curve c has a higher central scatter than a but otherwise has almost no subsidiary maxima.

Fig. 14 shows the time course of the scattering intensity at several angles. The top curve is the temperature, followed by the central scatter (b), the scatter at the position of the first maximum of the cold (ring-containing) solution (c) and the scatter at the position of the first maximum of the warm (microtubule-containing) solution (d).

The curves may be interpreted in the following way (Mandekow et al., 1980; Bordas et al., 1983). The central scatter rises after the temperature jump because microtubules are formed whose degree of polymerization is larger than that of the cold solution. However, the rise is preceded by a transient drop. From the difference pattern (see below) we conclude that this reflects the dissolution of rings prior to microtubule assembly. Most of the rise reflects the consumption of free subunits rather than the increase in average length of microtubules (see Fig. 7b). Following the reverse temperature jump the central scatter drops approximately to its initial value in a biphasic manner. A second cycle shows essentially the same features (brief undershoot, rise towards saturation, and biphasic decay). The absolute rates are retarded and the extent of polymerization is lower, indicating that a fraction of the protein does not participate in the reaction anymore, presumably due to the buildup of GDP.

Trace c is almost the mirror image of b in that it decreases following the temperature jump to 37°C and increases again in the cold. This is because the spacing chosen ($0.03 \ nm^{-1}$) is at a minimum of the microtubule pattern which develops upon microtubule polymerization. The mirror symmetry is not complete, however: during the undershoot of b the curve c decreases as well (see Fig. 13b). This means that there is an overall decrease of polymerization.

4.4. Difference intensity patterns
A solution-scattering pattern containing contributions from several species is often difficult to interpret. The problem is simplified by considering intensity differences. If a fraction Δx of subunits change from one state of aggregation to another, then

$$\Delta I = \Delta x \cdot (-p_1 i_1 + p_2 i_2) \tag{18}$$

If there is a large change in degree of polymerization then one of the two quantities in parentheses may be disregarded in first approximation, and the difference intensity can be interpreted in terms of a single species. This would be the case, for example, for the dissolution of rings into subunits, or for the association of subunits into microtubules.

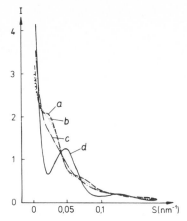

Fig. 13. Selected scattering traces observed during assembly. (a) Initial cold solution, (b) 45 sec after temperature jump, (c) 75 sec after temperature jump, (d) final warm solution. Traces b and c illustrate that a temporary drop in intensity occurs throughout the angular range shortly before the onset of microtubule assembly. The maxima of a arise from tubulin rings, those of d from microtubules. c cannot be interpreted in terms of rings or microtubules, nor is it a simple superposition of the two components. From Mandelkow et al. (1980).

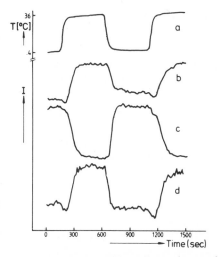

Fig 14. Time dependence of scattering at selected angles during two cycles of assembly and disassembly. (a) Temperature, (b) central scatter, (c) position of the first subsidiary maximum of the cold solution, (d) position of the first subsidiary maximum of the warm solution. Following the temperature jump there is a transient drop in intensity at all angles. This trend is reversed in b, d and e. Curve d lags behind b indicating that polymerization of microtubules is preceded by other intermediates. In c the initial drop continues in a biphasic manner since here the maximum of the cold solution is replaced by a minimum of the warm solution. Depolymerization after the reverse temperature jump is biphasic (see b), but no under- or overshoot is observed. The second polymerization is similar to the first except that the polymerized state is reached more slowly. This feature is explained in part by the production of GDP during the first cycle which acts as an inhibitor of assembly. From Mandelkow et al. (1980).

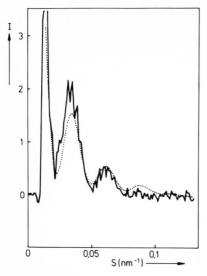

Fig. 15. Difference plot. The data are obtained by subtracting the intensity at the bottom of the under-shoot (see Fig. 14b) from the pattern of the initial cold solution, fitted against the computed scattering pattern of rings of mean diameter 37 nm. The plot illustrates that rings disappear during the undershoot. By varying the models one can show that rings actually break apart into fragments, rather than simply opening up. From Mandelkow et al. (1980).

Fig. 15 shows the difference between the initial cold solution and the pattern ob-tained at the bottom of the undershoot (see Fig. 14b). Superimposed on it is the theoretical scattering from a ring of diameter 37 nm. The plot illustrates that the in-itial event after the temperature jump is the dissolution of rings ($p = 14$ in the as-sumed model) into species of low degree of polymerization, presumably ring frag-ments and subunits. The data are not compatible with a simple opening up of the rings without loss of subunits, a fact which bears on possible assembly models (see below).

4.5. Intensity-versus-temperature plots: hysteresis

Fig. 16 shows the intensity plotted not against time, but against the temperature during a slow temperature scan. Initially we observe a biphasic undershoot between points A and C. After that the intensity rises, with a change in slope at D and a sat-uration point at E. The return curve (upper branch) is similar in shape except that there is no undershoot. The midpoints of the two branches are separated by 11°C so that the plot resembles a hysteresis curve. This behavior is common with enzymatic reactions when one step is slow (Frieden, 1970).

The region between points C and D is particularly interesting. Theoretical con-siderations (Oosawa and Kasai, 1962) and experimental evidence (Bryan, 1976; Engelborghs et al., 1976; Johnson and Borisy, 1977) indicate that microtubules are formed by a nucleation–condensation mechanism. The model predicts that when

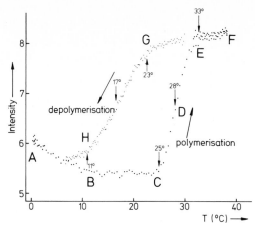

Fig. 16. Temperature dependence of central scatter during a slow temperature scan (1.2°C/min). This experiment was designed to allow the reactions to take place close to their thermodynamic equilibrium so that intermediate steps can be observed with better statistical accuracy. The initial drop in intensity due to the dissolution of rings takes place in two phases (A → B, B → C). It is followed by three phases of polymerization. We assign C → D to nucleation, D → E to microtubule growth, and E → F to some post-assembly events. During the reverse temperature scan the midpoint of disassembly is displaced to a lower temperature (17°C) which gives rise to hysteresis. The justification for regarding D as a transition point comes from the correlation analysis (Fig. 17). The data have been compiled during two successive runs; the gap between F and G is caused by the time required to dump the data from the local memory to the disk of the PDP 11/45.

a solution of subunits is brought into polymerizing conditions there is an initial lag phase during which nuclei are formed slowly. This is followed by the condensation of free subunits onto the ends of microtubules. The curve of Fig. 16 shows that the major part of the lag phase is devoted to the depolymerization of rings, rather than to the formation of nuclei. The period between C and D is characterized by an increase in polymerization without formation of microtubules. A difference plot between points D and C (not shown) indicates the objects formed are narrow elongated structures having a width compatible with a few tubulin protofilaments. In other words, the presumed nuclei of microtubules are probably associations of several short protofilaments. This is in good agreement with theoretical considerations based on the relative strengths of lateral and longitudinal bonds (Carlier and Pantaloni, 1978), and with electron microscope observations of polymorphic tubulin aggregates. Image processing of tubulin "hoops" suggests that a combination of three protofilaments plays a special role during nucleation (Mandelkow and Mandelkow, 1981).

4.6. Correlation analysis
In Fig. 16 we introduced the point D as a separate transition in the pathway of assembly although this is not very obvious from the figure. However, the various reaction steps are clearly resolved by cross-correlation of different parts of the scattering pattern. It can be shown that the change in intensity relative to some initial

166

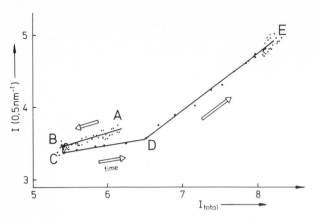

Fig. 17. Correlation analysis of a temperature scan experiment. The horizontal axis shows the total integrated intensity recorded by the detector, the vertical axis is the intensity at the first subsidiary peak of the warm solution ($S = 0.5\,nm^{-1}$). The plot reveals the different phases of the assembly pathway: A → B → C, biphasic disappearance of rings; C → D, nucleation; D → E, microtubule growth.

and final levels is independent of the scattering angle as long as there is only one reaction between two species involved. Therefore a plot of the intensity at one angle vs. that at another angle would yield a straight line. Conversely, a transition from one step in the reaction sequence to another is indicated by a change in slope of the correlation plot. This allows us to determine the time periods in which essentially only one reaction step takes place.

The correlation analysis is particularly valuable for temperature-scan experiments where all reactions occur slowly and close to equilibrium (Fig. 17). Denoting the initial and final states as A and E one finds three intermediate transitions at B, C and D. The undershoot prior to assembly extends from the initial state A to point C. It comprises two different reactions, roughly separated by point B. If the difference analysis is applied to each section independently it is seen that rings dissolve during both reactions. In the first phase the dissolving rings are smaller than in the second. The transition at D is marked clearly, in contrast to Fig. 16. It probably indicates the end of the nucleation phase which takes place between C and D.

4.7. Radius of gyration of tubulin
The structural interpretation of the intensity traces is based on a comparison with models whose scattering is calculated from Debye's formula (section 3.1). At limited resolution this can be greatly simplified by replacing the atoms by a continuous electron density, and by building the models up from identical spheres. This has the advantage that computing times become short enough to allow many trials.

The approach depends on a realistic choice for the diameter of the subunit spheres. For this reason we have measured the radius of gyration of tubulin. The protein was freed of MAPs by phosphocellulose chromatography (following Weingarten et al., 1975). The scattered intensity is plotted as log (I) vs. S^2, and the radius of gyration determined from the slope (Guinier and Fournet, 1955). The result of several measurements between 5 and 20 mg/ml is

$$r_g = 3.1 \pm 0.2 \text{ nm} \tag{19}$$

for the tubulin heterodimer. By comparison with other known data this value indicates that models based on spheres do not describe the tubulin dimer accurately. For instance, if one sphere of outer radius $r_o = \sqrt{5/3}\,r_g$ is filled with protein of partial specific volume 0.73 ml/g its molecular weight would be twice that of tubulin. The model can be improved by introducing a sphere for each monomer, separated by 4 nm (the axial repeat in microtubules) or by even more complicated variants.

All models based on the measured radius of gyration give similar scattering curves in the low and intermediate angular ranges, but discrepancies become apparent beyond $S = 1/(7 \text{ nm})$. Thus for higher resolution studies one will have to abandon the convenience of simple models and replace them by the actual shape of the tubulin dimer known from X-ray fiber diffraction (Mandelkow et al., 1977a).

4.8. Implications for models of microtubule assembly

A substantial amount of research has gone into the elucidation of the pathway of microtubule assembly over the past 10 years (see Kirschner, 1978; Roberts and Hyams, 1979). We will point out only a few issues in which X-ray scattering has been useful to solve certain questions.

One example is the role played by rings during microtubule polymerization. It is known that microtubule assembly in the presence of rings is much more efficient than in their absence. This is related to the fact that the MAPs which also act as promoters of assembly are mainly associated with the rings at low temperatures. As a result several assembly models have been proposed in which MAP-containing rings were depicted as nucleation centers. In some instances rings were also thought to contribute to microtubule elongation. The X-ray data show that rings disappear prior to microtubule assembly and are therefore unlikely to act either as nucleation centers or units of elongation. The dissolution of rings can be reconciled with the role of the ring-associated MAPs by noting that the rings may break apart into subunits or into oligomers containing several tubulin subunits, presumably complexed with MAPs. This is illustrated in Fig. 18. In this view the actual nucleation center is formed from ring fragments while intact rings are merely storage forms of tubulin and MAPs in the cold. It follows that chemical conditions which stabilize rings should act as inhibitors of microtubule assembly rather than as promoters, and this is indeed found in the presence of calcium or excess GDP.

Following the interpretation of rings as nuclei of assembly, several structural models of rings were proposed. Some authors considered rings as curved protofilaments while others thought them to be analogous to certain helices found in microtubules. X-Ray patterns of rings at higher resolution support the first view (data not shown), indicating that both rings and other assembly intermediates are based on a protofilament-type interaction between tubulin subunits.

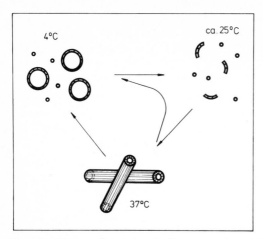

Fig. 18. Diagram illustrating the interconversion of rings and microtubules during temperature-induced assembly of microtubule protein in vitro. The initial cold solution contains tubulin subunits (α–β heterodimers) and rings (top left). During the temperature rise the rings break into smaller fragments (top right). These combine, probably in a sequence of sub-steps, to form the nucleation center which is then elongated by endwise addition of subunits to form microtubules (bottom). Dropping the temperature results in the endwise depolymerization of microtubules into tubulin subunits plus rings (top left) and/or ring fragments (top right) which later reassociate into rings. The difference between this scheme and others is that rings are not considered as direct intermediates of assembly. Rather, they are viewed as storage aggregates of tubulin at low temperatures which have to be broken down for assembly to occur. The microtubule-associated proteins which are tightly bound to rings are probably also bound to the ring fragments and thus promote the formation of nucleation centers. This would explain the observation that they both favor the formation of rings in the cold and of microtubules in the warm even though there is no direct interconversion between the two aggregates during assembly. From Mandelkow et al. (1980).

In its original form, the theory of nucleated assembly (Oosawa and Kasai, 1962) regarded assembly in terms of nucleation in which several subunits interact to form the starting helix, followed by elongation via subunits. The sequence found in the case of microtubules is more complicated since non-helical intermediates are formed which then associate into helices (an analogous situation has been found in the case of tobacco mosaic virus protein assembly; see Caspar, 1963). By slight variations in assembly conditions one can produce a variety of polymorphic tubulin aggregates which coexist with microtubules and whose structure probably reflects the bonding properties of the protein during the early stages of assembly (Mandelkow et al., 1977; Mandelkow and Mandelkow, 1979).

As shown by Fig. 16 the lag period preceding assembly (A–B–C) consists largely of a depolymerization process and this is not directly related to microtubule assembly. Other slow processes take place after the polymerization of the protein (E–F). Time-resolved X-ray scattering enables one to dissect the reaction into several main phases, and to distinguish between pre-nucleation events, nucleation and growth, and post-assembly events.

5. Conclusion

The history of the applications of synchrotron radiation in biology is short. Only a decade has elapsed since the first diffraction pattern from muscle using SR was reported (Rosenbaum et al., 1971). From the outset it was hoped that the technique could be applied to dynamical problems. However, time-resolved experiments became feasible only recently with the combination of a new X-ray camera (Hendrix et al., 1979), position-sensitive detector (Gabriel, 1977), and rapid data acquisition (Bordas et al., 1980) which are incorporated in the instrument X13 at EMBL Hamburg. The examples described in this review were drawn from our experience with microtubule assembly (Mandelkow et al., 1980; Bordas et al., 1983). In addition, the instrument has been applied to several biological problems, such as the contraction of muscle (Huxley et al., 1980, 1981, 1982), the equilibrium between subunits of the enzyme aspartate transcarbamylase (Moody et al., 1980), and structural changes in collagen (Nemetschek et al., 1977; Schilling et al., 1980). The technique has also been used to study problems of polymer chemistry (Koch et al., 1979; Elsner et al., 1981). In a parallel development it recently appeared feasible to monitor ligand binding in crystals of myoglobin using instrument X11 at EMBL (Bartunik et al., 1981). A survey of the research projects carried out at Hamburg (EMBL) and Orsay (DCI) has been published (Koch and Tardieu, 1982).

It is expected that the redesign of the storage ring DORIS in Hamburg will lead to an improvement in usable intensity which will be applied to a wide range of experiments in physics, materials science, and biology, in the new Hamburg Synchrotron Laboratory (HASYLAB) building (Beimgraben et al., 1981). Eventually it is hoped that all synchrotrons presently under construction will be surpassed by more advanced designs, such as the European Synchrotron Radiation Facility (ESF Proposal, 1979).

Acknowledgements

We would like to acknowledge the substantial contributions to the development of the techniques described here by our colleagues Christian Boulin, André Gabriel, Arnold Harmsen, Jules Hendrix, Michel Koch and Winfried Renner as well as the excellent support we have received from the technical staff at the EMBL Hamburg and the Max-Planck-Institute Heidelberg. Our special thanks go to our colleague Eva-Maria Mandelkow who was responsible for all biochemical aspects of this work both in Heidelberg and Hamburg, suggested many of the experiments described in section 4 and checked the results by a variety of complementary studies which were indispensable for their interpretation.

References

Amos, L.A. and Klug, A. (1974) Arrangement of subunits in flagellar microtubules, J. Cell Sci., 14, 523–549.

Barrington-Leigh, J. and Rosenbaum, G. (1974) A report on the application of synchrotron radiation to low-angle scattering, J. Appl. Cryst., 7, 117–122.

Bartunik, H.D., Jerzembek, E., Pruss, D. and Huber, G. (1981) Time-resolved three-dimensional study of ligand rebinding in carbonmonoxy myoglobin, Biophys. Struct. Mech., 7, 249.

Beimgraben, O., Graeff, W., Hahn, U., Knabe, J., Koch, E.E., Kunz, C., Materlik, G., Saile, V., Schmidt, W., Sonntag, B.F., Sprüssel, G., Weiner, E.W. and Zietz, R. (1981) Das neue Hamburger Synchrotronstrahlungslabor HASYLAB, Phys. Bl., 37, 2–10.

Berne, B.J. (1974) Interpretation of the light scattering from long rods, J. Mol. Biol., 89, 755–758.

Bordas, J. and Randall, J.T. (1978) Small-angle scattering and diffraction experiments in biology and physics employing synchrotron radiation and energy-dispersive techniques, J. Appl. Cryst., 11, 434–441.

Bordas, J., Munro, I.H. and Glazer, A.M. (1976) Small angle scattering experiments on biological materials using synchrotron radiation, Nature (London), 262, 541–545.

Bordas, J., Koch, M.H.J., Clout, P.N., Dorrington, E., Boulin, C. and Gabriel, A. (1980) A synchrotron radiation camera and data acquisition system for time resolved X-ray scattering studies, J. Phys. E: Sci. Instrum., 13, 938–944.

Bordas, J., Fourme, R. and Koch, M.H.J. (Eds.) (1982) Proceedings of EMBL Conference on Detectors, Hamburg 1980, Nucl. Instrum. Meth., 201, 1–280.

Borisy, G.G., Marcum, J.M., Olmsted, J.B., Murphy, D.B. and Johnson, K.H. (1975) Purification of tubulin and associated high molecular weight proteins from porcine brain and characterization of microtubule assembly in vitro, Ann. N.Y. Acad. Sci., 253, 107–132.

Bryan, J. (1976) A quantitative analysis of microtubule elongation. J. Cell Biol., 71, 749–767.

Carlier, M.F. and Pantaloni, D. (1978) Kinetic analysis of cooperativity in tubulin polymerization in the presence of guanosine di- or triphosphate nucleotides, Biochemistry, 17, 1908–1915.

Caspar, D.L.D. (1963) Assembly and stability of the tobacco mosaic virus particle, Adv. Protein Chem., 18, 37–121.

Cohen, C., Harrison, S.C. and Stephens, R.E. (1971) X-ray diffraction from microtubules, J. Mol. Biol., 59, 375–380.

Coss, R.A., Bamburg, J.R. and Dewey, W.C. (1981) The effects of X irradiation on microtubule assembly in vitro, Radiat. Res., 85, 99–115.

Czihak, G., Langer, H. and Ziegler, H. (Eds.) (1981) Biologie, Springer, Heidelberg.

DeBrabander, M. and DeMey, J. (Eds.) (1980) Microtubules and Microtubule Inhibitors 1980, Elsevier, Amsterdam.

Eigen, M. and DeMaeyer, L. (1974) Investigation of rates and mechanisms of reactions, in: A. Weissberger (Ed.), Techniques of Chemistry, Vol. 6, Wiley-Interscience, New York, pp. 63–146.

Elsner, G., Koch, M.H.J., Bordas, J. and Zachmann, H.G. (1981) Time-resolved small angle scattering during isothermal crystallization of unoriented poly(ethylene terephthalate) using synchrotron radiation, Makromol. Chem., 182, 1263–1269.

Engelborghs, Y., Heremans, K., de Maeyer, L. and Hoebeke, J. (1976) Effect of temperature and pressure on polymerisation equilibrium of neuronal microtubules, Nature (London), 259, 686–689.

European Science Foundation (ESF) proposal for the European Synchrotron Radiation Facility (ESRF) (1979) Suppl. I: Y. Farge and P.J. Duke (Eds.), The Scientific Case, Suppl. II: G. Thompson and M. Poole (Eds.), The Machine, Suppl. III: B. Buras and G.V. Marr (Eds.), Instrumentation, ESF, 1 quai Lezay-Marnésia, Strasbourg, France.

Fedorov, B.A. and Aleshin, V.G. (1967) Theory of low angle X-ray scattering by long rigid macromolecules in solution, Polymer Sci. U.S.S.R., 8, 1657–1666.

Fedorov, B.A., Shpungin, I.L., Gelfand, V.I., Rosenblat, V.A., Damaschun, G., Damaschun, H. and Papst, M. (1977) A study of microtubule structures in solution by small-angle X-ray scattering, FEBS Lett., 84, 153–155.

Finch, J.T. and Holmes, K.C. (1967) Structural studies of viruses, Methods Virol., 3, 351–474.

Frieden, C. (1970) Kinetic aspects of regulation of metabolic processes: The hysteretic enzyme concept, J. Biol. Chem., 245, 5788–5799.

Gabriel, A. (1977) Position sensitive X-ray detector, Rev. Sci. Instrum., 48, 1303–1305.

Godwin, R.P. (1969) Synchrotron radiation as a light source, Springer Tracts Mod. Phys., 51, 1–73.

Golding, F. (1982) CATY, a system for experiment control, data collection, data display and analysis, Nucl. Instrum. Meth., 201, 231–235.

Guinier, A. and Fournet, G. (1955) Small-Angle Scattering of X-Rays, Wiley, New York.

Haselgrove, J.C., Faruqi, A.R., Huxley, H.E. and Arndt, U.W. (1977) The design and use of a camera for low angle X-ray experiments with synchrotron radiation, J. Phys. E: Scient. Instrum., 10, 1035–1044.

Hendrix, J., Koch, M.H. and Bordas, J. (1979) A double focussing X-ray camera for use with synchrotron radiation, J. Appl. Cryst., 12, 467–472.

Holmes, K.C. and Barrington-Leigh, J. (1974) The effect of disorientation on the intensity distribution of non-crystalline fibers, I. Theory, Acta Cryst., A30, 635–638.

Huxley, H.E. and Brown, W. (1967) The low-angle X-ray diagram of vertebrate striated muscle and its behavior during contraction and rigor, J. Mol. Biol., 30, 383–434.

Huxley, H.E., Faruqi, A.R., Bordas, J., Koch, M.H.J. and Milch, J.R. (1980) The use of synchrotron radiation in time-resolved X-ray diffraction studies of myosin layer-line reflections during muscle contraction, Nature (London), 284, 140–143.

Huxley, H.E., Simmons, R.M., Faruqi, A.R., Kress, M., Bordas, J. and Koch, M.H.J. (1981) Millisecond time-resolved changes in X-ray reflections from contracting muscle during rapid mechanical transients, recorded using synchrotron radiation, Proc. Natl. Acad. Sci. (U.S.A.), 78, 2297–2301.

Huxley, H.E., Faruqi, A.R., Kress, M., Bordas, J. and Koch, M.H.J. (1982) Time-resolved X-ray diffraction studies of the myosin layer line reflections during muscle contraction, J. Mol. Biol., in press.

Isaacson, M.S. (1977) Specimen damage in the electron microscope, in: M.A. Hayat (Ed.), Principles and Techniques of Electron Microscopy, Vol. 7, Van Nostrand Reinhold, New York, pp. 1–78.

Johnson, K.A. and Borisy, G.G. (1977) Kinetic analysis of microtubule self-assembly in vitro, J. Mol. Biol., 117, 1–31.

Kam, Z., Koch, M.H.J. and Bordas, J. (1981) Fluctuation X-ray scattering from biological particles in frozen solution by using synchrotron radiation, Proc. Natl. Acad. Sci. (U.S.A.), 78, 3559–3562.

Kirschner, M.W. (1978) Microtubule assembly and nucleation, Int. Rev. Cytol., 54, 1–71.

Klug, A., Crick, F.H.C. and Wyckoff, H.W. (1958) Diffraction by helical structures, Acta Cryst., 11, 199–213.

Koch, M.H.J. and Bendall, P. (1981) INSCOM: An interactive data evaluation program for multichannel-analyzer type data, in: Proceedings of the Digital Equipment Computer User Society, Warwick, U.K., pp. 13–16.

Koch, M.H.J., and Bordas, J. (1983) X-Ray diffraction and scattering on disordered radiation, Nucl. Instrum. Meth., in press.

Koch, M.H.J. and Tardieu, A. (1982) Cinetix, La Recherche, in press.

Koch, M.H.J., Bordas, J., Schola, E. and Broecker, H.Ch. (1979) Kinetic study of the crystallization of stretched polyisobutylene using synchrotron radiation, Polym. Bull., 1, 709–714.

Koch, M.H.J., Stuhrmann, H.B., Tardieu, A. and Vachette, P. (1983) in: H.B. Stuhrmann (Ed.), Uses of Synchrotron Radiation in Biology, Academic Press, London, in press.

Kratky, O. (1963) X-ray small angle scattering with substances of biological interest in diluted solutions, Progr. Biophys., 13, 105–173.

Kunz, Chr. (Ed.) (1979) Synchrotron Radiation Techniques and Applications, Springer, Heidelberg.

Ledbetter, M.C. and Porter, K.R. (1963) A "microtubule" in plant cell fine structure, J. Cell Biol., 19, 239–250.

Luzzatti, V. (1960) Interpretation des mesures absolues de diffusion centrale des rayons X en collimation ponctuelle ou linéaire: Solutions de particules globulaires et de bâtonnets, Acta Cryst., 13, 939–945.

Luzzatti, V. and Tardieu, A. (1980) Recent developments in solution X-ray scattering, Annu. Rev. Biophys. Bioeng., 9, 1–29.

Mandelkow, E. and Mandelkow, E.-M. (1981) Image reconstruction of tubulin hoops, J. Ultrastruct. Res., 74, 11–33.

Mandelkow, E., Thomas, J. and Cohen, C. (1977) Microtubule structure at low resolution by X-ray diffraction, Proc. Natl. Acad. Sci. (U.S.A.), 74, 3370–3374.

Mandelkow, E., Harmsen, A., Mandelkow, E.-M. and Bordas, J. (1980) Microtubule assembly studied by time-resolved X-ray scattering, in: M. DeBrabander and J. DeMey (Eds.), Microtubules and Microtubule Inhibitors 1980, Elsevier, Amsterdam, pp. 105–117.

Mandelkow, E.-M. and Mandelkow, E. (1979) Junctions between microtubule walls, J. Mol. Biol., 129, 135–148.

Mandelkow, E.-M., Mandelkow, E., Unwin, P.N.T. and Cohen, C. (1977) Tubulin hoops, Nature (London), 265, 655–657.

Mandelkow, E.-M., Harmsen, A., Mandelkow, E. and Bordas, J. (1980) X-ray kinetic studies of microtubule assembly using synchrotron radiation, Nature (London), 287, 595–599.

Moody, M.F., Vachette, P., Foote, A.M., Tardieu, A., Koch, M.H.J. and Bordas, J. (1980) Stopped-flow X-ray scattering: The dissociation of aspartate transcarbamylase, Proc. Natl. Acad. Sci. (U.S.A.), 77, 4040–4043.

Nemetschek, Th., Jonak, R., Nemetschek-Gansler, H., Riedl, H., Niemann, E.G. (1977) Strahlenbiophysik, in: W. Hoppe, W. Lohmann, H. Markl and H. Ziegler (Eds.), Biophysik, Springer, Heidelberg, pp. 223–234.

Oosawa, F. and Kasai, M. (1962) A theory of linear and helical aggregations of macromolecules, J. Mol. Biol., 4, 10–21.

Powell, M.G.D. (1967) Curve fitting by cubic splines, AERE Report, Harwell, U.K., p. 367.

Renner, W., Mandelkow, E.-M., Mandelkow, E. and Bordas, J. (1983) Self-assembly of microtubule protein studied by time-resolved X-ray scattering using temperature jump and stopped flow, Nucl. Instrum. Meth., in press.

Roberts, K. and Hyams, J.S. (Eds.) (1979) Microtubules, Academic Press, London.

Rosenbaum, G. and Holmes, K.C. (1980) Small angle diffraction of X-rays and the study of biological structures, in: H. Winick and S. Doniach (Eds.), Synchrotron Radiation Research, Plenum, New York, pp. 533–564.

Rosenbaum, G., Holmes, K.C. and Witz, J. (1971) Synchrotron radiation as a source for X-ray diffraction, Nature (London), 230, 434–437.

Sayre, D., Kirz, J., Feder, R., Kim, D.M. and Spiller, E. (1977) Transmission microscopy of unmodified biological materials: comparative radiation dosages with electrons and ultrasoft X-ray photons, Ultramicroscopy, 2, 337–349.

Schilling, V., Bordas, J. and Koch, M.H.J. (1980) Evidence of forces produced by bilayers composed of long molecules, Naturwissenschaften, 67, 416–417.

Shannon, C.E. (1949) Communication in the presence of noise, Proc. Inst. Radio Eng., 37, 10–21.

Shelanski, M.L., Gaskin, F. and Cantor, C.R. (1973) Microtubule assembly in the absence of added nucleotides, Proc. Natl. Acad. Sci. (U.S.A.), 70, 765–768.

Sloboda, R.D., Dentler, W.L. and Rosenbaum, J.L. (1976) Microtubule-associated proteins and the stimulation of tubulin assembly in vitro, Biochemistry, 15, 4497–4505.

Stuhrmann, H.B. (1980) Small angle X-ray scattering of macromolecules in solution, in: H. Winick and S. Doniach (Eds.), Synchrotron Radiation Research, Plenum, New York, pp. 513–531.

Stuhrmann, H.B. and Miller, A. (1978) Small-angle scattering of biological structures, J. Appl. Cryst., 11, 325–345.

Webb, N.G., Samson, S., Stroud, R.M., Gamble, R.C. and Baldeschwieler, J.D. (1977) A focusing monochromator for small-angle diffraction studies with synchrotron radiation, J. Appl. Cryst., 10, 104–110.

Weingarten, M.D., Lockwood, A.H., Hwo, S.Y. and Kirschner, M.W. (1975) A protein factor essential for microtubule assembly, Proc. Natl. Acad. Sci. (U.S.A.), 72, 1858–1862.

Weisenberg, R.C. (1972) Microtubule formation in vitro in solutions containing low calcium concentrations, Science, 177, 1104–1105.

Zaremba, T.G. and Irwin, R.D. (1981) Effects of ionizing radiation on the polymerization of microtubules in vitro, Biochemistry, 20, 1323–1332.

Electrophoretic light scattering

B.R. WARE and DANIEL D. HAAS[1]

*Department of Chemistry, Syracuse University, Syracuse, NY 13210,
and [1]Kodak Research Laboratories, Eastman Kodak Company,
Rochester, NY 14650, U.S.A.*

Contents

1. Introduction 174
   1.1. How fast is ELS? 174
   1.2. Electrophoresis 175
   1.3. The Doppler shift 177
   1.4. Interference phenomena and the principle of beating 178
   1.5. Angular dependence 181
   1.6. Diffusion 183
2. Theory of the electrophoretic light-scattering spectrum 184
   2.1. Autocorrelation function 184
   2.2. Shot noise 185
   2.3. Scattered electric field 186
   2.4. Functional forms of data 188
3. Experimental methods 192
   3.1. Apparatus 193
   3.2. Sample heterogeneity 194
   3.3. Joule heating 195
   3.4. Electro-osmosis 196
   3.5. Transit-time broadening 197
   3.6. Chemical reaction 197
   3.7. Collection optics 198
   3.8. Multiple broadening causes 199
   3.9. Sample spectrum 199
4. Applications 200
   4.1. Applications to proteins 200
   4.2. Applications to polymers, viruses and vesicles 202
   4.3. Applications to particle electrophoresis 205
   4.4. Applications to living cells 207
   4.5. Conclusion 215
References 215

R.I. Sha'afi and S.M. Fernandez (Eds.), Fast Methods in Physical Biochemistry and Cell Biology
© 1983 Elsevier Science Publishers

174

1. Introduction

Electrophoretic light scattering (ELS) is a technique for the rapid measurement of electrophoretic drift velocities via the Doppler shifts of scattered laser light. Since its inception in 1971 (Ware and Flygare, 1971) ELS has been applied with success to the electrophoretic characterization of biological particles of all size ranges from small proteins to large living cells. The principal advantage of the technique is its ability to perform the electrophoretic characterization of many particles simultaneously.

Electrophoretic light scattering can be considered a special embodiment of both laser Doppler velocimetry (LDV) (Watrasiewicz and Rudd, 1976; Drain, 1980) and quasi-elastic light scattering (QELS) (Chu, 1974; Berne and Pecora, 1976). However, the application of an external electric field and the relatively slow particle velocities encountered in ELS often necessitate use of special light scattering configurations and sample chambers distinct from those normally employed in either LDV or QELS experiments. In addition, the mechanisms of spectral broadening in ELS measurements deserve special consideration.

In this chapter we discuss the principles and applications of ELS. Essential theoretical concepts from LDV, QELS, and electrophoresis will be described briefly with an emphasis on the physical principles which are essential for an understanding of ELS measurements. The methodology will be summarized generally with appropriate references to more detailed treatments. Finally we will present a comprehensive survey of the publications in this field which have appeared up to the time of this writing.

1.1. How fast is ELS?

The scattering of light by a particle, and hence the impression of a Doppler shift by a moving particle upon the light's frequency, are exceedingly rapid, essentially instantaneous events. However, the magnitudes of typical ELS Doppler shifts (10–100 Hz) place limits on the required duration of a single measurement. The precision of measurement of a spectral peak position is limited by the duration of observation of the sample, in accord with a fundamental restriction of the Fourier transform theorem which states that the smallest separation between resolvable points in the frequency spectrum is equal to the reciprocal of the longest time separation between any two data points acquired for that experiment, i.e., the time elapsed between the beginning and the end of the observation. For instance, if a particle is exposed to a sufficiently strong electric field so that the detected scattered light from the particle is Doppler-shifted by 100 Hz with respect to the frequency of the incident light, and a 0.5% precision (i.e., 0.5 Hz) is desired in the measurement of the shifted frequency, then that particle must be exposed to the electric field for at least $(0.5\,\mathrm{Hz})^{-1} = 2.0\,\mathrm{sec}$, and the consequent scattered light intensity from that particle must be recorded for that same period of time.

The rate at which data points must be acquired during this observation period is dictated by the highest frequency desired in the final spectrum. The Nyquist sampling theorem requires that at least two data points be obtained during each cycle of the highest frequency component included in the spectrum (Davenport and Root, 1958). In other words, the highest frequency which can be represented in the experimentally determined spectrum is limited to half the frequency of the data samplings of the signal wave form. If even higher frequencies are actually present in the original signal, those frequencies must be filtered out to prevent "aliasing" of their power into frequency bands within the range of the experimentally determined spectrum (Blackman and Tukey, 1958) (see Chapter 1).

The combined requirements of the Nyquist sampling theorem and the Fourier transform resolution theorem demand that all spectrum analyzers must retain a sufficient number of data points acquired at twice the Nyquist frequency to span the observation time required for the desired spectral resolution. For example, a 100-Hz-scale spectrum analyzer with 0.5 Hz resolution must collect at least $2 \times 100 \, \text{Hz} \times (0.5 \, \text{Hz})^{-1} = 400$ points of the signal wave form and the measurement will require at least 2 sec (5 msec per point).

1.2. Electrophoresis

ELS is fundamentally an electrophoresis technique, so we begin with a review of the classical principles of electrophoresis. An electric field can be applied to a sample solution by inserting two electrodes into the solution and attaching the exposed leads of these electrodes to the opposing-polarity terminals of a battery or other electrical power supply. Application of the electric field causes attraction of the charged sample particles towards the electrode of the opposite charge. The particles accelerate until the frictional force of the viscous drag due to their translation through the stationary solution matches the attractive force of the applied electric field; this process requires nanoseconds to microseconds (Haas, 1978) to attain the terminal electrophoretic velocity, which in the absence of shielding effects would be:

$$\vec{v} = \vec{E} Z / f \tag{1}$$

where Z is the particle charge, and E is the electric field strength and f is the particle's coefficient of viscous friction (Tanford, 1961; Velick, 1949; Davis and Cohn, 1939; Cohn and Edsall, 1943). Stoke's law stipulates that the hydrodynamic drag on a sphere of radius r produces a frictional drag coefficient of:

$$f = 6\pi\eta r \tag{2}$$

where η is the solution shear viscosity.

If the particle possesses substantial charge and the solution also contains small ions, ions of charge opposite in sign to that of the particle gather around the particle and reduce the effect of the applied electric field (Debye and Hückel, 1923a, b; Debye, 1924). The Debye–Hückel parameter characterizes the reciprocal of the exponential decay distance of this countercharge distribution gradient of the small ions away from the surface of each charged particle (Tanford, 1961):

$$\varkappa = \sqrt{8\pi N_A e^2 \Gamma / 1000 DkT} \tag{3}$$

where N_A is Avogadro's number (6.02×10^{23}); e is the elementary electric charge (4.8×10^{-10} esu); Γ is the ionic strength of the solution (to be discussed below); D is the dielectric constant at the low-frequency limit for the solution ($D_{water} = 81$); k is Boltzman's constant (1.38×10^{-16} erg/K); and T is the absolute temperature of the solution in Kelvin units (K). The electrophoretic mobility U of a particle is defined as the ratio of the particle's electrophoretic velocity to the applied electric field strength:

$$U = \vec{v}/\vec{E} \tag{4}$$

Taking into account shielding of the particle of interest by the counterion cloud, the observed electrophoretic mobility of a sphere of radius r is predicted by Henry's law (Tanford, 1961) to be:

$$U = \frac{Ze}{6\pi\eta r} \frac{\chi(\varkappa r)}{1 + \varkappa r} \tag{5}$$

Henry's function $\chi(\varkappa r)$ is a sigmoid curve whose value monotonically increases from unity for $\varkappa r$ less than 0.1 to a maximum of 1.5 for $\varkappa r$ greater than 1000.

For practical purposes, there are only two general classes of electrophoresing particles which need to be considered: particles small with respect to the Debye–Hückel screening distance, and particles large compared to $1/\varkappa$ of the sample solution. In the case of small particles ($\varkappa(\varkappa r)\approx 1$), Henry's law reduces to:

$$U = \frac{Ze}{6\pi\eta \{r(1 + \varkappa r)\}} \tag{6}$$

where the equation is written in a form emphasizing that a particle with radius smaller than the Debye–Hückel distance $1/\varkappa$ moves with the electrophoretic mobility of an unshielded particle of the same charge but larger effective hydrodynamic radius.

Henry's law for large particles ($\varkappa(\varkappa r)\approx 1.5$ and $\varkappa r \gg 1$) gives:

$$U = \frac{Ze}{4\pi\eta\varkappa r^2} \tag{7}$$

Since the charge of a large particle is expected to reside on the particle's surface, the surface charge density:

$$s = \frac{Ze}{4\pi r^2} \tag{8}$$

can be used to recast the expression for electrophoretic mobility:

$$U = \frac{s}{\eta \varkappa} \tag{9}$$

with the result that the electrophoretic mobility of a large particle is independent of its radius.

The most important component of a solution for electrophoretic analysis is obviously the particles of interest whose electrophoretic behavior is to be studied. The solution pH, and buffer chosen to control the hydrogen ion concentration, is the second ranking concern since pH is often the principal determinant of macromolecular charge in solution. The third most important constituents of a sample solution are the small ions of the added salts, whose significance is often overlooked.

The ionic strength of a solution is defined as:

$$\Gamma = \tfrac{1}{2} \sum_i C_i Z_i^2 \tag{10}$$

with the summation extending over each mobile ion of molar concentration C_i and charge Z_i free in the solution. The direct dependence of the Debye–Hückel parameter and resultant electrophoretic mobility upon ionic strength were pointed out in the previous section. The ionic strength, and the identity of ions which constitute it, influence the observed electrophoretic mobility both by shielding effects and by direct effects on the electrical charge of the particles.

1.3. The Doppler shift

In ELS one measures electrophoretic velocities through the Doppler shifts of scattered laser light. By analogy with sound waves from a source moving towards the detector at speed v, the Doppler shift of the frequency for either light or sound is given by:

$$\Delta v = \frac{v}{c} \, v_0 \tag{11}$$

where v_0 is the original frequency in the source's inertial frame of reference and c is the characteristic speed of the wave (i.e., the speed of light for light scattering). If the source is moving at some angle θ with respect to the detector, the magnitude of the Doppler shift is diminished by a factor of $\sin \theta$.

The fact that the Doppler shift impressed upon the scattered light by a moving particle is directly proportional to that particle's velocity may give the misleading impression that the light-scattering spectrum will simply be a histogram of the sample particles' instantaneous velocities. In addition to the electrophoretic drift, the particles undergo the jostling motions of random Brownian diffusion. These random components have velocities which are orders of magnitude greater than the electrophoretic drift velocities at the usual electric field strength. However, in actual practice, the electrophoresis causes a well-characterized spectral peak at the proper Doppler shift frequency, with a linewidth dependent upon the particle's diffusion constant D. This primary dependence of the spectral shift upon electrophoresis rather than upon the abrupt thermal motions can be reconciled with the Doppler shift explanation by interpreting the Doppler shift in classical terms as a continuously increasing phase shift impressed upon light of the original source frequency by the particle's motion, causing an apparent frequency shift in the observed light. Only those motions which carry the particle a substantial fraction of the wavelength of the illuminating light can cause a noticeable change in the time required for successive crests of the scattered light wave to reach the detector, thereby affecting the perceived frequency of scattered light seen by the detector. Individual steps in the random walk executed by the sample particle are much too small (only a few Å) to be detectable with light of optical wavelengths. The cumulative effects of many steps in a particle's diffusion are only noticeable as a secondary contribution to the electrophoresis peak in an ELS spectrum because diffusion broadens the distribution in times required for sample particles to migrate roughly a wavelength of light while being forced through solution by an applied electric field.

1.4. Interference phenomena and the principle of beating

Although in theory the Doppler shift produced by an electrophoresing particle could be detected directly, this shift is so small (about 100 Hz shift on a carrier frequency of about 10^{15} Hz) that no optical filters of sufficient resolution are available. The only practical way to detect the shift is by optically beating the Doppler-shifted scattered light with a portion of the original illuminating light. This situation can be analyzed by simply considering the simultaneous observation of light scattered from two particles: (1) the test particle which is moving under the influence of the electric field and might also be executing random Brownian movement or any other motion of interest; and (2) a stationary particle, referred to as the local oscillator, which might be rendered immobile by attachment to the wall of the sample chamber. Most of the important effects seen in light-scattering experiments can be explained with the aid of this simple system.

The oscillating electric field of the illuminating light forces the electron distributions of each of the molecules constituting these two model particles to oscillate slightly about their normal motions with respect to their associated atomic nuclei. In accordance with the classical Drude model (Drude, 1907), the frequency of the electrons' oscillations is the same as the driving frequency of the illuminating light seen in the rest frames of each particle; and the phase of the electrons' oscillations with respect to the light's electric field oscillations is determined by the proximity of the light's frequency to the resonance frequencies of the electrons (i.e., the absorption frequencies of the particles). These electron oscillations in turn cause the particles to appear as secondary sources radiating light of the same frequency as the incident light, although with amplitude and phase shift with respect to the illuminating light as prescribed by the particles' positions and constituent matter.

If the two particles are separated by a distance d, as depicted in Fig. 1, then their far-field intensity pattern is a series of interference maxima and minima at scattering angles determined by the interparticle separation, d, and the wavelength of the illuminant, λ. The interference pattern is caused by the phase difference in the superposition of the oscillating electric fields emanating from the two particles. Maxima of the interference pattern are located at observation angles θ for which the distance the light travels from the source by way of one particle to the detector is precisely an integral number, m, of illuminant wavelengths greater than the distance from the source via the other particle to the detector:

$$\sin\left(\theta_{\substack{\text{intensity}\\\text{maximum}}}\right) = \frac{m\lambda}{d} \tag{12}$$

This criterion for the location of interference maxima is derived in detail in any elementary physics textbook describing wave behavior (Jenkins and White, 1957; Halliday et al., 1970; Sears and Zemansky, 1960).

Suppose the test particle moves slightly farther away from the stationary local oscillator, perpendicular to the direction of the illumination, as depicted in Fig. 2. This increase in d, the interparticle spacing, causes a decrease in the angular separation of the far-field interference maxima and minima. Upon sufficient test particle motion, a photodetector originally placed to observe an interference maximum (Fig. 2a) would be receiving a minimum of interference intensity due to the overall contraction of the interference pattern (Fig. 2b). Monitoring the photodetector's

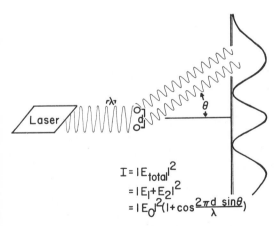

Fig. 1.   Diagram of the scattering of coherent light of wavelength λ by two particles separated by a distance d. The occurrence of an intensity maximum or minimum at the detector depends upon λ, d, and θ, the scattering angle.

180

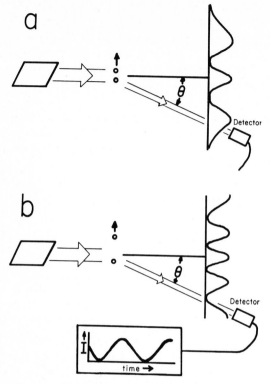

Fig. 2.   Diagram of the condition shown in Fig. 1 when the two particles are in relative motion. The motion produces a time-varying intensity at the detector, because positions of the interference maxima and minima change with respect to the detector due to the increasing separation between the particles.

output discloses a regular sinusoidal oscillation in photocurrent with time as the test particle moves at constant velocity away from the local oscillator, and successive interference maxima and minima of scattered light intensity sweep across the photosensitive region of the detector. This oscillation of the detector's output is exemplified in the photocurrent-versus-time plot at the base of Fig. 2. It probably comes as no surprise that the frequency of photocurrent oscillation is identical to the Doppler shift appropriate for the test particle's velocity; but note that the stationary local oscillator was needed to implement this scheme of "beating" or "heterodyning" the test particle's scattered light. Presence of the local oscillator converts the Doppler-shifted light of constant amplitude from the moving test particle into a total scattered light flux of varying intensity at the detector, indirectly rendering the miniscule Doppler shift detectable.

A particle moving towards the bottom of the page in Fig. 2 cannot be distinguished from a particle moving up by the interference method, because the signal modulation is generated simply by the particle going from positions producing scattering maxima to positions producing scattering minima for either direction of travel. However, direct determination of the Doppler shift would readily show an increased light frequency observed for a particle moving down toward the detector, but decreased frequency for a particle moving upward. The insensitivity of the "beating" method to the sign of direction of the observed particle's motion is called the Doppler ambiguity, from a similar problem encountered in radar applications. If necessary, the Doppler ambiguity can be resolved in ELS by moving the local oscillator particle at a known speed in a predetermined direction (Stevenson, 1970). These efforts generally are not necessary in ELS experiments since the sign of the charge is usually known or can easily be observed or inferred. ELS experiments can exploit the Doppler ambiguity. Since spectra are identical for applied electric fields of the same strength but opposite polarity, the applied field direction can be reversed to prevent accumulation of the charged particles or to reverse electrode reactions.

1.5. Angular dependence

According to Eqn. 12, a detector receiving light scattered perpendicular to the illumination direction ($\theta = 90°$) would observe an interference maximum replacing a minimum for $\lambda/2$ translation of the particle perpendicular to the illumination direction. But note that for the same detector placement to receive 90° scattering, *no* change in interference intensity is observed for particle motion at 45° to the illumination direction (from upper left to lower right). This is because, for any scattering angle, the loci of points of equal optical path length from the illuminator's phase front via the point of interest to the detector's phase front constitute lines (actually, planes in three-dimensional space) which bisect the scattering angle. Motions which carry a particle from one point to any other within the same plane of constant optical path length will not alter the relative phases of the scattered light intensity reaching the detector from the particle and the local oscillator. Only the component of particle motion perpendicular to these planes will cause modulation of the observed scattering intensity and result in a detectable signal change. If the test particle is originally in a plane causing an interference maximum at the detector, translocation of the particle by:

$$\text{scattering plane spacing} = \frac{\lambda}{2 \sin (\theta/2)} \tag{13}$$

normal to the original plane will carry the particle to the next plane for maximum scattering intensity as depicted in Fig. 3; and the signal at the detector will pass from a maximum through a minimum to another maximum as a result of such motion. Note that the separation between planes of equal scattering intensity is determined only by the wavelength of the illuminant and by the scattering angle selected for observation by the arrangement of the detector with respect to the illumination beam, not by any properties of the scattering particles. Eqn. 13 is a statement of the familiar Bragg law condition (Bragg, 1912, 1968).

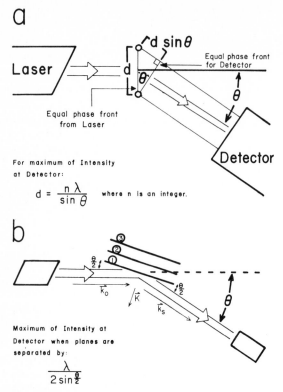

Fig. 3. Diagram of the condition shown in Figs. 1 and 2 with the required geometrical constructs for determining interference intensity dependence upon scattering angle.

The periodic spacing of these planes permits them to be succinctly characterized by a vector, called the K vector or scattering vector, whose amplitude is inversely proportional to the spacing of its associated planes (Eqn. 13) and whose direction is normal to that set of planes:

$$\vec{K} = \frac{4\pi}{\lambda} \sin(\theta/2)\hat{K} \tag{14}$$

where $\hat{K}$ is the unit vector directed normal to the planes (see Fig. 3). The K vector is simply the difference between vectors describing the illuminating beam, $\vec{k}_0$, and scattered light, $\vec{k}_s$:

$$\vec{K} = \vec{k}_0 - \vec{k}_s \tag{15}$$

whose directions coincide with the propagation direction of their respective light waves (i.e., normal to their respective wave fronts) and whose amplitudes are inversely proportional to their respective wavelengths:

$$\vec{k}_0 = \frac{2\pi}{\lambda_0} \hat{k}_0 \qquad (16)$$

$$\vec{k}_s = \frac{2\pi}{\lambda_s} \hat{k}_s \qquad (17)$$

in the same fashion as the K vector is defined by the wavelength (*i.e.*, periodic spacing) and normal of its planes of equal scattering intensity. As mentioned earlier, the Doppler wavelength shift induced by electrophoresis of most sample particles is only one part in ten billion, so the wavelength of the scattered light is typically considered to be identical to the illuminant's wavelength for most purposes:

$$\lambda_0 \cong \lambda_s \qquad (18)$$

Since the illuminant and scattered light wavelengths and k vector amplitudes are nearly equal, the amplitude of the K vector is determined solely by the angular difference of the constituent light vectors which gives rise to the sin $(\theta/2)$ factor of Eqn. 14.

1.6. Diffusion

Up to this point in the analysis, the test particle has been assumed to be moving at constant speed with respect to the local oscillator, so that the detected signal indicates intensity modulated at only a single frequency given by the appropriate Doppler shift. But in reality all matter possesses thermal energy which is manifested by submicron particles suspended in water as random Brownian motion. This diffusional motion superimposes a random walk on the persistent directed motion of an electrophoresing particle. An ensemble of electrophoresing particles initially located on one plane of maximum scattering intensity will require a distribution of times to reach the neighboring plane of maximum scattering intensity, resulting in a diffusion-broadened spectral peak at the Doppler shift. The peak width caused by diffusion exhibits an angular dependence proportional to the square of the amplitude of the scattering K vector (K^2 dependent) whereas the Doppler-induced peak shift is only linearly proportional to $|K|$ (K dependent). Thus the analytical resolution of the technique may be improved, when diffusion is a significant contribution to the linewidth, by working at low scattering angle (Ware and Flygare, 1971).

Although these angular dependences will be derived mathematically in the next section, they can be rationalized via the previous discussion of the scattering planes' separation. Diffusion is characterized by short, jerky motions in random directions which seldom carry the particle very far away from its starting position. Spectra obtained at small scattering angles (small $|K|$, large plane spacing) respond primarily to the long-range directed motion of electrophoresis and are only slightly affected by the random Brownian motion. But at large scattering angles (large $|K|$, small plane spacing), these random motions of diffusion are increasingly likely to carry a particle from one plane to the next, making the diffusion broadening larger in comparison to the electrophoretic Doppler shift.

2. Theory of the electrophoretic light-scattering spectrum

The observable quantity of an electrophoretic light-scattering experiment is the photodetector's output current in response to the light intensity incident on its photosensitive surface. Exposed to an instantaneous light intensity, $I(t)$, a typical photomultiplier produces a current of photoelectrons, $i(t)$, at a rate of:

$$i(t) = I(t)A\gamma Me/h\nu \tag{19}$$

where A is the photosensitive area of the detector's photocathode upon which the light impinges; γ is the probability of ejection of a photoelectron from the photocathode upon absorption of an incident photon, i.e., the photomultiplier's quantum efficiency; M is the average number of electrons produced at the end of the amplifying dynode string for each electron emitted from the photocathode, i.e., the photomultiplier gain; e is the charge of an electron; h is Planck's constant; and ν is the frequency of light received by the photomultiplier. The light intensity incident on the photodetector is proportional to the square of the light's instantaneous electric field, $\varepsilon(t)$, averaged over the photosensitive area of the photomultiplier, according to Poynting's theorem (Jackson, 1962):

$$I(t) = c|\varepsilon(t)|^2/8\pi = c\varepsilon^*(t)\varepsilon(t)/8\pi \tag{20}$$

where c is the speed of light and the asterisk denotes the complex conjugate.

2.1. Autocorrelation function
The modulations impressed on the scattered light intensity by the sample particles' motions can be inferred from the photocurrent fluctuations. The fluctuations can be quantified by determining either the photocurrent's power spectrum, $S_i(\omega)$, or its autocorrelation function, $C_i(\tau)$, which are directly interconvertible as Fourier transforms of one another according to the Weiner–Khinchine theorem (Kittel, 1958; Berne and Pecora, 1976) (see Chapter 1):

$$S_i(\omega) = FT\{C_i(\tau)\} = \int_{-\infty}^{\infty} C_i(\tau) \exp(-j\omega\tau)d\tau \tag{21}$$

where ω is the modulation frequency in angular units (radians/sec); the term j occurring in the argument of the exponential is equal to $\sqrt{-1}$; the i subscripts refer to the photocurrent; and τ is the lag time between two events. The autocorrelation function, $C_i(\tau)$, is the ensemble average of the product of the photocurrent at any instant multiplied by the photocurrent at an instant τ later (McQuarrie, 1976; Chu, 1974):

$$C_i(\tau) = <i(t)i(t + \tau)> = \lim_{T \to \infty} \frac{1}{2T} \int_{-T}^{T} i(t)i(t + \tau)dt \qquad (22)$$

Qualitative description of autocorrelation functions for a few situations may provide some intuition about the behavior of $C_i(\tau)$. If the sample particles are not moving, the scattered light intensity and photocurrent will remain constant with time; and the autocorrelation function will simply be a constant of amplitude $I^2(0)$ because the photocurrent at any instant has the same value as the photocurrent at any other instant, regardless of the time τ separating any two observations of the immobile particles. If the particles are moving, the autocorrelation will exhibit changes dependent upon lag time τ because the photocurrent amplitude at any instant is distinctly different from the photocurrent value occurring τ later. In the context of electrophoresis, the photocurrent oscillations from the concerted, directed electrophoretic motion are sinusoids whose phases depend on the particles' positions at initiation of recording the scattered light; but the autocorrelation function is always a cosine of the same frequency as the photocurrent, and of amplitude $I^2(0)/2$, regardless of the phase of the sinusoidal photocurrent. The photocurrent due to diffusive motion is a randomly varying quantity, and the corresponding autocorrelation function is an exponential which decays monotonically with τ from a maximum amplitude of $I^2(0)$. In the case of simultaneous electrophoresis and diffusion, the autocorrelation function is expected to be a damped cosine whose frequency is related to the electrophoretic drift velocity of the particles and whose damping constant is related to their diffusion coefficient.

Substitution of Eqn. 20 into Eqn. 19 and 22 results in a photocurrent autocorrelation function which is dependent on the ensemble average of the product of four different values of the scattered electric field. Siegert's relation simplifies this complicated dependence to three sets of products of pair-wise expectation values (Siegert, 1943; Whalen, 1971). One set is recognized as the autocorrelation function of the scattered electric field multiplied by its complex conjugate. Another set is identified as the mean square scattered electric field amplitude multiplied by its complex conjugate; this set can be equated to the squared magnitude of the electric field autocorrelation function evaluated for lag time $\tau = 0$. The final set vanishes because it is the product of time averages of rapidly oscillating functions. The photocurrent autocorrelation function's dependence on the autocorrelation function of the scattered electric field becomes:

$$C_i(\tau) = (A\gamma Mec/8\pi h\nu)^2 \{|C_\varepsilon(0)|^2 + |C_\varepsilon(\tau)|^2\} \qquad (23)$$

2.2. Shot noise

At this point, account must be taken of the fact that the photocurrent is composed of discrete photoelectrons which are released at random times from the photocathode. The temporal correlation of each photoelectron with any other is governed by the scattered electric field as expressed in Eqn. 23. But the intrinsic *pulsed* nature of the quantized photocurrent gives rise to an additional term of $Mei_0 \, \delta(\tau)$ in the photocurrent autocorrelation function (Kittel, 1958), called the "shot noise". The Dirac delta function $\delta(\tau)$ reflects the fact that each photocurrent pulse is of very short duration, and is perfectly correlated with itself but not well correlated with any other randomly occurring photocurrent pulse. The photocurrent autocorrelation function becomes:

$$C_i(\tau) = Mei_0\delta(\tau) + i_0^2 + i_0^2 \frac{|C_\varepsilon(\tau)|^2}{|C_\varepsilon(0)|^2} \tag{24}$$

where i_0 is the average value of the photocurrent as related to intensity I through Eqn. 19. Many of the constants have been subsummed into the normalizing factor $|C_\varepsilon(0)|^2$ which is proportional to the average intensity by virtue of the autocorrelation function definition and Poynting's theorem. Only the last term conveys interesting information because it depends on the time evolution of the sample particles' positions.

2.3. Scattered electric field

The electric field at the detector due to scattering from the sample is the sum of the individual electric fields, $\varepsilon_m(t)$, of light traveling via the time-dependent locations, $r_m(t)$, of the N particles in the sample:

$$\varepsilon_s(t) = \sum_{m=1}^{N} \varepsilon_m(t) = \sum_{m=1}^{N} A_{s,m} \exp\left(j\vec{K} \cdot \vec{r}_m(t)\right) \exp\left(-j\omega_0 t\right) \tag{25}$$

where K is the scattering vector probed by the apparatus; $A_{s,m}$ is the light-scattering amplitude of the mth particle, and ω_0 is the carrier frequency in radians/sec of the illuminating light. Also, the electric field from the local oscillator, ε_{lo}, must be incident on the detector simultaneously with the sample's scattered light in order to obtain "beating" and impart sensitivity to directed motions of the sample particles:

$$\varepsilon_{lo}(t) = A_{lo} \exp\left(-j\omega_0 t\right) \tag{26}$$

This local oscillator is either collected from unmodulated light scattered by the glass–solution and glass–air interfaces at the illumination beam entrance and exit windows of the sample chamber or from a separately directed beam from the same laser. Thus, the total electric field of the scattered light reaching the detector is:

$$\varepsilon(t) = \varepsilon_{lo}(t) + \varepsilon_s(t) \tag{27}$$

Expanding the autocorrelation function of the total electric field received by the detector in terms of ε_{lo} and ε_s gives rise to four terms: two terms are recognized as the autocorrelation functions of the local oscillator and of the scattering particles' electric fields, respectively; and the two cross-product terms vanish because the net scattered field from the particles may have any arbitrary phase with respect to the local oscillator electric field.

$$C_\varepsilon(\tau) = C_{\varepsilon_{lo}}(\tau) + C_{\varepsilon_s}(\tau) \tag{28}$$

By use of relation 26, the autocorrelation of the local oscillator electric field is simply:

$$C_{\varepsilon lo}(\tau) = |A_{lo}|^2 \exp(j\omega_0\tau) \tag{29}$$

Determination of the autocorrelation function of the electric field scattered by the moving particles is more complex. Substitution of relation 25 into the constitutive relation for the autocorrelation function, and matching of the carrier-wave frequency terms, gives:

$$C_{\varepsilon_s}(\tau) = \exp(j\omega_0\tau) < \{\sum_{m=1}^{N} A_{s,m}{}^* \exp(-j\vec{K} \cdot \vec{r}_m(t))\}$$

$$\times \{\sum_{q=1}^{N} A_{s,q} \exp(j\vec{K} \cdot \vec{r}_q(t + \tau))\} > \tag{30}$$

This autocorrelation function can be divided into two contributions: (1) the cross-correlations of the field scattered by one particle with the field scattered by another particle at either the same or different instants, which all average to zero for every possible pairing of different scatterers; and (2) the sum of correlations of each particle's scattered field with itself at two instants.

The scattered field autocorrelation function reduces to the sum over all particles of the self-correlation in position for each particle:

$$C_{\varepsilon_s}(\tau) = \exp(j\omega_0\tau)\sum_{m=1}^{N} |A_{s,m}|^2 < \exp(j\vec{K} \cdot \{\vec{r}_m(t + \tau) - \vec{r}_m(t)\}) > \tag{31}$$

The expectation value in Eqn. 31 accounts for the likelihood that a particle will move from one position at one instant to another position at time τ later.

By stationarity, no time or position in the sample volume is preferred over any other, so the probability of any particle moving from point r at time t to point $r+R$ at time $t+\tau$ is just the same as the probability of traveling a distance R away from an arbitrary origin in time τ. This probability function describing the devolution of a particle away from its starting point is denoted $G(R, \tau)$; and the integration over all space of $G(R, \tau)$ multiplied by an interference weighting term $\exp(-jK \cdot R)$ is equivalent to the expectation value of Eqn. 31:

$$< \exp (j\vec{K} \cdot \{\vec{r}_m(t + \tau) - \vec{r}_m(t)\}) > = \int_{-\infty}^{\infty} G(\vec{R},\tau) \exp (-j\vec{K} \cdot \vec{R})d\vec{R} \qquad (32)$$

Because the particles are identical, $A_{s,m} = A_s$, and the summation of the scattered electric field can be performed over the N particles:

$$C_{\varepsilon_s}(\tau) = N|A_s|^2 \exp (j\omega_0\tau) \int_{-\infty}^{\infty} G(\vec{R},\tau) \exp (-j\vec{K} \cdot \vec{R})d\vec{R} \qquad (33)$$

The space–time probability density function for a particle which starts at the origin at time zero and moves under the combined influence of diffusion and a constant electric field E applied in the x direction obeys the following partial differential equation:

$$\frac{\partial}{\partial \tau} G(\vec{R},\tau) = D\nabla^2 G(\vec{R},\tau) + UE \frac{\partial}{\partial x} G(\vec{R},\tau) \qquad (34)$$

where D is the diffusion coefficient of the particle and U is its electrophoretic mobility. For the initial condition of the particle starting at the origin $R = 0$ at $\tau = 0$:

$$G(\vec{R},\tau) = \delta(\vec{R}) \qquad (35)$$

the solution of this partial differential equation in time and three spatial dimensions is (Haas, 1978):

$$G(\vec{R},\tau) = (4\pi D|\tau|)^{-3/2} \exp (-R^2/4D|\tau|) \exp - \frac{UE}{4D} \{UE|\tau| + x\} \qquad (36)$$

2.4. Functional forms of data

The dependence on absolute value of lag time ($|\tau|$) arises because the space–time probability density is expected to be symmetric with respect to the lag time origin, according to the fluctuation–dissipation theorem:

$$G(\vec{R},\tau) = G(\vec{R},-\tau) \qquad (37)$$

Performing the integral over all space of this space–time probability density weighted by the interference factor as required by Eqn. 33 (which is identical to computing the spatial Fourier transformation of $G(R, \tau)$) allows computation of the autocorrelation function of the scattered electric field:

$$C_{\varepsilon_s}(\tau) = N|A_s|^2 \exp\left(-DK^2|\tau| + jUEK_x\tau + j\omega_0\tau\right) \tag{38}$$

where K^2 is the square of the magnitude of the scattering vector K, and K_x is the projection of the scattering vector along the X direction. Summing the expressions for the autocorrelation functions of the local oscillator and sample particle scattered electric fields (Eqns. 29 and 38, respectively) provides the total scattered field autocorrelation function, $C_\varepsilon(\tau)$. Squaring the absolute magnitude of $C_\varepsilon(\tau)$ and substituting into Eqn. 24 completes the computation of the photocurrent autocorrelation function expected from a heterodyne ELS experiment:

$$C_i(\tau) = Mei_0\delta(\tau) + i_0^2 + \frac{i_0^2}{(|A_{lo}|^2 + N|A_s|^2)^2}\left\{|A_{lo}|^4 + N^2|A_s|^4\exp\left(-2DK^2|\tau|\right)\right.$$

$$\left. + 2N|A_{lo}|^2|A_s|^2\exp\left(-DK^2|\tau|\right)\cos\left(UE_x\tau\right)\right\} \tag{39}$$

The final term in Eqn. 39 is the anticipated damped cosine oscillation caused by simultaneous electrophoresis and diffusion.

The photocurrent power spectrum is the Fourier transform with respect to time of the photocurrent autocorrelation function, as stipulated by the Weiner–Khinchine theorem (Eqn. 21):

$$S_i(\omega) = Mei_0 + i_0^2\left\{1 + \frac{|A_{lo}|^4}{(|A_{lo}|^2 + N|A_2|^2)^2}\right\}\delta(\omega)$$

$$+ i_0^2\frac{N^2|A_s|^4}{(|A_{lo}|^2 + N|A_2|^2)^2}\left\{\frac{4DK^2}{\omega^2 + (2DK^2)^2}\right\} \tag{40}$$

$$+ i_0^2\frac{2N|A_{lo}|^2|A_s|^2}{(|A_{lo}|^2 + N|A_2|^2)^2}\left\{\frac{DK^2}{(\omega - UEK_x)^2 + (DK^2)^2}\right.$$

$$\left. + \frac{DK^2}{(\omega + UEK_x)^2 + (DK^2)^2}\right\}$$

where K is the magnitude of the scattering vector defined in Eqn. 14; θ is the scattering angle between the illuminating laser beam and the direction toward the detector as measured in the scattering medium; n is the index of refraction of the scattering medium; λ_0 is the wavelength of the illuminant in vacuum, and the wavelength of light in the scattering solution is $\lambda = \lambda_0/n$. For an applied electric field perpendicular to the direction of propagation of the illuminating laser beam:

$$K_x = \vec{K}\cos\frac{\theta}{2} = \frac{2\pi n}{\lambda_0}\sin\theta \tag{41}$$

The first term in the photocurrent power spectrum in Eqn. 40 is the shot noise which produces constant power at all frequencies. The second term is a DC signal due to the average intensity from the local oscillator and particles. The third term corresponds to the scattered light from the particles interfering with itself; this "homodyne" or "self-beat" term is a peak of half-width at half-maximum:

$$\Delta v_{1/2}^{homo} = \frac{2DK^2}{2\pi} \tag{42}$$

centered at zero frequency, and is present even without a local oscillator signal. This homodyne term is insensitive to net directed motions of the scatterers. The fourth term is the "heterodyne" or "cross-beat" term which produces the Doppler-shifted peak, conveying the most interesting information available in an ELS experiment. The principal component of the heterodyne term is a Lorentzian-shaped peak centered at frequency:

$$v_{shift}^{het} = \left| \frac{UEK \cos (\theta/2)}{2\pi} \right| \tag{43}$$

of half-width at half-maximum:

$$\Delta v_{1/2}^{het} = \frac{DK^2}{2\pi} \tag{44}$$

plus a minor component consisting of the tail of this peak's power spectral density which extends beyond the spectrum origin and overlaps onto the low-frequency range.

The heterodyne shift is predicted to be proportional to the applied electric field and to the sample particles' electrophoretic mobility, while the heterodyne width is proportional to the sample's diffusion coefficient. The heterodyne signal dominates the homodyne term in the photocurrent power spectrum if the local oscillator amplitude significantly exceeds the total scattering from the sample particles. The diffusion coefficient and electrophoretic mobility of the sample particles can be determined by fitting the photocurrent power spectral density data to a baseline plus heterodyne-curve-shaped function using the known values of λ, θ, E, and n which were controlled during the experiment. The absolute scale of the amplitude of the power spectrum is immaterial to this determination of the parameters of interest.

The upper limit of useful local oscillator strength is the value at which the total intensity of detected light causes the photodetector and analysis electronics to respond non-linearly, degrading the detection system's sensitivity to sample-motion-induced modulation of the scattered intensity.

From the dependence of K upon θ (Eqn. 14), the heterodyne peak is expected to broaden with increasing scattering angle, θ:

$$\Delta v_{1/2}^{\text{het}} \propto \left(\sin \frac{\theta}{2} \right)^2 \tag{45}$$

faster than it shifts:

$$v_{\text{shift}}^{\text{het}} \propto \sin \frac{\theta}{2} \tag{45}$$

Fig. 4 illustrates this trend. Actual ELS spectra (Figs. 5 and 6) collected with the chamber described by Haas and Ware (1976) for 100-μM solutions of hemoglobin clearly exhibit the features predicted by the functional dependence of Eqn. 40.

1. The spectral shift measured for one scattering angle is linearly proportional to the applied electric field (Fig. 5).

2. A well-shifted peak obtained with large applied electric field is a symmetrical Lorentzian of width similar to the zero-frequency-centered heterodyne peak obtained with no field at the same scattering angle (Fig. 5).

3. Peak shift is linearly proportional to sin ($\theta/2$) for constant applied electric field (Fig. 6).

4. Peak width increases faster than peak shift for greater scattering angle (Fig. 6).

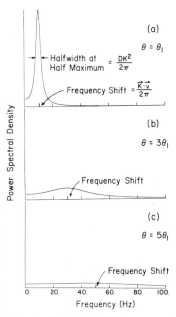

Fig. 4.   Theoretical ELS spectara showing the dependence upon scattering angle for an experiment with constant electrophoretic mobility and electric field strength. The linewidth increases as the square of the sine of half the scattering angle, while the shift increases only as the first power. Thus, for diffusion-broadened spectra, the ratio of shift to width decreases with increasing scattering angle, as was first derived by Ware and Flygare (1971).

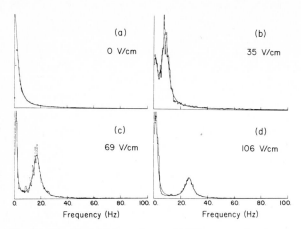

Fig. 5. Electric field dependence of ELS spectra of carboxyhemoglobin (100 μM heme) in a glycine–NaOH–NaCl buffer of pH 9.6 and ionic strength 12 mM, observed at $\theta = 2°$ scattering angle. The stepped lines represent the experimental data, and the smooth lines are computer fits of the data to a shifted Lorentzian peak with appropriate diffusion width. (From Haas, 1978.)

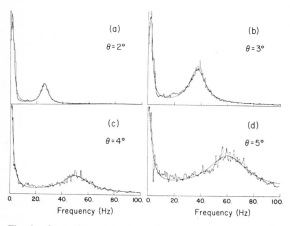

Fig. 6. Scattering angle dependence of ELS spectra of carboxyhemoglobin (100 μM heme) in a glycine–NaOH–NaCl buffer of pH 9.6 and ionic strength 12 mM. For all spectra the electric field strength was 106 V/cm. The stepped lines and smooth lines are as described in Fig. 5. (From Haas, 1978).

3.   Experimental methods

Detailed descriptions of ELS apparatus and methodology have appeared in the recent literature (Smith and Ware, 1978; Uzgiris, 1981a, b). We will therefore summarize the experimental details very briefly and deal primarily with issues of experimental design in this section.

3.1. Apparatus

A schematic diagram of an ELS apparatus is shown in Fig. 7. Light from a continuous-wave laser (low power He-Ne lasers are generally adequate) is split into two beams, one of which illuminates the sample and the other of which acts as the local oscillator. An electric field is created in the chamber by the passage of a current provided by a constant-current power supply. Light from the moving particles is mixed with the local oscillator at the phototube surface to produce the oscillating photocurrent, which is amplified and then passed into a spectrum analyzer for determination of the ELS spectrum.

The central feature of an ELS apparatus is the sample chamber, which must provide for application of the electric field as well as entry and exit of the incident and scattered radiation. Two fundamentally different configurations can be distinguished. Ware and Flygare (1971) employed a Tiselius-like configuration in which separated electrodes are spaced by a narrow channel which includes the illuminated scattering volume. This configuration was later modified (Haas and Ware, 1976; Smith and Ware, 1978) to reduce the volume by employing hemicylindrical electrodes spaced by a narrow channel. In a different approach, Uzgiris (1972, 1981a, b) has employed a small set of parallel-plate electrodes spaced by approximately 1 mm. The two approaches lead to a very different set of experimental problems. The channel configuration suffers the drawback of possible interference from electro-osmosis (to be discussed later) and has a higher thermal time constant. The parallel-plate configuration introduces possible error in the electric field strength

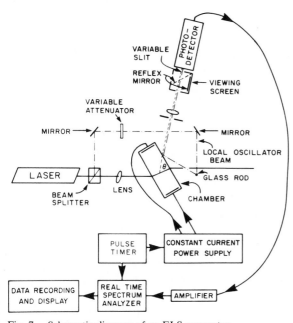

Fig. 7. Schematic diagram of an ELS apparatus.

and uniformity due to edge effects or imperfect parallelism of the electrodes, and may suffer a number of possible problems that result from electrode reactions very near the scattering region. In an attempt to reduce the latter problems, the field direction is generally reversed at a frequency higher than the resolution frequency; these modulations produce harmonic sidebands or broadening of the spectrum (Bennett and Uzgiris, 1973; Haas, 1978). Spurious broadening of ELS spectra is highly undesirable, since the ability to produce a true electrophoretic histogram is one of the major advantages of the technique.

The two types of ELS chamber configuration differ also in their susceptibility to thermal convection. The narrow-channel approach suppresses all convection until the solution becomes unstable by the Rayleigh–Bénard criterion (Schmidt and Saunders, 1938). The parallel-plate configuration is subject to convection at much lower current densities because it has no vertical boundary, although the convection at low power densities may be a regular vertical flow which contributes little breadth to the spectrum until the Rayleigh–Bénard instability is reached.

A final important consideration in chamber design is the size of the illuminated volume. The parallel-plate configuration restricts the choice of scattering angle and the size of the scattering volume, since the opaque electrodes must be close to the scattering volume. The narrow-channel approach permits selection of a wider range of scattering angles and, since the sides of the scattering volume are not masked, allows inclusion of a much wider scattering volume. This advantage is particularly important in the applications to cells and large particles at low number density, for which inclusion of a statistical number of particles in the scattering volume is the primary limitation. A broad scattering volume can also be important for reducing transit-time broadening (to be discussed later).

Choice of a chamber configuration is a critical decision for the new investigator. Of the two available commercial ELS instruments, one uses only the channel-type design, while the other provides a choice of either design. Obviously an objective evaluation should be made by comparing the quality of published spectra and the number and scope of successful applications in the literature at the time of the decision.

Efforts to minimize spectral broadening often dictate experimental practices. A basic cause of ELS spectral linewidths is the random thermal motion of the sample particles, whose diffusion effects have been described in the previous section and whose magnitude is determined by the choice of scattering angle. In addition there are a number of other effects which can lead to line broadening or spectral degradation. We now consider several of these effects.

3.2. Sample heterogeneity

A sample may be heterogeneous in two important respects: (1) size (polydispersity); and (2) electrophoretic mobility heterogeneity. A polydisperse sample of small particles may exhibit peak broadening due to the dependence of electrophoretic mobility upon particle radius (Eqn. 6). A sample of large particles with identical surface charge densities may not show any direct evidence of polydispersity in a single ELS spectrum because all of the particles would be expected to electrophorese at the same velocity regardless of size according to Eqn. 9. Comparison of ELS spectra collected at various angles may permit some inference about the behaviors of different-sized particles in a size-polydisperse sample, because particles much smaller than the illuminant's wavelength scatter an equal amount of light in all directions about the oscillation axis, whereas light scattered from different portions of a large particle will destructively interfere, causing the scattered intensity from large particles to taper off at higher scattering angles.

The spectrum from an electrophoretically heterogeneous sample is a superposition of Lorentzians centered at various frequencies, corresponding to the electrophoretic mobilities of each of the sample particles. Due to the small diffusion coefficients of large particles such as blood cells, their ELS spectra can usually be interpreted as histograms of the cells' electrophoretic mobilities; and surface-charge-density polydispersity is the principal cause of peak width, rather than diffusion. If the difference in electrophoretic mobilities is insufficient to produce individually resolvable peaks, apportionment of a single inhomogeneously broadened peak into its constituent population contributions is exceedingly difficult. Measuring peak width for several scattering angles is a good test for deciding whether a peak is broadened by diffusion or by electrophoretic mobility dispersion since width due to the latter is linearly proportional to K (i.e., sin $(\theta/2)$). Also, a linear increase in peak width with greater applied electric field may be an indication of electrophoretic heterogeneity.

Sample purification, especially dust-particle removal from aqueous solutions, poses the most troublesome stumbling block to performance of successful ELS experiments on submicron-sized particles. Since the scattering intensity depends on the square of the molecular weight, a single dust particle may scatter as much light as all of the molecules of a dilute protein solution in the observed scattering region. A peculiar effect seen in some preliminary ELS experiments on macromolecules is a spectral peak exhibiting the anticipated shift, but *too narrow* width! This situation may be due to the presence of dust particles in the sample with accretions of the sample macromolecules on their surfaces. The relatively large dust particles dominate the scattering spectrum and produce narrow lines in accordance with their small diffusion coefficients. The macromolecules adsorbed to the dust-particle surfaces impart the appropriate charge density to produce an electrophoretic mobility of the dust which is similar to that expected for the macromolecules alone. Acquisition of shifted spectral peaks slightly wider than diffusion-broadened can be a criterion for dust-free samples in ELS experiments on small particles.

3.3. Joule heating

Neophytes to the practice of electrophoresis are often unaware of the fact that a constant electric field can be maintained in a conductive solution only by a continuous flow of electrical current through the solution. The electric field E at any point in the solution is a function of the conductivity σ of the solution at that point, and of the electrical current density I/B flowing through that region, as predicted by the microscopic version of Ohm's law:

$$\vec{E} = \vec{I}/\sigma B \tag{47}$$

where I is the ionic electrical current and B is the cross-sectional area through which the current flows in the vicinity of the point of interest. The inevitable Joule heating generated by regions conducting electrical current will cause the solution temperature to increase. A temperature rise of approximately 2°C was expected to occur during the 2-sec application of the strongest electric field employed for Figs. 5 and 6. Four broadening phenomena can be engendered by a temperature increase.

1. The solution viscosity decreases, thereby increasing the particles' electrophoretic mobilities during the course of the observation.

2. Some regions of the solution are close to thermally conductive chamber walls and remain cooler than regions in the center of the solution; solution conductivity increases with temperature, so the electric field in the center of the solution rises faster than the field near the walls and is time-dependent.

3. Sufficiently large vertical temperature gradients in the solution can cause convection which superimposes its net directed movement on the particles' other motions (Schmidt and Saunders, 1938).

4. Decreasing viscosity increases the diffusion coefficients of the particles during the ELS data collection period.

The first-cited phenomenon can cause the greatest Joule-heating-related broadening, but it can be minimized by establishing the electric field with a constant-current power supply. A constant-voltage power supply creates constant electric field in the scattering region regardless of the temperature increase, but the particles' speed will increase during the data collection period due to the dropping viscosity, resulting in a spuriously broadened spectrum. If a constant-current power supply is used instead of a constant-voltage supply, the increasing conductivity creates a decreasing electric field strength in the solution which compensates for the increasing sample electrophoretic mobility. The particles move at constant speed during the course of the ELS experiment when using a constant-current source, avoiding peak broadening due to heat-generation-induced peak shift slewing.

The second and fourth phenomena are second-order effects which are expected to be small, while the likelihood of the onset of convection can be minimized by proper sample chamber design which avoids large vertical free-solution volumes and large negative vertical temperature gradients near the observed scattering region.

3.4. Electro-osmosis

The walls of an ELS sample chamber generally possess some net charge. The wall charge attracts oppositely charged ions to the mobile solution region near the walls. Application of an electric field to the solution drives these mobile charges, and the water that these charged particles drag along by viscous friction, in one direction near the wall. Since the chamber is closed, a compensating backflow passes through the center of the solution in order to maintain zero net solution flow through any cross-section bounded by chamber walls. This electro-osmosis superimposes a position-dependent parabolic velocity profile upon the sample particles in the chamber, with highest velocity in one direction near the walls and an extremum of velocity in the opposite direction at the solution center (Beniams and Gustavson, 1942; Cornish, 1928; Currie, 1931; Stoltz et al., 1969). Electro-osmosis can affect the apparent electrophoretic mobility of a sample particle by imparting an electric-field-strength-dependent increment to an electrophoresing particle's velocity. The parabolic velocity distribution can also broaden the Doppler-shifted peak if the observed scattering region is large enough to encompass regions of different electro-osmotic velocities. Problems of electro-osmosis can be handled in three ways.

1. Use a narrow illumination beam to avoid sampling a large range of velocities; obtain the highest resolution spectra for detailed analysis with the observed scattering region placed at the center of the chamber where the extremum of the parabolic velocity profile produces the least change in velocity with position; measure the electro-osmotic velocity profile as a function of position in the scattering chamber, then adjust the calculated mobilities for its effects.

2. Coat the chamber walls with substances which create the least surface charge under the solution conditions to be used in the ELS experiments (Smith, 1977; Haas and Ware, 1976; Smith and Ware, 1978).

3. Design the chamber so that the applied electric field at the walls is much less than the applied electric field in the observed scattering region (Uzgiris, 1974).

3.5. Transit-time broadening

The frequency resolution of a spectrum determined for a wave form is inversely proportional to the length of time that wave form is sampled, as mentioned previously in the context of Fourier transforms. This limitation imposes a lower bound on the peak width observed for a particle which passes through the observed scattering region in less time than the duration of data collection for one electric field application (Edwards et al., 1971). The magnitude of this transit-time broadening is independent of scattering angle, and can be reduced by enlarging the observed scattering region in the direction of travel of the electrophoresing sample particles. But since the best signal-to-noise ratio is obtained with a sharply focused illuminating beam, a compromise can be struck by using cylindrical lenses to collimate the laser along the direction of the applied electric field but to focus the laser in the perpendicular direction, thereby creating a long, narrow, brightly lit scattering region about 1 mm wide and 100 μm high (Josefowicz, 1975; Haas, 1978).

3.6. Chemical reaction

If an electrophoresing particle spontaneously undergoes a chemical reaction which changes that particle's charge or size, the particle would abruptly change its velocity to a value appropriate for its new electrophoretic mobility. Alternatively, the chemical reaction might produce a particle with different scattering power than the previous form. In either case, the change in the scattered intensity wave form could be interpreted as an annihilation of the first wave form and creation of the second wave form during the course of a single data-collection period. As with transit-time broadening, the fact that each of these wave forms can not be sampled for the full data-collection period implies that the Doppler-shifted peaks corresponding to reactant and product must each be broadened by the Fourier-transform-mandated limitation in proportion to the reciprocal of the characteristic length of time, $\tau_{reaction}$, that the particle spends in each distinct form:

$$\Delta v_{reaction} \approx \frac{1}{2\pi\tau_{reaction}} \tag{48}$$

This broadening is angle-independent.

The reversible binding and release of hydrogen ions (H^+) by almost any electrophoresing macromolecule might be considered as a candidate for reaction-induced spectral broadening. The molecules in a typical protein sample should possess a distribution of net charge with a standard deviation of 1 or 2 charges about the mean value of the population (Linderstrøm-Lang and Nielsen, 1959; Cohn and Edsall, 1943). However, work of Eigen (1964) indicates that hydrogen ions are exchanged with the protein sites at a rate of $10^{10} - 10^{12}$ M^{-1} sec^{-1}. Therefore, a typical protein molecule exhibits many different values of net charge before it moves sufficient distance to cause modulation of the detected scattered light intensity. Over the time scale of resolution of ELS experiments, all of the protein molecules appear to be moving at the electrophoretic velocity appropriate to the mean of their charge distribution, and no perceptible broadening results from these charge fluctuations. Thus, a chemical reaction will only perturb an ELS spectrum if its time scale is on the order of the spectral time scale. To date this effect has not been observed.

3.7. Collection optics

In an ill-considered attempt to obtain greater signal, the aperture stop of the detection optics might be enlarged to capture more light. This endeavor is not productive for two reasons.

1. The signal-to-noise ratio of the light-scattering spectrum does not increase once the aperture is larger than one coherence area, which is the area of the central diffraction maximum produced at the aperture by a coherent source of the same spatial dimensions and position as the observed scattering region:

$$coherence\ area = \frac{(R\lambda)^2}{hl} \tag{49}$$

where R is the distance from the observed scattering region to the aperture; λ is the light wavelength; h is the verticle height and l is the horizontal length of the aspect of the observed scattering region as viewed from the aperture (Pusey and Vaughan, 1975; Benedek, 1969; Dubin, 1972).

2. The detector receives light from a larger range of scattering angles, broadening the measured spectrum since it is a superposition of the constituent spectra for the range of observed angles.

3.8. Multiple broadening causes

Typically, an ELS experiment is subject to several of these sources of broadening. Some of these phenomena individually cause less than 5% broadening, and the effects of others can be minimized by the experimental practices previously described. Numerical simulation has demonstrated that the cumulative broadening due to simultaneous occurrence of these cited broadening sources adds approximately in quadrature:

$$\Delta v_{1/2,\ \text{total}}^{\text{het}} = \sqrt{(\Delta v_{\text{diffusion}})^2 + \sum_{p} (\Delta v_{\text{cause p}})^2} \tag{50}$$

as long as each of the individual broadening effects is less than the inevitable diffusion component. Uzgiris (1981a, b) has pointed out that, if one considers all broadening sources other than diffusion to be angle-independent, an optimum angle for the measurement can be defined in terms of the shift-to-width ratio if the diffusion coefficients can be estimated. For some measurements, however, it is of value to maintain a significant diffusion width in order to be able to detect aggregation or other changes in the sample.

3.9. Sample spectrum

We conclude this section with a sample ELS spectrum for the case of suspended large particles. Red blood cells in physiological ionic strength media often serve as electrophoretic standards. In Fig. 8 we show an ELS spectrum of fresh human red blood cells in phosphate-buffered saline. The peak Doppler shift corresponds to a mobility, corrected to 25°C, of 1.33×10^{-4} cm^2/sec·V. The ratio of the shift to full

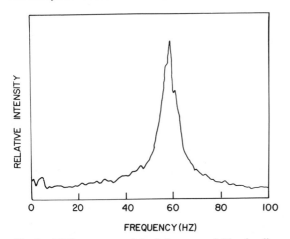

FREQUENCY (HZ)

Fig. 8. ELS spectrum of fresh human red blood cells suspended in phosphate-buffered saline at physiological pH and ionic strength. For this spectrum $\theta = 30°$, $\varepsilon = 23$ V/cm, and $T = 28°C$. The peak mobility, corrected to 25°C, is 1.33×10^{-4} cm^2/sec·V and the width is ±3.5%, in agreement with the best classical measurements. Spectrum taken by James Klein. This spectrum is an amplitude spectrum (square root of the power spectrum).

width is 8.6, indicating a dispersion of ±3.5% in mobility of the cells. This spread in mobility values represents the true electrophoretic distribution of red cells (Seaman, 1975) and diffusion broadening is slight, as verified by the linear dependence of the half-width on the scattering angle. When performed properly, ELS measurement should produce the theoretically predicted spectrum with no significant experimental artifacts. The most important hurdles are design of a good experiment and preparation of good samples.

4. Applications

In principle electrophoretic light scattering is applicable to the study of the electrokinetic properties of any charged species in solution or suspension, but in practice the application of the technique, and particularly its efficacy vis-à-vis other electrophoretic techniques, must be considered separately for different particle-size domains. ELS is very difficult to apply to particles of low molecular weight ($\leq 10\,000$) because of the low light scattering cross-section. In the molecular weight range of most proteins, ELS is feasible but difficult, largely because of the substantial contribution of diffusion to the ELS linewidth. In this range ELS is superior to the moving boundary technique for low-ionic-strength applications but probably inferior for applications at physiological ionic strength and above. In cases for which electrophoretic resolution is the desired result and a determination of the absolute magnitude of the electrophoretic mobility is not essential, the combined electrophoretic–chromatographic separation techniques using gels and other supporting media are clearly superior to ELS or moving boundary methods. For the electrophoretic characterization of large particles ($> 1~\mu$m) such as living cells, the classical technique is the optical cytopherometer, in which the experimenter selects individual particles in the field of view of a microscope and visually clocks their motion in a known electric field. ELS has proved to be superior to the optical cytopherometer, largely because of the ability to analyze many particles simultaneously and with comparable or superior accuracy. ELS is particularly well suited to investigations of particles in the intermediate size range ($0.1–1\mu$m) for which the application of either moving boundary or microscope techniques is difficult. We shall discuss the biological applications of ELS in order of increasing particle size, beginning with protein solutions, proceeding to systems such as nucleic acids and vesicles which are in the intermediate range, and finishing with a review of the applications of ELS to living cells.

4.1 Applications to proteins

The first successful application of electrophoretic light scattering was reported by Ware and Flygare in 1971. Using an autocorrelator plug-in of a laboratory computer, they were able to make a simultaneous determination of the electrophoretic mobility (from the period of the cosine oscillation of the autocorrelation function) and the diffusion coefficient (from an estimate of the time constant of the damped cosine) of bovine serum albumin (BSA) in solution. The solutions were prepared by dissolving crystalline BSA to a weight concentration of 5% and titrating to pH 9.2 with n-butylamine. Conductivity measurements performed on these solutions indicated that the ionic strength was about 10 mM (Ware, 1972). The measured electrophoretic mobility of $u_{20} = 1.8 \pm 0.2 \times 10^{-4}$ cm^2/sec·V agrees well with classical measurements under the same conditions (Schlessinger, 1958), although incorrect assumptions of the ionic strength have led to erroneous comparisons in the literature (Caflisch et al., 1980; Uzgiris, 1981a). The diffusion coefficient was reported to be $D_{20,\omega} = 6 \pm 1 \times 10^{-7}$ cm^2/sec, also within the correct range (Dubin et al., 1967).

In a subsequent paper Ware and Flygare (1972) reported further ELS measurements on bovine serum albumin solutions. For these experiments the solutions were dialyzed against a buffer of pH 9.4 and ionic strength 0.004 M. The electrophoretic mobility at the lower ionic strength, corrected for viscosity to 20°C, was $3.5 \pm 0.2 \times 10^{-4}$ cm^2/sec·V. The Fourier transforms of the autocorrelation functions revealed the presence of a second peak at about 30% lower mobility, which these authors attributed to dimers of BSA. Weissman and Ware (1978) have discussed the application of fluctuation transport theory to such experiments at low ionic strength and have suggested an alternative interpretation of the data involving fluctuations in the monomer : dimer ratio. The primary significance of these early experiments was the demonstration of the feasibility of the technique, including the demonstration of the theoretically predicted dependence on scattering angle and electric field strength. Ware and Flygare (1972) also demonstrated explicitly the applicability of ELS to multi-component systems by studying mixtures of BSA and fibrinogen and achieving clear resolution of the two species.

Three other groups have measured the electrophoretic mobility of BSA using laser Doppler techniques. Mohan et al. (1976) reported an electrophoretic mobility for BSA of 3.0×10^{-4} cm^2/sec·V at pH 8.6 and ionic strength 0.005. More recently Caflisch et al. (1980) studied two different lots of BSA from different manufacturers. One lot had a mobility exactly equal to the value reported by Ware and Flygare under the same conditions, but the second lot had a mobility that was about 25% lower. Caflisch et al. conclude that BSA is not a good mobility standard, though for reasons of availability and stability it is often used as a reference or marker protein. Drifford et al. (1981) have studied the mobility of BSA as a function of pH at low ionic strength in an effort to compare methods for determination of the apparent electrical charge. Their data are reported to be consistent with the data of Ware and Flygare (1971) and of Mohan et al. (1976).

The electrophoretic analysis of blood serum or plasma is an important clinical test for which improvement in speed, resolution or accuracy could be of major significance. Several laser Doppler groups have attempted to characterize blood plasma, but all have reported a lack of reproducibility (Ware, 1972, 1974; Flygare et al., 1976; Mohan et al., 1976; Smith and Ware, 1978; Uzgiris, 1981a, b). The few published spectra have been taken for samples which were dialyzed to low ionic strength and high pH and subsequently filtered, a procedure which is likely to decompose and remove the plasma lipoproteins. These spectra show an identifiable albumin peak and a number of lower mobility peaks which are presumably due to immunoglobulins and fibrinogen. No one has yet reported the definite assignment of any of these peaks nor the type of reproducibility that would be necessary for such an assignment. At present it seems unlikely that ELS will become competitive with current clinical methods for plasma analysis.

Hemoglobin has been studied extensively by many hydrodynamic techniques because of the great interest in the relationship of the tertiary and quaternary state of the hemoglobin tetramer to its oxygen-binding capacity. Haas and Ware (1976, 1978) have used ELS and photon correlation spectroscopy to study the electrophoretic mobilities and diffusion coefficients of hemoglobin at high pH. These studies represent the only published ELS spectra to date for which the data have been fit to the predicted Lorentzian form for diffusion broadening consistent with the known diffusion coefficient. Examples are shown in Figs. 5 and 6. The photon correlation spectroscopy results confirmed the dissociation of hemoglobin from tetramers to dimers above pH 10 and provided new estimates of the dissociation equilibrium constants in this range. The ELS data revealed that the electrophoretic mobilities of tetramers and dimers are indistinguishable in this range to within at most 7%, implying an increase in the electrical charge of the dimer of at least 2.8 – 4.4 net negative charges upon dissociation.

Future applications of ELS to protein solutions will probably focus on studies of dissociating or aggregating systems such as hemoglobin or will be directed toward the investigation of fundamental physical questions such as those raised by Weissman and Ware (1978). The application of ELS to electrophoretic resolution of multi-component protein solutions is not an anticipated growth area unless there are major breakthroughs in the low-angle experimental methodology required.

4.2. Applications to polymers, viruses and vesicles

Synthetic and biological polymers, viruses and vesicles are generally in the size range of 0.01–1 μm and larger. They are difficult or impossible to visualize in an optical microscope, and they often do not maintain the stable concentration gradients which are essential in the moving boundary technique. Applications of ELS in this area have been quite successful. ELS spectra for these particles are generally broadened both by diffusion and by electrophoretic heterogeneity, so that both charge density and size effects (aggregation, changes in conformation) can be studied simultaneously. As stated earlier, diffusion and electrophoretic heterogeneity broadening can be separated experimentally by taking advantage of their different dependences upon scattering angle and electric field strength.

Application of ELS to solutions of non-globular polyelectrolytes began with the report by Hartford and Flygare (1975) of the electrophoretic behavior of calf-thymus DNA. They found that the mobility was in linear inverse proportion to the square root of the ionic strength. They also found that the electrophoretic mobility of denatured DNA was about 15% lower than the mobility of native DNA. Although this difference is consistent with reports of previous workers (Costantino et al., 1964; Olivera et al., 1964), the mobility magnitudes reported by Hartford and

Flygare are 20–30% higher. A recent report by Drifford et al. (1981) confirms the values of Hartford and Flygare as well as the salt dependence, though one of their lower molecular weight (3×10^5) samples showed a mobility about 10–15% lower than the values reported by Hartford and Flygare.

Schmitz (1979) has studied DNA solutions using ELS with a sinusoidal electric field. The ELS spectra consisted of several peaks which were at multiples of the driving frequency. Variation of the overtone amplitude distribution with the fundamental frequency led to the determination of a reasonable value for the electrophoretic mobility of the DNA. In a previous paper, Schmitz (1976) had predicted that the technique of employing sinusoidal driving fields could be used to obtain precise values for the number of relaxation modes present and their characteristic relaxation times; to date that possibility has not been realized experimentally.

Electrokinetic behavior of synthetic polyelectrolytes has been studied extensively by Drifford and coworkers using ELS, dynamic light scattering, and tracer techniques. In the most comprehensive study (Magdelénat et al., 1979), the transport properties of chondroitin sulfate in aqueous solutions were measured as a function of the concentration of mono-, di- and trivalent cations. An effective charge was calculated from the ratio of the electrophoretic mobility to the diffusion coefficients and the results were compared with the counterion condensation theory of Manning (1978, 1979a). Their results showed that monovalent cations are displaced by trivalent cations, but in both cases at concentrations of the substituting ion that are lower than the theoretical prediction. However, the hydrodynamic transition, reflected in an increased polymer diffusion coefficient as trivalent cations replaced divalent cations, occurred at a charge density which was in good agreement with theory. These data were interpreted in terms of a sequential series of cation substitution steps; and the presence of a "second layer" of counterions about the polyelectrolyte was proposed. In a related study, Meullenet et al. (1979) examined the electrophoretic behavior of solutions of a linear alternating copolymer of maleic acid and ethyl vinyl ether in aqueous NaBr solutions. They observed that the electrophoretic mobility was essentially independent of molecular weight but was a sensitive function of the degree of neutralization (charge density), a result which has also been reported by Cha et al. (1980) using ELS to study solutions of hydrolyzed polyacrylamide. The apparent valences calculated from the electrophoretic mobilities and diffusion coefficients of Meullenet et al. were surprisingly low. Although these workers suggested that the results may indicate an invalidity of the Debye–Hückel approximation in these systems, Manning (1979b) commented that a more likely explanation is an invalidity of the calculation of the apparent charge from the ratio of the electrophoretic mobility to the diffusion coefficient, a result of the fact that the effective friction coefficients for electrophoresis and diffusion are not the same. In a recent review of the data and theories of this issue, Drifford et al. (1981) express the concurring opinion that intramolecular frictional effects are the central issue in the understanding of the electrophoretic behavior of polyelectrolytes.

Two groups have reported ELS studies of viruses. Hartford and Flygare (1975) measured the electrophoretic mobility of tobacco mosaic virus as a function of solution ionic strength and observed that these particles are highly charged at pH 7. Rimai et al. (1975) studied the electrophoretic behavior of several RNA tumor viruses and concluded that, in terms of the density of surface charged groups, the RNA tumor virus membrane is not very different from the erythrocyte membrane. Since the surface chemistry of viruses is an important element in the mechanism of attachment and invasion of the host cell, the electrophoretic characteristics of viruses under different solution conditions or conditions of preparation and treatment may be of considerable interest, and it is reasonable to suppose that this will be an expanding area of application for ELS in the future.

Vesicles of synthetic and natural origin are the subjects of intense investigation in a number of areas of biological research. In many cases the density of electrically charged groups on the vesicle exterior may be a parameter of interest, as for example in the case of secretory vesicles, which must approach and fuse with the like-charged membrane of the host cell in order to accomplish exocytotic release of their contents. The first ELS study of secretory vesicles was reported by Siegel et al. (1978), who measured the effects of Ca^{2+} and Mg^{2+} on the electrophoretic mobility of chromaffin granules from the bovine adrenal medulla. Sample data from one of these experiments are shown in Fig. 9. The presence of 0.5 mM Ca^{2+} produced a measurable decrease in the average electrophoretic mobility and in the width of the ELS spectrum. However, these workers observed a similar effect with Mg^{2+}, which does not induce secretion. Their results were in contradiction with earlier microelectrophoresis experiments (Matthews et al., 1972; Dean and Matthews, 1974) showing a selectively higher affinity for Ca^{2+}, which had been taken as supporting evidence for the "electrostatic hypothesis" of secretory exocytosis (Dean, 1975).

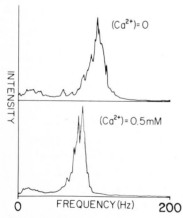

Fig. 9. ELS spectra of bovine chromaffin granules in the absence and presence of 0.5 mM Ca^{2+}. The medium contained 0.30 M sucrose to maintain isotonicity and was buffered at pH 6.9 and ionic strength 15 mM.

The conclusion of Siegel et al. (1978), that the Ca^{2+}-specific induction of secretion can not be attributed to the electrostatic effects of Ca^{2+} binding, was extended to synaptic transmission in a study of synaptic vesicles and synaptosomal membranes (Siegel and Ware, 1980). In this case a somewhat higher affinity of Ca^{2+} over Mg^{2+} was observed, but the difference was negligible in the physiologically relevant range. Matthews and O'Connor (1978) have reported an ELS measurement of the electrophoretic mobility of zymogen granules in a preliminary note, but no conclusions are drawn from their data. Petty et al. (1980a) have measured the electrophoretic mobilities of mast cell granules in conjunction with a study of electrokinetic alterations of mast cells after activation, to be discussed later.

Schlieper et al. (1981a) have used ELS to study the surface charge properties of (Na^+,K^+)-ATPase-containing microsomal vesicles derived from guinea-pig kidney. From the pH and ionic strength behavior of the electrophoretic mobility, they inferred that there was one net negative charge per protein unit with a pK of 3.9. They also determined the association constants for Mg^{2+} and ATP. In a separate study the same group (Schlieper et al., 1981b) reported the electrokinetic effects of propranolol, tetracaine, lidocaine and procaine on phosphilipid vesicles prepared from charged and neutral phospholipids. All four cationic drugs decreased the mobility of negatively charged vesicles and increased the (positive) mobility when added to previously neutral vesicles. The drugs are listed above in descending order of activity by these criteria.

Yen et al. (1981) have reported ELS measurements on vesicles prepared from human red cell membranes. These vesicles are formed with either the normal membrane sidedness (right-side-out vesicles, ROV) or with inverted sidedness (inside-out vesicles, IOV). The proportions of ROV/IOV can be manipulated by variation of the conditions of preparation. One then makes use of the fact that the electrophoretic properties of these vesicles are determined exclusively by the outward membrane exposed to solution, so that ROV reflect the characteristics of the red cell exterior membrane surface and IOV reflect the characteristics of the inside (cytoplasmic) surface of the red cell membrane. Yen et al. showed that at neutral pH, ROV had a (~25%) higher mobility than IOV, and that the two peaks could be resolved in the ELS spectrum to provide a quantitative estimate of the IOV/ROV ratio. The ROV peak coincided with the mobility of fresh red cells and of resealed ghosts. These workers performed a pH titration and a series of enzymatic treatments to reveal the differences in the chemistry of the two sides of the red cell membrane (Yen et al., to be published).

4.3. Applications to particle electrophoresis

Particles in the size range above 1 μm are particularly attractive specimens for electrophoretic analysis by laser Doppler methods. The high scattering cross-section provides an ample signal from a dilute suspension, and contamination by dust is usually negligible by comparison. Moreover, greater electrophoretic resolution is achievable, since the diffusion width of the spectrum is inversely proportional to the particle radius. The latter fact makes it possible to work at higher scattering angles, for which the Doppler shifts for a given velocity are higher; this in turn means that lower velocities can be detected with sufficient accuracy, so that experiments at higher salt concentrations can be performed satisfactorily. It is thus not surprising that large particles have been the most common specimens for ELS application.

Polystyrene latex spheres are a common test sample for a number of techniques. Their electrophoretic mobilities are very dependent on sample history and solution conditions, so they are not a good standard for absolute mobility, but they are a convenient test sample for instrumental development. Application of laser Doppler spectroscopy to particle electrophoresis was reported independently in 1972 by Yoshimura et al. and by Uzgiris. Both groups reported simultaneous measurement of the Doppler shift linewidth for suspensions of polystyrene latex spheres and verified the correct relationship between Doppler shift and applied electric field. In subsequent publications Yoshimura et al. (1975a, b) reported the effects of particle size on electrophoretic properties measured by ELS. They also observed (Yoshimura et al., 1975b) a very interesting coupling of the electrophoretic velocities of these charged particles; at low ionic strength ($<10^{-4}$ M) latex particles of different size, which have different mobilities, moved with a single velocity. This apparent coupling could be removed to give a bimodal spectrum characteristic of two species either by raising the ionic strength or lowering the particle concentration. These very interesting observations seem to be related to electrostatic repulsions of the like-charged particles, except that, as has been pointed out by Uzgiris (1981a), the average interparticle spacing is much greater than the Debye shielding length. It is also puzzling that the limiting spectra at the higher ionic strengths ($\sim 2 \times 10^{-4}$ M) are broader and more greatly different in mobility than would be expected for the individual particles. It is possible that the fluctuation mobilities seen in ELS experiments differ from the tracer particle mobilities under these low-salt conditions (Weissman and Ware, 1978).

Goff and Luner (to be published) have studied the electrophoretic properties of particles of a number of different types of latex materials and have classified these materials on the basis of the respective ionic strength dependences of their electrophoretic mobilities. Their data illustrate the potential of ELS for characterization of particles which are employed in coating processes, particularly for the paper industry.

Uzgiris and Costaschuck (1973) studied the adsorption of a cationic polymer onto negatively charged polystyrene latex spheres and observed the anticipated charge reduction and reversal as increasing amounts of polycation were added. Flocculation was observed at the charge reversal point, and an even greater agglomeration was produced by large excesses of the polycation. Uzgiris (1976) later used polystyrene spheres to develop a laser Doppler antibody assay for the antigen–antibody reaction. The principle of this method is that antigen-coated spheres will exhibit a lower mobility when reacted with antibody, since the antibodies have a lower charge density and, upon binding to the spheres, extend the shear plane of the spheres to a surface of lower potential. The sensitivity of the technique was demonstrated to be of the order of 10 ng antibody per ml. The primary disadvantage of the technique is that non-specific binding of antibody to the sphere surface limits the specificity of the assay and introduces a requirement for repeated washing of the treated spheres prior to electrophoretic analysis. To date there have been no further reports of refinement of this method or of its application to immunological research, but the study of surface reactions is a major area for development of ELS application.

Josefowicz and Hallett (1975a) used polystyrene latex spheres as the test sample for the demonstration of electrophoretic detection by a crossed-beam (sometimes called "differential Doppler") method. Although this method has certain advantages for extremely dilute samples (of the order of one particle in the laser beam at any given time), its use has not been pursued for ELS applications, for which the simultaneous analysis of many particles is a motivating objective. Harvey et al. (1976) also used polystyrene spheres as a test sample to demonstrate that the frequency sidebands which accompany the square-wave modulation of electric field commonly employed with parallel-plate chambers can be reduced greatly by sweeping over a range of modulation frequencies.

4.4. Applications to living cells

Electrophoretic analysis of living cells has been an established technique for the study of the cell surface for 5 decades (Abramson, 1934; Seaman, 1975). Despite the great breadth of application and the extensive literature of cell electrophoresis, the scientific yield of these endeavors has been comparatively meager. Three fundamental aspects of the approach are severely limiting. The first is the simple fact that the electrophoretic mobility is a single parameter and thus is inadequate to characterize a system as complex as the surface of a living cell. The second point is that the electrophoretic mobility is a function of a number of physical characteristics of the cell surface (number and distribution of charged groups; detailed structure of the cell surface; concentration, size and electrovalence of counterions in the supporting electrolyte and their interaction with charged groups on the cell surface) and that it is not possible to calculate accurately these more fundamental properties of the system from the electrophoretic data. The third unfortunate limitation is that biological particles do not vary greatly in electrophoretic mobility. Most cell types range within a factor of 2 in mobility for a given set of solution conditions; with instrumental accuracies of typically a few per cent, the analytical resolution of the technique is thus severely limited. Because of these limitations, it is a considerable challenge to design cell electrophoresis experiments which can lead to meaningful interpretation; but certainly there are cases for which the charge density on the cell surface, and particularly the distribution of the surface charge density among a population of cells, can be measured to discover new phenomena and to answer significant scientific questions. Even in these cases, cell electrophoresis has been limited by the tedious, time-consuming, subjective, and statistically insufficient methodology of the optical cytopherometer. There has therefore been considerable optimism that the laser Doppler approach would become an improved technique for cell electrophoresis and that the particular advantages of ELS (principally the result of the ability to measure many cells simultaneously) would permit the design and execution of more meaningful cell electrophoresis experiments. We now review the substantial evidence that this objective has been realized, proceeding essentially in chronological order.

The first applications of the laser Doppler principle to cell electrophoresis were largely for the purpose of demonstration and development of the analytical capabilities of the technique. Uzgiris (1972) measured the electrophoretic mobilities of bacteria and erythrocytes in the initial demonstration of his methodology, and Uzgiris and Kaplan (1974, 1976a) studied the electrophoretic mobilities of both erythrocytes and lymphocytes as a function of pH and ionic strength. At pH 7 the ionic strength dependence of the two cell types was found to be different, so that the electrophoretic distinction of the two was substantially enhanced at low ionic strength. They also observed some bimodal character in the lymphocyte spectra and speculated on its biological significance, which will be elaborated further.

Josefowicz and Hallett (1975b) used ELS to study the electrokinetic alterations produced by incubation of rat-thymus lymphocytes with pokeweed mitogen. They found that incubation for 15 min with 50 μg/ml mitogen decreased the average electrophoretic mobility of the lymphocytes to less than half the original value. Kaplan and Uzgiris (1975) reported much smaller decreases of electrophoretic mobility when human peripheral lymphocytes were incubated with the mitogens phytohemagglutinin (PHA) or concanavalin A (Con A). Using incubation times of from 40 min to 90 h, they observed mobility reductions of only a few per cent at pH 7, even for lectin concentrations as high as 800 μg/ml. Somewhat larger differences were seen at lower pH, and, in the case of Con A, the mobility reductions could be further enhanced by the addition of antibody to Con A. Perhaps most surprising was the observation that these alterations could be completely reversed by washing the cells.

Uzgiris and Kaplan (1976b) observed an alteration of the electrophoretic mobility distribution of human peripheral lymphocytes from former tuberculosis patients after the cells were exposed to the tuberculin antigen purified protein derivative (PPD). In 15 out of 20 cases, the distribution showed a new high-mobility population which was not elicited by exposure of lymphocytes from any of 9 control donors to PPD. Although the reaction seems to be specific, the mechanism for a mobility change of a substantial fraction of lymphocytes in response to a specific antigen remains unexplained. It was suggested that PPD-specific lymphocytes may have produced a soluble factor that influenced the electrophoretic mobility of a larger subpopulation of lymphocytes.

Kaplan and Uzgiris (1976) pursued the issue of lymphocyte electrophoretic heterogeneity by selective enrichment of populations using nylon columns and erythrocyte-rosetting techniques. In agreement with previous workers using classical techniques, they found that T lymphocytes have a higher electrophoretic mobility than B lymphocytes. They also reported an intermediate population which (like B cells) did not pass through a nylon column, but (like T cells) formed rosettes with sheep erythrocytes and did not have surface immunoglobulins or complement receptors. These cells were presumed to be a subpopulation of T cells. More recently, Kaplan et al. (1979) demonstrated an association between the number of high-affinity erythrocyte rosette-forming cells and the intensity of the high mobility peak in the spectrum. Thus the two subpopulations of T cells observed in their laser Doppler spectra were associated with differing affinities for erythrocytes, though there were consistently more low-mobility cells than could be accounted for by low-affinity erythrocyte rosette-forming cells. Uzgiris et al. (1978, 1979) found that in cancer patients the fraction of cells falling in the high-mobility class decreased by 20 ± 11% of total cells, but they observed no significant decrease of high-affinity erythrocyte rosette-forming cells. It was concluded that the correspondence between high affinity and high mobility found for normal donor lymphocytes may not hold for a subset of lymphocytes found in individuals with cancer.

Smith et al. (1976) used ELS to compare the electrophoretic distributions of human peripheral blood mononuclear white cells from normal subjects to those from patients with acute lymphocytic leukemia (ALL). Whereas the normal cells had a bimodal distribution characteristic of lymphocytes, the leukemic cells had a monomodal, relatively narrow distribution with a mode mobility ranging from 7 to 28% below the mobility of normal cells. They also found that optimized cryopreservation techniques have no significant electrokinetic effect on either normal or leukemic cells, a fact which permits case history analyses of individuals for whom sequential aliquots of cells have been cryopreserved. Subsequently Smith et al. (1978) reported a comparison of the electrophoretic distinctions of T and B lymphocytes and ALL cells under conditions of varying ionic strength. Enhanced populations of T and B cells were prepared by selective rosette-depletion procedures. In agreement with previous reports, it was found that T cells have a higher mobility than B cells. The fraction of each was estimable from the bimodal ELS spectra, and good agreement was obtained by comparison with rosetting identification for a group of normal donors whose T/B ratio varied from 1.4 to 4.0. An example is shown in Fig. 10. New information about these cell types was obtained by reaction with the enzyme neuraminidase, which cleaves charged sialic acid groups from the cell surface. After neuraminidase treatment the order of mobility for T and B cells was reversed and the differences were even more distinct, indicating that T cells have a substantially greater amount of available sialic acid on their surfaces. Leukemic cells, whose mobility at 0.015 M ionic strength is in the range of B-cell mobilities, also have a substantial surface density of accessible sialic acid; after neuraminidase treatment they exhibit a very narrow mobility distribution with a mode mobility just below that of T cells. An illustration of the comparison after neuraminidase treatment is shown in Fig. 11. The ionic strength dependence of the mobility of leukemic cells is also distinct; at physiological salt concentration, the leukemic cells overlap with the higher (T-cell) mobilities of the normal populations. Thus leukemic cells are electrokinetically distinct from and presumably differ in cell surface structure from either T or B cells.

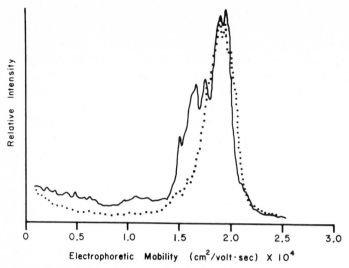

Fig. 10. A comparison of the electrophoretic mobility distributions (at 0.015 M ionic strength) of a fresh human mononuclear white blood cell sample before (solid line) and after (dotted line) EAC rosette depletion (which removes principally B cells). The whole sample contained 63% T cells and 16% B cells as determined by rosette assays. From these data and similar experiments it was concluded that T and B cells can be distinguished on the basis of their electrophoretic mobility, T cells having a higher mobility than B cells. (From Smith et al., 1978.)

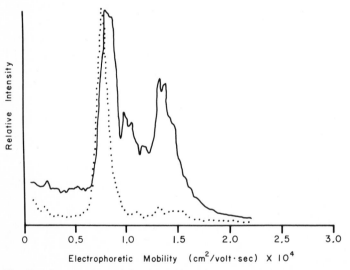

Fig. 11. A comparison of the electrophoretic mobility distributions for normal (solid line) and acute lymphocytic leukemic (dotted line) human mononuclear blood cells after treatment with neuraminidase. Rosetting experiments demonstrated that the low-mobility peak in the normal cell distribution is attributable to T cells and the higher-mobility peak to B cells. Leukemic cells had a very narrow mobility distribution after neuraminidase treatment, with a mode mobility which was slightly, but reproducibly, below the peak for T cells. (From Smith et al., 1978.)

Josefowicz et al. (1977) measured the ELS spectra of fractions of mouse-thymus lymphocytes which had been separated by velocity sedimentation in a density gradient. Faster-sedimenting (larger) cells were shown to have a higher mean mobility, and the presence of physically distinct subpopulations was evident from the structure of the ELS spectra for the individual velocity fractions.

Red blood cells have been studied extensively by classical electrophoresis techniques (Seaman, 1975). The red blood cell electrophoretic mobility varies by as much as a factor of 2 among different species, but within a species is remarkably constant and insensitive to pathological conditions. There have, however, been widely accepted reports that the oldest circulating human erythrocytes have an electrophoretic mobility up to 30% lower than the youngest cells, separated on the basis that older cells are denser (Danon and Marikovsky, 1961; Yaari, 1969). The reduced charge density was proposed as a mechanism for the recognition of senescent cells by the spleen. However, in a joint publication from three research groups including an ELS group (Josefowicz and Ware), this proposition was laid to rest by the demonstration that circulating human erythrocytes separated on the basis of density do not differ in electrophoretic mobility, although, because of membrane loss, they do differ in sialic acid content (Luner et al., 1977).

Furcinitti and Hunter (1978) used ELS to characterize quail femoral medullary cavity cells according to their electrophoretic heterogeneity prior to separation by preparative cell electrophoresis. Their preliminary results indicated that subpopulations of these endosteal cells were sufficiently distinct electrophoretically that electrophoretic separation should be readily feasible. The use of ELS as a rapid diagnostic technique for pre-characterization of cells and particles prior to time-consuming preparative separations is an excellent idea which should find extensive future application.

Most of the very recent literature in ELS has come from the research group of Ware and has dealt with the application of ELS to reactions at the cell surface. The remainder of this article will summarize those reports.

Petty et al. (1979) observed the electrokinetic alterations which resulted from reaction of macrophages and eosinophils from the guinea-pig peritoneal cavity with soluble IgG immune complexes (insulin plus anti-insulin). Macrophage electrophoretic mobilities decreased uniformly by about 60%, whereas eosinophils showed a bimodal response. As illustrated in Fig. 12, one population of eosinophils showed a negligible eletrophoretic alteration, while the major population decreased in mobility by about a factor of 3. The clear distinction of these two subpopulations of eosinophils demonstrates the power of the ability of ELS to generate a simultaneous determination of the complete electrophoretic mobility distribution of a sample of electrokinetically diverse, suspended cells.

Petty and Ware (1979) observed the response of peritoneal macrophages to the tetrameric lectin concanavalin A (Con A). Although binding of Con A to macrophage surfaces had no electrophoretic effect at the concentrations employed, subsequent incubation of the washed cells for 90 min produced a greatly altered electrophoretic distribution, whose mode mobility was about 15% lower but whose

212

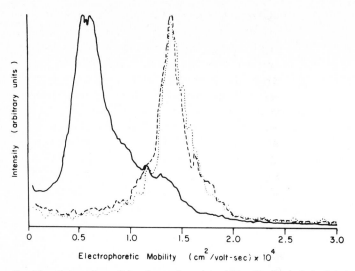

Fig. 12.   Comparison of the electrophoretic mobility distributions (at 0.015 M ionic strength) of guinea-pig resident peritoneal eosinophils treated with (a) bovine insulin (---), (b) guinea-pig antibovine insulin (···), and (c) bovine insulin and guinea-pig antibovine insulin (——).

width increased 3-fold. This effect was attributed to cross-linking of cell surface receptors by the following experiment: reaction of the same cells with dimeric succinyl Con A (which has similar metabolic effects but induces little surface cross-linking) produced no electrokinetic alteration, but a dramatic electrokinetic alteration of the same form could be regenerated by incubation with anti-Con A, which would cross-link the succinyl Con A dimers on the cell surface. These electrokinetic effects on resident peritoneal macrophages were then correlated with alterations of surface morphology (Petty, 1980). Con A was shown to decrease the number of surface folds and ruffles and to be internalized via endocytic vesicles. Succinyl Con A caused no morphological change nor was it internalized unless it was cross-linked by anti-Con A. Similarly is was shown that anionic sites, as visualized by cationized ferritin labeling, were redistributed on the macrophage surface following treatment with Con A or succinyl Con A plus anti-Con A, but not succinyl Con A alone, corroborating the ELS results.

Petty and Ware (1981a) have recently performed a similar series of experiments on inflammatory (oil-elicited) guinea-pig peritoneal macrophages. Surprisingly, the electrokinetic effects observed previously for resident cells were completely absent for inflammatory cells. Scanning electron micrographs showed that the morphological change in the surface folds of the membrane was essentially identical, and transmission electron micrographs using ferritin-conjugated Con A showed vesicular internalization of the same form as for resident cells. However, the anionic site redistribution observed in the resident cells did not occur upon cross-linking of the Con A receptors in the inflammatory cells, in corroboration of the ELS results. It appears that resident macrophages specifically partition anionic groups into en-

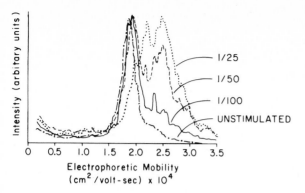

Fig. 13. Effect of immunologic activation upon rat mast cell electrophoretic mobility distribution. Dose response of activation is indicated by showing ELS spectra of mast cells incubated at 37°C with various dilutions (1/100, 1/50, 1/25) of rabbit anti-rat F(ab')$_2$ antiserum. Apparently exocytotic secretion adds new highly charged membrane to the cell surface, creating a population of cells with substantially higher electrophoretic mobility. (From Petty et al., 1980a.)

docytic vesicles, whereas inflammatory macrophages do not. This difference in endocytic mechanisms is one of several newly discovered distinctions between these two cell types; its fundamental significance is not yet clear, but it may be related to a difference in digestive mechanism in the phagolysosome whereby the presence of a higher surface charge density on the interior vesicle surface serves to stabilize a lower pH or steeper ionic distribution in the resident cells. Petty and Ware (1981b) have performed a similar series of ELS experiments on human granulocytes incubated with Con A and succinyl Con A and have found an electrokinetic effect of the same form and magnitude as observed in the resident macrophage.

The inverse of endocytosis, generally called exocytosis, is a fundamental aspect of certain types of vesicular secretion. Petty et al. (1980a) have used ELS to characterize the electrophoretic alterations accompanying stimulated secretion by rat serosal mast cells. They observed a dramatic increase in electrophoretic mobility of stimulated cells that went through a bimodal distribution suggestive of an all-or-none process. Sample data are shown in Fig. 13. The interpretation of these experiments (with associated controls) was that the interior surface of the secretory vesicles (granules) bears a high net negative charge, which adds to the negative surface charge density of the mast cell when the exocytic process transforms the mast cell granule interior to become part of the mast cell exterior. The selective asymmetry of the granule membrane, with a higher interior charge density, is the same as that inferred for the endocytic vesicles of the resident macrophage and circulating granulocyte.

Circulating lymphocytes, in response to immunological stimulation, release a number of soluble factors (lymphokines) that are thought to mediate the expression of cellular immunity. Of these factors the migration inhibitory factor (MIF) and leukocyte inhibitory factor (LIF) have received great attention because of their apparent ability to concentrate cells in an affected area by inhibiting the migration of macrophages (MIF) or polymorphonuclear leukocytes (LIF). Modulation of the surface charge density has been proposed as a plausible mechanism for this activity. Petty et al. (1980b) used ELS to show that LIF has no effect on polymorphonuclear leukocyte surface charge. Two distinct lymphocyte supernatant species in the molecular weight ranges 10–20 K and 30–60 K were found to decrease the mobility significantly (~10%), but neither fraction showed migration-inhibitory activity. In a similar study by the same research groups (Petty et al., 1980c), it was shown that MIF does not affect the electrophoretic mobility of guinea-pig macrophages. In fact, unfractionated supernatants from stimulated human lymphocytes did not alter the electrophoretic mobility distributions of guinea-pig macrophages. Supernatants from antigen-stimulated guinea-pig lymphocytes did reduce the electrophoretic mobilities of guinea-pig macrophages, but the factor responsible was shown to be in the apparent molecular weight range 10–25 K and thus distinct from MIF. The most general conclusion from these studies is that lymphokine migration inhibitory activity is not the result of alteration of the surface charge density of the target cell.

Other aggregation and cell adhesion phenomena have been attributed to electrostatic interactions with the cell surface and thus to cell surface charge density. Two recent studies using ELS have addressed this issue. Wright et al. (1980) found no correlation between electrophoretic mobility and rate of spontaneous aggregation in either normal or transformed fibroblasts. Using the ultrastructural marker polycationized ferritin, they did find that cells which have a high net rate of spontaneous aggregation also show rearrangement of anionic sites on their surface membrane. The important conclusion is that the rate of spontaneous cell aggregation (which is a relevant phenomenon for cancer metastasis and other processes) is affected not by the density of charged groups on the cell surface, but rather by their lateral mobility in the cell membrane. In a related study by the same two research groups (Hoover et al., 1980), the process of adhesion of leukocytes to endothelium, an important initial event in the acute inflammatory reaction, was investigated. Although several chemotactic agents, including divalent cations, showed activity both in promoting cell adhesion and in reducing cell electrophoretic mobilities, there was no correlation between the magnitudes of the two activities. Moreover, certain control factors were active in reducing the electrophoretic mobilities without inducing increased adhesion. The conclusion from these studies must be that electrostatic models of cell–cell interactions are not good predictors of actual biological behavior.

Petty et al. (1981) have recently reported a study of the electrophoretic properties of circulating cells from patients with chronic lymphocytic leukemia (CLL) and hairy cell leukemia (HCL), including a comparison with their normal cell type counterparts. Hairy cells were electrophoretically indistinguishable from normal monocytes; but after treatment with neuraminidase, the two cell types showed very different electrophoretic profiles (see Fig. 14). CLL cells overlapped with the lower portion of the normal lymphocyte distribution, corresponding to B cells, but were closer to the T-cell peak after treatment with neuraminidase. This behavior is very similar to that of cells from patients with acute lymphocytic leukemia (Smith et al.,

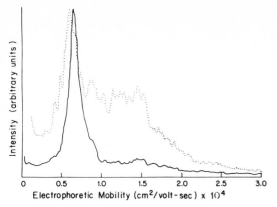

Fig. 14. Electrophoretic mobility distributions of neuraminidase-treated chronic lymphocytic leukemia lymphocytes (——) and normal lymphocytes (···). CLL cells are reduced to a uniform distribution (compare to Fig. 11). (From Petty et al., 1981.)

1978). Thus HCL and CLL cell types can be distinguished from normal monocytes and lymphocytes by ELS with the inclusion of neuraminidase treatment. In addition, HCL and CLL cells are distinguishable from each other using ELS (see Fig. 14), a fact which could be of clinical diagnostic significance.

4.5. Conclusion

Successful applications of electrophoretic light scattering to biological systems in all relevant size ranges have been documented. Use of laser Doppler detection introduces experimental complications which should be offset by the increased power of the method. ELS measurements are fundamentally electrophoresis experiments. The most important aspect of experimental design is to choose a system for which electrophoretic mobility magnitudes or comparisons can provide pertinent information for significant biological or medical questions.

References

Abramson, H.A. (1934) Electrokinetic Phenomena and their Application to Biology and Medicine, Chemical Catalog Co., New York.

Benedek, G.B. (1969) Polarisation, Matière, et Rayonnement, Presses Universitaires de France, Paris, pp. 49–84.

Beniams, H. and Gustavson, R.G. (1942) Theory and application of a two path rectangular microelectrophoresis cell, J. Phys. Chem., 46, 1015–1023.

Bennett, A.J. and Uzgiris, E.E. (1973) Laser Doppler spectroscopy in an oscillating electric field, Phys. Rev. A, 8, 2662–2669.

Berne, B.J. and Pecora, R. (1976) Dynamic Light Scattering, Wiley-Interscience, New York.

Blackman, R.B. and Tukey, J.W. (1958) The Measurement of Power Spectra, Dover, New York, pp. 31–33, 117–120, 125.

Bragg, W.L. (1912) Diffraction of short electromagnetic waves by a crystal, Proc. Cam. Phil. Soc., 17, 43–57.

Bragg, W.L. (1968) X-Ray crystallography, Scient. Amer., 219 (July), 58–70.

Caflisch, G.B., Norisuye, T. and Yu, H. (1980) Electrophoretic light scattering of bovine serum albumin, J. Colloid Interface Sci., 76, 174–181.

Cha, C.Y., Folger, R.L. and Ware, B.R. (1980) Electrophoretic light scattering of water-soluble polyelectrolyte, J. Polymer Sci., Polymer Phys. Ed., 18, 1853–1858.

Chu, B. (1974) Laser Light Scattering, Academic Press, New York.

Cohn, E.J. and Edsall, J.T. (1943) Proteins, Amino Acids, and Peptides, Hafner, New York.

Cornish, R.J. (1928) Flow in a pipe of rectangular cross-section, Proc. Roy. Soc., A120, 691–700.

Costantino, L., Liquori, A.M. and Vitagliano, V. (1964) Influence of thermal denaturation on the electrophoretic mobility of calf thymus DNA, Biopolymers, 2, 1–8.

Currie, B.W. (1931) Electro-endoosmosis in closed cylindrical tubes of large diameter, Phil. Mag., 12, 429–438.

Danon, D. and Marikovsky, Y. (1961) Différence de charge électrique de surface entre érythrocytes jeunes et âgés, C.R. Acad. Sci. (D), 253, 1271–1272.

Davenport Jr., W.B. and Root, W.L. (1958) Random Signals and Noise, McGraw-Hill, New York.

Davis, B.D. and Cohn, E.J. (1939) The influence of ionic strength and pH on electrophoretic mobility, J. Am. Chem. Soc., 61, 2092–2098.

Dean, P.M. (1975) Exocytosis modeling: an electrostatic function for calcium in stimulus-secretion coupling, J. Theor. Biol., 54, 289–308.

Dean, P.M. and Matthews, E.K. (1974) Calcium-ion binding to the chromaffin granule surface, Biochem. J., 142, 637–640.

Debye, P. (1924) Osmotic equation of state and the activity of strong electrolytes in dilute solutions, Physik Z., 25, 97–107.

Debye, P. and Hückel, E. (1923a) Lowering of freezing point and related phenomena, Physik Z., 24, 185–206.

Debye, P. and Hückel, E. (1923b) The limiting law of electrical conductivity, Physik Z., 24, 305–325.

Drain, L.E. (1980) The Laser Doppler Technique, Wiley, New York.

Drifford, M., Menez, R., Tivant, P., Nectoux, P. and Dalbiez, J.P. (1981) Diffusion quasiélastique de la lumière sous champ électrique: mobilité électrophorétique et charge apparente des macromolécules, Rev. Phys. Appl., 16, 19–33.

Drude, P. (1907) The Theory of Optics, C.R. Mann, and R.A. Millikan, (Transl.), Longmans, Green, and Co., New York, pp. 382–399.

Dubin, S.B. (1972) Methods Enzymol. 26, 119.

Dubin, S.B., Lunacek, J.H. and Benedek, G.B. (1967) Observation of the light scattered by solutions of biological macromolecules, Proc. Natl. Acad. Sci. (U.S.A.), 57, 1164–1171.

Edwards, R.V., Angus, J.V., French, M.J. and Dunning Jr., J.W. (1971) Spectral analysis of the signal from the laser Doppler flowmeter: time-independent systems, J. Appl. Phys., 42, 837–850.

Eigen, M. (1964) Proton transfer, acid-base catalysis, and enzymatic hydrolysis: Part I: Elementary processes, Angew. Chem., Int. Edn., Engl., 3, 1–19.

Flygare, W.H., Ware, B.R. and Hartford, S.L. (1976) Electrophoretic light scattering, in: C.T. O'Konski, (Ed.), Molecular Electro-Optics, Dekker, New York, pp. 321–366.

Furcinitti, P.S. and Hunter, S.J. (1978) Electrophoretic heterogeneity of avian femoral medullary cavity cells, in: N. Catsimpoolas (Ed.), Electrophoresis '78, Elsevier, New York, pp. 373–381.

Haas, D.D. (1978) Electrophoretic Light Scattering of Dilute Protein Solutions, Ph.D. Thesis, Harvard University, Cambridge, Ma.

Haas, D.D. and Ware, B.R. (1976) Design and construction of a new electrophoretic light-scattering chamber and applications to solutions of hemoglobin, Anal. Biochem., 74, 175–188.

Haas, D.D. and Ware, B.R. (1978) Electrophoretic mobilities and diffusion coefficients of hemoglobin at high pH, Biochemistry, 17, 4946–4950.

Halliday, D., Resnick, R., Edwards, F. and Merrill, J. (1970) Fundamentals of Physics, Wiley, New York, pp. 1068–1080.

Hartford, S.L. and Flygare, W.H. (1975) Electrophoretic light scattering on calf thymus deoxyribonucleic acid and tobacco mosaic virus, Macromolecules, 8, 80–84.

Harvey, J.D., Walls, D.F. and Woolford, M.W. (1976) Electrophoretic investigations by laser light scattering, Optics Commun., 18, 367–370.

Hoover, R.L., Folger, R., Haering, W.A., Ware, B.R. and Karnovsky, M.J. (1980) Adhesion of leukocytes to endothelium: roles of divalent cations, surface charge, chemotactic agents, and substrate, J. Cell Sci., 45, 73–86.

Jackson, J.D. (1962) Classical Electrodynamics, Wiley, New York, pp. 189–194, 202–205, 611–621.

Jenkins, F.A. and White, H.E. (1957) Fundamentals of Optics, 3rd edn., McGraw-Hill, New York, pp. 232–239.

Josefowicz, J.Y. (1975) Electrophoretic Light Scattering and its Application in Studies of Cellular and Viral Systems, Ph.D. Dissertation, University of Waterloo, Waterloo, Ont., Canada.

Josefowicz, J. and Hallett, F.R. (1975a) Homodyne electrophoretic light scattering of polystyrene spheres by laser cross-beam intensity correlation, Appl. Opt., 14, 740–742.

Josefowicz, J. and Hallett, F.R. (1975b) Cell surface effects of pokeweed observed by electrophoretic light scattering, FEBS Lett., 60, 62–65.

Josefowicz, J.Y., Ware, B.R., Griffith, A.L. and Catsimpoolas, N. (1977) Physical heterogeneity of mouse thymus lymphocytes, Life Sci., 21, 1483–1487.

Kaplan, J.H. and Uzgiris, E.E. (1975) The detection of phytomitogen-induced changes in human lymphocyte surfaces by laser Doppler spectroscopy, J. Immunol. Methods, 7, 337–346.

Kaplan, J.H. and Uzgiris, E.E. (1976) Identification of T and B cell subpopulations in human peripheral blood: electrophoretic mobility distributions associated with surface marker characteristics, J. Immunol., 117, 1732–1740.

Kaplan, J.H. and Uzgiris, E.E. (1980) in: H.E. Nieburgs (Ed.), Prevention and Detection of Cancer, Vol. II, Marcel Dekker, New York, p. 2385.

Kaplan, J.H., Uzgiris, E.E. and Lockwood, S.H. (1979) Analysis of T cell subpopulations by laser Doppler spectroscopy and the association of electrophoretic mobility differences with differences in rosette-forming ability, J. Immunol. Methods, 27, 241–255.

Kittel, C. (1958) Elementary Statistical Physics, Wiley, New York, pp. 133–140.

Lastovka, J.B. and Benedek, G.B. (1966) Spectrum of light scattered quasielastically from a normal liquid, Phys. Rev. Lett., 17, 1039–1042.

Linderstrøm-Lang, K. and Nielsen, S.O. (1959) in: M. Bier (Ed.), Electrophoresis, Academic Press, New York, pp. 60–61.

Luner, S.J., Szklarek, D., Knox, R.J., Seaman, G.V.F., Josefowicz, J.Y. and Ware, B.R. (1977) Red cell charge is not a function of cell age, Nature (London), 269, 719–721.

Magdelénat, H., Turr, P., Tivant, P., Chemla, M., Menez, R. and Drifford, M. (1979) The effect of counter-ion substitition on the transport properties of polyelectrolyte solutions, Biopolymers, 18, 187–201.

Manning, G.S. (1978) The molecular theory of polyelectrolyte solutions with applications to the electrostatic properties of polynucleotides, Quart. Rev. Biophys., 11(2), 179–246.

Manning, G.S. (1979a) Counterion binding in polyelectrolyte theory, Accounts Chem. Res., 12, 443–449.

Manning, G.S. (1979b) Comments on "Electrophoretic light scattering of linear polyelectrolyte aqueous salt solutionsT, J. Phys. Chem., 84, 1059.

Matthews, E.K. and O'Connor, M.D.L. (1978) Photon correlation spectroscopy and laser Doppler anemometry; a new method for determining secretory particle electrophoretic mobilities, J. Physiol., 278, 1–2P.

Matthews, E.K., Evans, R.J. and Dean, P.M. (1972) Electrokinetic properties of isolated chromaffin granules, Biochem. J., 130, 825–832.

McQuarrie, D.A. (1976) Statistical Mechanics, Harper and Row, New York.

Meullenet, J.P., Schmitt, A. and Drifford, M. (1979) Electrophoretic light scattering of linear polyelectrolyte aqueous salt solutions, J. Phys. Chem., 83, 1924–1927.

Mohan, R., Steiner, R. and Kauffmann, R. (1976) Laser Doppler spectroscopy as applied to electrophoresis in protein solutions, Anal. Biochem., 70, 506–525.

218

Olivera, B.M., Baine, P. and Davidson, N. (1964) Electrophoresis of the nucleic acids, Biopolymers, 2, 245–257.

Petty, H.R. (1980) Response of the resident macrophage to concanavalin A, Alterations of surface morphology and anionic site distribution, Exp. Cell Res., 128, 439–454.

Petty, H.R. and Ware, B.R. (1979) Macrophage response to concanavalin A: effect of surface crosslinking on the electrophoretic mobility distribution, Proc. Natl. Acad. Sci. (U.S.A.), 76, 2278–2282.

Petty, H.R. and Ware, B.R. (1981a) The inflammatory macrophage–concanavalin A interaction: A thin-section and scanning electron microscopy and laser Doppler electrophoretic investigation of surface events, J. Ultrastruct. Res., 75, 97–111.

Petty, H.R. and Ware, B.R. (1981b) Electrokinetic response of granulocytes to concanavalin A and succinyl-concanavalin A, Cell Biophys., 3, 19–28.

Petty, H.R., Folger, R.L. and Ware, B.R. (1979) Electrokinetic study of the reactions of peritoneal macrophages and eosinophils with IgG immune complexes, Cell Biophys., 1, 29–37.

Petty, H.R., Ware, B.R. and Wasserman, S.I. (1980a) Alterations of the electrophoretic mobility distribution of rat mast cells after immunologic activation, Biophys. J., 30, 41–50.

Petty, H.R., Smith, B.A., Ware, B.R. and Rocklin, R.E. (1980b) Alterations of polymorphonuclear leukocyte surface charge by stimulated lymphocyte supernatants, Cell Immunol., 54, 435–444.

Petty, H.R., Ware, B.R., Remold, H.G. and Rocklin, R.E. (1980c) The effects of stimulated lymphocyte supernatants on the electrophoretic mobility distribution of peritoneal macrophages, J. Immunol., 124, 381–387.

Petty, H.R., Ware, B.R., Liebes, L.F., Pelle, E. and Silber, R. (1981) Electrophoretic mobility distributions distinguish hairy cells from other mononuclear blood cells and provide evidence for the heterogeneity of normal monocytes, Blood, 57, 250–255.

Pusey, P.N. and Vaughan, J.M. (1975) Dielectric and Related Molecular Processes, Vol. 2, The Chemical Society, Burlington House, London, pp. 48–105.

Rimai, L., Salmeen, I., Hart, D., Liebes, L., Rich, M.A. and McCormick, J.J. (1975) Electrophoretic mobilities of RNA tumor viruses, Studies by Doppler-shifted light scattering spectroscopy, Biochemistry, 14, 4621–4627.

Schlessinger, B.S. (1958) Electrophoretic and titration study of bovine plasma albumin, J. Phys. Chem., 62, 916–920.

Schlieper, P., Mohan, R. and Kauffmann, R. (1981a) Electrokinetic properties of (Na^+,K^+)-ATPase vesicles as studied by laser Doppler spectroscopy, Biochim. Biophys. Acta, 644, 13–23.

Schlieper, P., Medda, P.K. and Kauffmann, R. (1981b) Drug-induced zeta potential changes in liposomes studied by laser Doppler spectroscopy, Biochim. Biophys. Acta, 644, 273–283.

Schmidt, R.J. and Saunders, O.A. (1938) Motion of a fluid heated from below, Proc. Roy. Soc. A, 165, 216–228.

Schmitz, K.S. (1976) Quasielastic light scattering by flexible polymers in the presence of a sinusoidal driving field: internal relaxation modes, Chem. Phys. Lett., 42, 137–140.

Schmitz, K.S. (1979) Quasielastic light scattering by biopolymers, Center-of-mass motion of DNA in the presence of a sinusoidal electric field, Chem. Phys. Lett., 63, 259–264.

Seaman, G.V.F. (1975) Electrokinetic behavior of red cells, in: D. Surgenor (Ed.), The Red Blood Cell, Vol. II, Academic Press, New York, pp. 1135–1229.

Sears, F.W. and Zemansky, M.W. (1960) College Physics, 3rd edn., Addison-Wesley, Reading, MA, pp. 902–909.

Siegel, D.P. and Ware, B.R. (1980) Electrokinetic properties of synaptic vesicles and synaptosomal membranes, Biophys. J., 30, 159–172.

Siegel, D.P., Ware, B.R., Green, D.J. and Westhead, E.W. (1978) The effects of Ca^{2+} and Mg^{2+} on the electrophoretic mobility of chromaffin granules measured by electrophoretic light scattering, Biophys. J., 22, 341–346.

Siegert, A.J.F. (1943) MIT Rad. Lab. Report 465.

Smith, B.A. (1977) The Study of Cell Surface Charge by Electrophoretic Light Scattering, Ph.D. Thesis, Harvard University, Cambridge, Ma.

Smith, B.A. and Ware, B.R. (1978) Apparatus and methods for laser Doppler electrophoresis, in: D.M. Hercules, G.M. Hieftje, L.R. Snyder and M.A. Evenson (Eds.), Contemporary Topics in Analytical and Clinical Chemistry, Vol. 2, Plenum, New York.

Smith, B.A., Ware, B.R. and Weiner, R.S. (1976) Electrophoretic distributions of peripheral blood mononuclear white cells from normal subjects and from patients with acute lymphocytic leukemia, Proc. Natl. Acad. Sci. (U.S.A.), 73, 2388–2391.

Smith, B.A., Ware, B.R. and Yankee, R.A. (1978) Electrophoretic mobility distributions of normal human T and B lymphocytes and of peripheral blood lymphoblasts in acute lymphocytic leukemia: Effects of neuraminidase and of solvent ionic strength, J. Immunol., 120, 921–926.

Stevenson, W.H. (1970) Optical frequency shifting by means of a rotating diffraction grating, Appl. Opt., 9, 649–652.

Stoltz, J.F., Stoltz, M., Peters, A., Collin, F., Larcan, A. and Schneider, M. (1969) Method for measuring the electrophoretic mobility of colloidal particles in suspension, Theory and comparison of various electrophoretic cells, J. Chim. Phys. Physiochim. Biol., 66(5), 922–928.

Tanford, C. (1961) Physical Chemistry of Macromolecules, Wiley, New York.

Uzgiris, E.E. (1972) Electrophoresis of particles and biological cells measured by the Doppler shift of scattered laser light, Optics Commun., 6, 55–57.

Uzgiris, E.E. (1974) Laser Doppler spectrometer for study of electrokinetic phenomena, Rev. Sci. Instrum., 45, 74–80.

Uzgiris, E.E. (1976) A laser Doppler assay for the antigen–antibody reaction, J. Immunol. Methods, 10, 85–96.

Uzgiris, E.E. (1981a) Laser Doppler methods in electrophoresis, Progr. Surface Sci., 10, 53–164.

Uzgiris, E.E. (1981b) Laser Doppler spectroscopy: Applications to cell and particle electrophoresis, Adv. Colloid Interface Sci., 14, 75–171.

Uzgiris, E.E. and Costaschuk, F.M. (1973) Investigation of colloid stability in polyelectrolyte solutions by laser Doppler spectroscopy, Nature Phys. Sci., 242, 77–79.

Uzgiris, E.E. and Kaplan, J.H. (1974) Study of lymphocyte and erythrocyte electrophoretic mobility by laser Doppler spectroscopy, Anal. Biochem., 60, 455–461.

Uzgiris, E.E. and Kaplan, J.H. (1976a) Laser Doppler spectroscopic studies of the electrokinetic properties of human blood cells in dilute salt solutions, J. Colloid Interface Sci., 55, 148–155.

Uzgiris, E.E. and Kaplan, J.H. (1976b) Tuberculin-sensitized lymphocytes detected by altered electrophoretic mobility distributions after incubation with the antigen PPD, J. Immunol., 117, 2165–2170.

Uzgiris, E.E., Kaplan, J.H., Cunningham, T.J., Lockwood, S.H. and Steiner, D. (1978) Laser Doppler spectroscopy in experimental and clinical immunology, in: N. Catsimpoolas (Ed.), Electrophoresis '78, Elsevier, New York, pp. 427–440.

Uzgiris, E.E., Kaplan, J.H., Cunningham, T.J., Lockwood, S.H. and Steiner, D. (1979) Association of electrophoretic mobility with other cell surface markers of T cell subpopulations in normal individuals and cancer patients, Eur. J. Cancer, 15, 1275–1280.

Velick, S.F. (1949) Interaction of enzymes with small ions, I. An electrophoretic and equilibrium analysis of aldolase in phosphate and acetate buffers, J. Phys. Coll. Chem., 53, 135–149.

Ware, B.R. (1972) The Invention and Development of Electrophoretic Light Scattering, Ph.D. Thesis, The University of Illinois at Urbana, Champaign, Il.

Ware, B.R. (1974) Electrophoretic light scatttering, Adv. Colloid Interface Sci., 4, 1–44.

Ware, B.R. and Flygare, W.H. (1971) The simultaneous measurement of the electrophoretic mobility and diffusion coefficient in bovine serum albumin solutions by light scattering, Chem. Phys. Lett., 12, 81–85.

Ware, B.R. and Flygare, W.H. (1972) Light scattering in mixtures of BSA, BSA dimers, and fibrinogen under the influence of electric fields, J. Colloid Interface Sci., 39, 670–675.

Watrasiewicz, B.M. and Rudd, M.J. (1976) Laser Doppler Measurements, Butterworths, London.

Weissman, M.B. and Ware, B.R. (1978) Applications of fluctuation transport theory, J. Chem. Phys., 68, 5069-5076.

220

Whalen, A.D. (1971) Detection of Signals in Noise, Academic Press, New York, pp. 96–99.

Wright, T.C., Smith, B., Ware, B.R. and Karnovsky, M.J. (1980) The role of negative charge in spontaneous aggregation of transformed and untransformed cell lines, J. Cell Sci., 45, 99–117.

Yaari, A. (1969) Mobility of human blood cells of different age groups in an electric field, Blood, 33, 159-163.

Yen, W., Mercer, R.W., Ware, B.R. and Dunham, P.B. (1981) Electrophoretic light scattering of red cell membrane vesicles: a study of the electrical charges of the cytoplasmic membrane surface, in: S.-H. Chen, B. Chu and R. Nossal (Eds.), Proceedings, NATO Advanced Study Institute on Scattering Techniques Applied to Supramolecular and Nonequilibrium Systems, Plenum, New York.

Yoshimura, T., Kikkawa, A. and Suzuki, N. (1972) Spectroscopy of light scattered by suspended charged particles, Jpn. J. Appl. Phys., 11, 1797–1804.

Yoshimura, T., Kikkawa, A. and Suzuki, N. (1975a) Measurements of electrophoretic movements with an optical beating spectrometer, Jpn. J. Appl. Phys., 14, 1853–1854.

Yoshimura, T., Kikkawa, A. and Suzuki, N. (1975b) The spectral profile of light scattered by particles in electrophoretic movement, Optics Commun., 15, 277–280.

CHAPTER 9

Time-resolved fluorescence spectroscopy

SALVADOR M. FERNANDEZ

Department of Physiology, University of Connecticut Health Center,
Farmington, CT 06032, U.S.A.

Contents

1. Introduction 222
2. Experimental approaches 223
   2.1. Pulsed methods 223
   2.2. The phase shift method: sinusoidally modulated excitation 225
   2.3. The cross-correlation method: randomly fluctuating excitation 227
3. Technical considerations 228
   3.1. Time-correlated single photon (TCSP) detection 228
        3.1.1. Instrumentation 228
               3.1.1.1. Timing electronics 229
               3.1.1.2. Dead time and data collection rates 231
               3.1.1.3. Light sources 233
               3.1.1.4. Photomultiplier tube response 239
        3.1.2. Deconvolution of fluorescence decay curves 240
               3.1.2.1. Method of moments 241
               3.1.2.2. Laplace transformation 243
               3.1.2.3. Iterative convolution 244
               3.1.2.4. Other methods 245
               3.1.2.5. Criteria for goodness of fit 246
               3.1.2.6. Comparative performance of various methods 247
   3.2. Phase shift method 248
        3.2.1. Instrumentation 248
               3.2.1.1. Modulations of the exciting light 249
               3.2.1.2. Phase-sensitive detection 251
               3.2.1.3. Photomultiplier tube response 252
4. Fluorescence lifetimes 253
   4.1. Technical considerations 254
   4.2. Representative biological applications 254
        4.2.1. Photophysics of aqueous tryptophan 254
        4.2.2. Dynamics of protein conformation revealed by fluorescence
               of tryptophan residues 255

R.I. Sha'afi and S.M. Fernandez (Eds.), Fast Methods in Physical Biochemistry and Cell Biology
© *1983 Elsevier Science Publishers*

222

5. Time- and phase-resolved emission spectra 256
  5.1. Time-resolved emission spectra (TRES) 257
      5.1.1. Representative applications 258
  5.2. Phase-resolved spectra 259
6. Time-resolved fluorescence anisotropy measurements 260
  6.1. Technical considerations 261
  6.2. Information provided by TRF anisotropy measurements 262
  6.3. Representative applications 264
      6.3.1. Structural dynamics of biological membranes 264
      6.3.2. Dynamics of biopolymers 266
7. Differential polarized phase fluorometry 267
  7.1. Information provided by the differential polarized phase method . 267
  7.2. Applications 269
8. TRF spectroscopy in the study of resonance energy transfer 269
  8.1. Biochemical applications 271
9. Application of TRF spectroscopy to cell biology 272
References 275

1. Introduction

The simplest item of information that can be obtained from a time-resolved fluorescence (TRF) measurement is the fluorescence lifetime of the fluorphore. With additonal effort, TRF spectroscopy provides a method to monitor spectral and fluorescence polarization changes which occur during the excited state lifetime.

Quantum mechanics show that in the ultraviolet and the visible range of the spectrum fluorescence lifetimes are of the order of $10^{-7} - 10^{-10}$ sec. Thus, TRF spectroscopy enjoys access to a time scale in which a broad assortment of physicochemical processes of fundamental interest in biochemistry, and cell and molecular biology take place. While fluorescence lifetimes extend into the $10^{-11} - 10^{-12}$ sec range, albeit less commonly, the study of phenomena which occur in this time scale belongs to the realm of picosecond spectroscopy. Since this is the topic of the following chapter, discussion of such fast processes will not be presented here.

Excited state interactions such as proton transfer, excimer and exciplex formation, and dipolar relaxation processes which may include solvent relaxation or reorientation of a biopolymer matrix around the excited state dipole moment of a fluorophore are but a few of the processes that can be studied with TRF spectroscopy. Rotational diffusion of fluorophores either covalently bound to a biomolecule or thermodynamically partitioned into macromolecular assemblies such as biomembranes can also be investigated by examining the time dependence of the fluorescene emission anisotropy. The rotational behavior of a covalently bound probe can supply information about the size, shape, flexibility and conformation of the biopolymer to which it is attached; membrane-partitioned fluorophores can provide knowledge about structural fluctuations and other dynamic properties of such assemblies that are thought to play important functional roles.

Additionally, TRF measurements can reveal information about inter- and intramolecular distances of the order of 10–100 Å since fluorescene lifetimes offer a direct means of calculating resonance energy transfer efficiencies between donor and acceptor fluorphores. Likewise, through the effect of other types of dynamic quenching processes on fluorescence lifetimes, the diffusion of solutes over distances of the order of 10–100 nm can be investigated. Thus, intelligence about the dynamics of the immediate and extended microenvironment of a fluorophore can be readily obtained from measurement of its fluorescent lifetime.

With proper experimental design, the general phenomena discussed above can be utilized to obtain knowledge about an enormous variety of specific biochemical, biophysical, and cellular processes. In what follows an attempt will be made to illustrate the versatility and power of TRF spectroscopy by examining representative recent applications of biological interest. The reader should be aware, however, that the focus of this chapter is on methodology. The inclusion of specific examples, therefore, will be guided by their ability to illustrate the use of a particular technique, and the overall selection of material is heuristic in purpose. Citations to the literature, consequently, are neither comprehensive nor historical.

For a more complete survey of the literature, especially earlier applications, the reader is referred to the excellent reviews by Yguerabide (1972) and Rigler and Ehrenberg (1973). More recently, in an article which is somewhat more limited in scope, Lakowicz (1980) has elegantly discussed the application of TRF spectroscopy to several phenomena of biological interest.

2. Experimental appoaches

In general terms, time-resolved fluorescence measurements are performed by determining the kinetic response of a fluorescent sample to excitation by light of time-varying intensity. Short pulses of light or continuous illumination with a sinusoidally modulated source are the most common modes of excitation, although light of randomly fluctuating intensity can also be employed. Each of these approaches can be implemented in practice in a variety of ways.

2.1. Pulsed method

In this approach the fluorescent sample is excited with a short pulse of light and the decay of intensity of the emitted fluorescence is determined as a function of time after excitation (Lewis et al., 1973). The observed decay curves, however, are distorted by the finite duration of the excitation pulse and by the limited frequency response of the detection system. Thus, the measured fluorescence decay curve $F(t)$ is given by the convolution integral (see Chapter 1)

$$F(t) = L(t) * f(t) * I(t) \tag{1}$$

where $L(t)$ represents the time profile of the exciting light flash, $f(t)$ is the true fluorescence response of the sample; i.e., the response to a delta function excitation, and $I(t)$ is the impulse response function of the apparatus (photo-multiplier tube and associated electronic signal processing system).

The measured pulse shape $E(t)$ of the exciting light pulse $L(t)$ is similarly given by

$$E(t) = L(t) * I(t) \tag{2}$$

By virtue of the associative and commutative properties of convolution, it follows from Eqns. 1 and 2 that the observed fluorescence decay, $F(t)$ can be treated as the convolution of an effective apparatus function $E(t)$ with the true fluorescent response $f(t)$:

$$F(t) = E(t) * f(t) \tag{3}$$

In practice $F(t)$ and $E(t)$ are experimentally determined and $f(t)$ is obtained by deconvolution of Eqn. 3, a procedure which is usually accomplished with a computer. The excitation function $E(t)$ ideally should be narrow relative to the fluorescence lifetime being measured, thus the width of $E(t)$ should be of the order of 1 nsec or less.

The insensity profiles $E(t)$ and $F(t)$ can be determined with a variety of techniques. The most common of these are streak camera detection (Bradley, 1977), pulse sampling techniques (Steingraber and Berlman, 1963) and time-correlated single photon detection (Bollinger and Thomas, 1961; Isenberg, 1975).

Streak camera detection may be advantageous when excitation of the sample results in a large number of photons at the detector (Walden and Winefordner, 1980). In this method the entire intensity decay can be recorded from a single shot of the exciting source. This method possesses the advantages of speed and high resolution ($\sim$ 10 psec). For further details on this approach the reader is referred to Chapter 10 of this volume.

In sampling techniques the fluorescence intensity is usually determined with a so-called boxcar integrator (Badea and Georghiou, 1976) or with a transient digitizer (Docchio et al., 1981; Crosby and MacAdam, 1981). In the former, the insensity is sampled at a series of sequential time intervals which are variably delayed with respect to the excitation. These acceptance windows are generated by either gating the detector gain or sampling the PMT anode current with an oscilloscope. Each PMT pulse accepted is only sampled once; this information can then be digitized and stored in the memory of a multichannel analyzer (MCA)) where each channel corresponds to a sampling time point on the horizontal sweep of the oscilloscope. A transient digitizer, on the other hand, samples many points on a single wave form. A drawback of this device, however, is a relatively slow sampling time which may limit its resolution. Sampling times of 2 nsec have been reported recently (Specht, 1980). Both methods are amenable to long-range signal averaging but are sensitive to drift. Their main advantages are low expense, simplicity, and fast data collection times for intense emitters.

In the time-correlated single photon (TCSP) detection method, the time lag between excitation (an invariant point on the intensity profile of the exciting flash) and the subsequent arrival of an individual fluorescence photon at the detector is measured. The sample is excited repetitively, and after a sufficient number of time lags have been collected, a distribution function of the number of photons which arrive at the detector per unit time as a function of time lag after excitation can be reconstructed. This provides the time profile, $F(t)$, of the emitted light intensity. The advantage of TCSP detection include exceptionally high sensitivity, low noise, stability against drift and high time resolution ($\sim$ 50 psec). Technical aspects of this technique are considered in greater detail below.

Note that while a streak camera directly records the intensity decay resulting from a single excitation flash, sampling and photon-counting methods are indirect in that the time dependence of the emission is reconstructed from the results of many individual excitation events.

2.2. The phase shift method: sinusoidally modulated excitation

In this approach the fluorescent sample is excited by a continuous source of light whose intensity is sinusoidally modulated at a single frequency. Under these conditions, the emitted fluorescence is modulated at the same frequency but demodulated and phase shifted. From measurements of either the phase retardation or the demodulation of the emission, the fluorescence lifetime can be calculated as follows (Bailey and Rollefson, 1953; Spencer and Weber, 1970). Consider a homogeneous population of fluorophores which decay exponentially with lifetime τ. The modulated exciting light $E(t)$ can be represented as

$$E(t) = A + B \cos \omega t \tag{4}$$

where $\omega = 2\pi f$ is the angular modulation frequency. Then, the fluorescence $F(t)$ is given by

$$F(t) = A + B \cos \delta \cos (\omega t - \delta) \tag{5}$$

The phase shift, δ, is related to the fluorescence lifetime by Eqn. 6,

$$\tan \delta = \omega \tau \tag{6}$$

and the demodulation is given by

$$\cos \delta = [1 + (\omega \tau)^2]^{-\frac{1}{2}} \tag{7}$$

Values of the lifetime calculated from the phase lag (Eqn. 6) and from the de-modulation (Eqn. 7) agree only if the decay follows a single exponential. In a heterogeneous emitting population the lifetime measured from the degree of modulation is generally longer than the weighted average of the component lifetimes, whereas lifetimes determined from the phase lag will always be shorter than the weighted average. This provides a means to analyze non-exponential behavior.

Considering the time scale of typical fluorescence phenomena (10^{-9} sec), the theory of Fourier transforms dictates that the modulation frequency, f, be of the order of 100 MHz. This is the frequency domain counterpart of the pulse width re-quirements ($\sim$ 1 nsec) discussed in the previous section.

To determine the phase shift experimentally the modulated exciting beam is split into two components. One of these becomes a reference beam which is detected di-rectly with a PMT, and the other is used to excite the sample; the resulting fluores-cence signal is detected by a second PMT. The phase difference between the refer-ence and fluorescence signals is then measured with a phase-sensitive detector such as a lock-in amplifier. In order to avoid problems characteristic of RF signal pro-cessing, the RF modulated fluorescence and reference signals are often not de-tected directly at the anode of the PMT. Instead, a heterodyne procedure is employed to translate the information from the original high frequency to a low fre-quency (< 100 Hz). This is accomplished by modulating the gain of the PMTs at a dynode with an RF signal of frequency $F + \Delta f$ ($\Delta f < 100$ Hz). The resulting low-frequency cross-correlated anode signals (Δf Hz) which contain the phase and modulation information of the original high frequency signals are then measured with the phase sensitive detector.

An optical delay line technique rather than electronic means can be employed to determine the magnitude of the phase shift (Menzel and Popovic, 1978). In this case, the modulated reference beam is directed through an optical delay onto the same PMT that detects the sample fluorescence. The reference beam is delayed op-tically with a movable mirror and suitably attenuated to be equal in amplitude and opposite in phase to the fluorescence. When this condition is met, the detector photocurrent is a direct current with no ac component at the modulation frequency; determination of the ac null is accomplished by connecting the PMT to a radio re-ceiver tuned to the modulation frequency, or preferably with a spectrum analyzer. Once the ac null is established, the same procedure is repeated with the sample re-placed by a scattering medium. The optical path change Δx between the two optical delays yields the phase shift,

$$\delta = \frac{\omega \Delta x}{c} \tag{8}$$

where c is the speed of light. This optical approach simplifies the instrumentation required for phase shift measurements.

Advantages of the phase shift method include relatively simple instrumentation, rapid data collection, high precision, straightforward data analysis which obviates the need for deconvolution, and subnanosecond time resolution. Phase fluorometers possessing picosecond resolution have been described (Haar and Hauser, 1978). A disadvantage of this approach, however, is that unlike pulse methods, it does not produce a complete time decay curve which can be useful for analyzing heterogeneous emissions and other sources of non-exponential relaxation. To generate information equivalent to that obtained by the pulsed method would require a Fourier analysis of lifetimes measured as a function of modulation frequency. Phase fluorometers with continuously variable frequency have been described (Hauser and Heidt, 1975). Complex decays can be studied, however, by modulating the exciting light at several discrete frequencies, albeit with a restricted range of resolution.

2.3. The cross-correlation method: randomly fluctuating excition

In contrast to the two approaches to TRF measurements described above which require either a pulsed or a sinusoidally modulated excitation source, an alternative technique can be employed which utilizes continuous excitation from a cw laser. This approach is based on the fact that intensity fluctuations arise in a free running inhomogeneously broadened laser from interference between longitudinal modes which have random phase relationships (Chapter 10). The principle of this method is based on linear response theory (Hieftje and Haugen, 1981). In essence, a time decay curve equivalent to that obtained in a pulsed excitation experiment is obtained by cross-correlating the randomly fluctuating cw excitation with the sample emission (Dorsey et al., 1979).

   The cross-correlation method can be implemented in practice with an intensity interferometer. In this apparatus the exciting laser output is split into two beams; one of these is used to excite the sample whose fluorescence is detected by a PMT, and the other is directed through a variable optical delay onto a reference PMT. The output signals from the two PMTs are processed through a double balanced mixer and detected with a lock-in amplifier.

   When both detectors view the exciting light (a scattering medium in the sample compartment), both detectors record the signal $E(t)$ (assuming identical detector responses), which as we have already seen is the convolution of the light source $L(t)$ and the detector response $I(t)$ (Eqn. 2). The correlation of the response of the two detectors, C_{EE}, as a function of time lag Δt introduced by the optical delay is given by (see Chapter 1)

$$C_{EE}(\Delta t) = E(t) \star E(t) \qquad (9)$$

   To determine the fluorescence decay, $f(t)$, the scattering medium is replaced with the sample. The fluorescence signal $F(t)$ produced at the detector is then given as before by Eqn. 3. The cross-correlation, C_{EF}, between the excitation response $E(t)$ and the fluorescence signal $F(t)$ can then be obtained and is given by

$$C_{EF}(\Delta t) = E(t) \bigstar F(t) \tag{10}$$

The cross-correlation C_{EF} is expected to persist at longer delay times than the autocorrelation C_{EE} due to the finite fluorescence lifetime of the sample. From general properties of convolutions and correlations (Brigham, 1974) it is easy to show that

$$C_{EF}(\Delta t) = C_{EE}(\Delta t) * f(\Delta t) \tag{11}$$

That is, the cross-correlation is given by the convolution of the autocorrelation with the true fluorescence decay. Thus, once $C_{EF}(\Delta t)$ and $C_{EE}(\Delta t)$ are experimentally determined, $f(t)$ can be extracted from Eqn. 11 by any suitable deconvolution technique.

As it is readily apparent from comparison of Eqns. 3 and 11, the relationship between the auto- and cross-correlations is analogous to that between the scattered light and the fluorescence signal of a pulsed experiment. An obvious advantage of this approach is that it eliminates the need to mode-lock the laser source (vide infra) or for high frequency modulation of the exciting light. The instrumentation thus becomes considerably simpler and less expensive. A serious disadvantage, however, is the limited sensitivity of this approach (Dorsey et al., 1979). In addition, complex non-exponential decays might be difficult to analyze with this method.

3. Technical considerations

3.1. Time-correlated single photon (TCSP) detection

The TCSP method is probably the most widely employed technique for the performance of TRF measurements. It is generally the method of choice when maximum sensitivity is a major concern and it is especially useful when sample photolability dictates that excitation be carried out at low average power. The excellent signal-to-noise ratio achievable with this technique endows it with a large dynamic range which makes possible collection of decay curves over many orders of magnitude of intensity. At present, this method achieves its maximum potential when a mode-locked or cavity-dumped laser is employed as the excitation source. The narrow light pulses ($\sim$ 10 psec), stable pulse shapes, and high repetition rates (1–100 MHz) of such lasers result in superior time resolution and high data collection rates. Technical aspects of the TCSP method have been previously reviewed by Isenberg (1975).

3.1.1. Instrumentation
The TCSP technique requires three important components: a source of short light pulses of stable shape at high repetition rate; a fast timing system, and a detector

with high gain and linearity and good single photon resolution. Fig. 1 shows a conventional arrangement which was designed for making TRF measurements from single cells (Fernandez, 1982). The light source is a mode-locked synchronously pumped cavity-dumped tunable dye laser which will be described below. The heart of the timing system consists of a time-to-amplitude converter (TAC) and a multichannel pulse height analyzer (MCA). The operation and performance of the various system components is discussed below.

3.1.1.1. Timing electronics. The TAC is the fast clock of the system. It is a device which accepts two temporally displaced input pulses, a start and a stop pulse, and generates for every pair of these a conversion output pulse of analog amplitude proportional to the time interval between the start and stop pulses. The conversion output of the TAC is sent to the MCA which performs a pulse height analysis. In addition to the start and stop signal inputs, the TAC has two control inputs (gate and strobe) and two status outputs (true start and busy), the use of which will become apparent as this discussion progresses.

The overall system operates as follows: A small fraction of the exciting light pulse is split-off and directed onto a fast photodiode (PD) (rise time < 50 psec). The output pulse of the PD is processed through a constant fraction discriminator which generates a timing logic pulse at some invariant time point on the profile of the PD

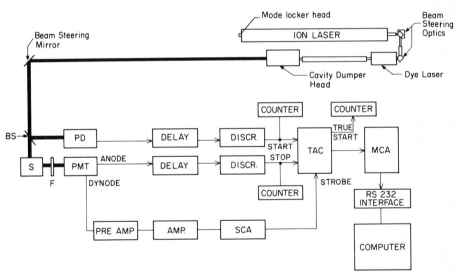

Fig. 1.   Block diagram of a time-resolved fluorimeter employing the TCSP technique. The pulsed light source is a Spectra Physics mode-locked synchronously pumped cavity-dumped tunable dye laser which consists of a krypton ion laser (model 171), a dye laser (model 375) and a cavity dumper (model 344). The start signal for the TAC is derived from a fast photodiode (PD) (Spectra Physics 403B) and the stop signal is picked off the anode of an RCA 8850 PMT. Both discriminators are Elscint STB-N-2 models, the TAC is an Ortec 457, and the MCA is a Canberra Series 40. BS, bean splitter; S, sample; SCA, single channel analyzer; F, optical filter and/or polarizer (or monochromator).

output. This logic pulse is fed into the start input of the TAC and provides the reference timing signal for excitation of the sample. The timing of arrival of an individual fluorescence photon at the detector is derived from the anode signal of the PMT which, after being processed by a second constant fraction discriminator, is connected to the stop input of the TAC. Leading edge timing may be employed in the PD channel since PD pulses have relatively constant amplitude; PMT anode pulses, on the other hand, possess a large amplitude dynamic range and thus require constant fraction discrimination to minimize timing walk (Gedcke and McDonald, 1968; Leskovar et al., 1976). The TAC conversion output which contains the timing information coded in the form of an analog voltage amplitude is routed to the MCA for analysis. After a suitably large number of excitation cycles, an amplitude histogram is reconstructed in the MCA which represents the probability $F(t)$ that the first photoelectron of the PMT is released in the time interval $(t, t + dt)$ after a time origin related to the excitation event.

Since the TAC can only process one stop pulse for each valid start pulse, the MCA histogram is a direct analog of the decay under observation only when a single photon reaches the detector during each cycle of measurement. If two or more photons were to reach the detector during a measurement time span, the second and subsequent photons are not recorded and the resulting time distribution would be distorted; i.e., biased toward shorter times. This pulse pile-up problem can be solved by applying a theoretical correction to the data (Coates, 1968) or by operating the system at sufficiently low count rates that the probability of two-photon events is negligible. The former method assumes Poisson statistics which may not hold at high count rates when source polarization and coherence effects may become significant; the latter method results in unduly long data collection times. A third more convenient approach is to allow a higher photon flux and to reject multiphoton events electronically; a procedure which makes no assumptions about the nature of the statistics.

Hardware rejection of multiphoton events can be accomplished with a pulse-pair discriminator which inhibits the TAC when two photons are detected within the measuring time span (Davis and King, 1970), or by energy windowing (Schuyler and Isenberg, 1971). In this latter approach a signal is picked off a dynode of the PMT and processed with a preamplifier and amplifier whose combined time constant is sufficiently long that dynode pulses separated by less than the TAC measuring range are integrated into a single pulse of larger amplitude. The amplified dynode pulses are fed to a single channel analyzer (SCA) set to reject pulses which do not meet the amplitude requirements of single photon events, and the output of the SCA is used to strobe the information out of the TAC onto the MCA. If no strobe pulse arrives within a certain time period, the TAC resets and the conversion output is discarded. A disadvantage of the strobe method to reject multiphoton events is that it introduces a non-linearity between the integrated TAC output and both the fluorophore concentration and the fluorescence intensity. As the intensity increases, the probability of occurrence of a multiphoton event becomes greater and so does the fraction of events that is rejected.

3.1.1.2. Dead time and data collection rates. Optimization of data collection rates, while always an experimental goal, becomes crucially important when studying samples of low stability or when the sample is a dynamic entity such as a living cell. The rate at which data can be processed is limited by the measuring system dead time: the time during which the system is busy processing information and thus cannot receive new input. Understanding the sources of dead time is important to optimize data collection rates, particularly when employing high repetition rate sources (1–100 MHz) such as mode-locked and cavity-dumped lasers. If the laser repetition rate is increased above a certain frequency, the event period (interval between excitation pulses) becomes shorter than the system dead time and therefore a stop pulse may cause the next event to be ignored. As the repetition rate is increased, a greater fraction of events is disregarded. A thorough analysis of dead time in the TCSP method has been presented recently (Haugen et al., 1979).

The primary sources of dead time in the TCSP technique are the TAC conversion time, t_c, and the time it takes the MCA to perform a pulse height analysis. Unlike pulse-sampling techniques where the system dead time is fixed, in the TCSP method the overall dead time is variable and depends on the exeperimental conditions. This can be readily seen by considering the operation of the TAC with reference to the timing diagram shown in Fig. 2: a start pulse triggers the generation of a linear ramp voltage within the TAC; arrival of a stop pulse interrupts this process and causes the generation of a conversion output pulse of amplitude equal to the

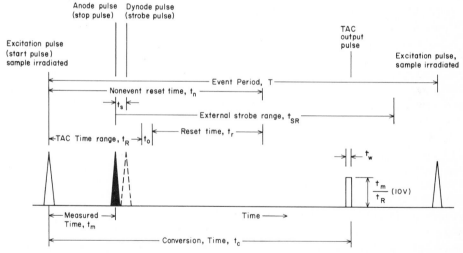

Fig. 2. Timing diagram for an externally strobed TAC in the conventional configuration. The reset time t_r, output pulse width t_w, and overrange t_o are the internal dead times of the TAC. The interval between excitation pulses is T, and t_{SR} represents the time after a stop pulse during which a strobe pulse will be accepted. t_s is the time elapsed between the arrival of the stop and strobe pulses, and t_c and t_n correspond to the conversion times in the presence or absence of a photoelectron event respectively. (Redrawn from Haugen et al., 1979.)

ramp voltage. Before this information is sent out to the MCA, however, the TAC waits for a period t_{SR} for the external strobe signal. After this, a second time interval, the reset time, t_r, elapses before the TAC is ready to receive new input. The conversion time, t_c, is then the sum of the measured time, t_m, plus a dead time which depends on t_s, the arrival time of the strobe pulse. If a given conversion is not strobed, then $t_c = t_m + t_{SR} + t_r + t_w$; for no conversion (i.e., no stop pulse) the non-event reset time $t_n = t_R + t_o + t_r$, where t_o is a fixed overrange characteristic of the TAC. Typically, $t_r \sim 5\,\mu sec$, $t_w \sim 2\,\mu sec$, $t_{SR} \sim 5\,\mu sec$, and $t_R \sim 0.5\,\mu sec$, resulting in a non-strobed conversion time of about $12\,\mu sec$. Therefore, the relationship between the true start rate versus the laser repetition rate will be linear with unity slope up to about $80\,kHz$ ($1/t_c$) where it would start to roll off. It is apparent that in this system configuration there is no advantage in running the laser above a frequency equal to $1/t_n$, and that the frequency producing the most true starts must be empirically determined.

A disadvantage of this conventional system configuration is that not every laser pulse produces a true start even when the TAC is operating at well below its maximum rate. This can be remedied by configuring the system in the so-called interactive mode in which the busy output of the TAC is used to control the laser (Haugen et al., 1979). In this configuration the sample is never excited unless the TAC can process the data; therefore, every excitation is a true start and the linear range of the true start versus laser repetition rate curve is extended. To obtain intensity or fluorophore concentration information, however, the integrated TAC output must be normalized to the number of excitation events.

A further drawback of the conventional configuration, interactive or otherwise, is that if a stop pulse is not sensed, the total non-event recovery time, t_n, is spent without acquiring data. A more efficient configuration for operating the TAC would be that in which only single-photon events are allowed to start a conversion. This is accomplished with the so-called inverted configuration (Taylor and Becker, 1972) in which the PMT monitoring the fluorescence provides the start pulse, and the PD output, delayed by an interval equal to the TAC range, t_R, provides the stop pulse (Fig. 3). To reject multiphoton events, the dynode control signal is sent to the TAC gate input operated in the coincidence mode. In this mode of operation the start pulse is not accepted unless it overlaps in time the control pulse at the TAC input. This, by the way, requires that the anode pulse be delayed by an interval equal to the integration time of the dynode channel. Thus, timing requirements in this case become more stringent than when the strobe option is employed. In the inverted configuration, events detected immediately after excitation generate large amplitude conversions, while those detected at later times generate small amplitude conversions; thus, the MCA reconstructs the decay curve reversed in time. Unlike in the strobed mode, the single photon check does not affect the dead time since the control signal is used to gate the start pulse before it triggers the TAC. This feature, coupled with the fact that only single-photon events trigger conversions, results in high data collection rates. Disadvantages of this approach are that (a) the integrated TAC output is never linear with intensity or sample concen-

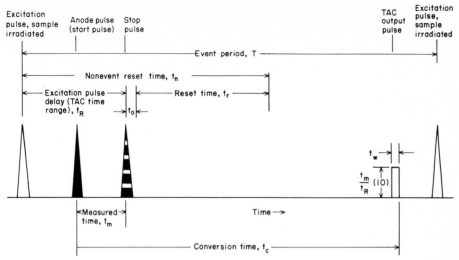

Fig. 3. Timing diagram for inverted TAC configuration. The excitation pulse is delayed by a time equal to the TAC range and is used as a stop pulse. (Redrawn from Haugen et al., 1979.)

tration, unless the average anode pulse separation is greater than the dead time and (b) whenever the event period, T, is less than the average dead time, the fluorescence decay will be distorted because the resetting of the TAC will be random with respect to excitation of the sample.

As in the conventional configuration, the inverted mode of operation can be made interactive with the laser source by employing the busy output signal of the TAC to control the laser. In this case, every true start results in a conversion and the maximum possible data rate is attained. Furthermore, linear operation is always achieved and no statistical distortion of the decay can occur when $T < t_{Dt}$, because time zero is always well defined. This is achieved at the expense of more complex electronics; and, of course, to preserve intensity information the integrated output of the TAC must be normalized to the number of excitation events.

Finally, the MCA contribution to the overall system dead time must be considered. The MCA dead time consists primarily of a fixed ADC dead time, a variable channel address digitizing time and the memory cycle time. MCAs can have dead times that range from a few to tens of microseconds and thus may limit further the data rate of the TAC alone.

3.1.1.3.   Light sources. Since its first introduction in the early 1960's (Bollinger and Thomas, 1961), the TCSP method has undergone two periods of major technological advance. During the first decade of its use, progress in detector and amplifier technology resulted in significanctly improved resolution and sensitivity; during the last decade a drastic revolution in the methods available to generate ultrashort light pulses has further enhanced the power of this technique.

The chief attributes of a pulsed light source are pulse duration, pulse peak power, and repetition rate. The pulse full width at half-maximum should be at least as short as the relaxation time of the phenomena being investigated; the shorter the pulse width the more accurate deconvolution procedures become. Pulse peak power should be sufficiently high to generate large signals but not so extreme that it would create non-linear effects or photodamage the sample, and finally, the repetition rate should be high and variable to permit either maximizing data collection rates or minimizing sample degradation by maintaining low average power. Below, we examine a variety of pulsed sources and investigate to what extent they meet these requirements.

Gas discharge spark-gap flash lamps either free running or thyratron-driven (D'Alessio et al., 1964; Yguerabide, 1965; Lewis et al., 1973) have been the most commonly employed light sources for pulse fluorimetry. A number of shortcomings, however, render these flash lamps far less than ideal for the purposes of TCSP detection. Their major drawbacks are low pulse peak power ($\sim 10^{-6}$ W), low repetition rate (~ 40 kHz) and poor pulse shape stability. Pulse widths typically range from 1 to 3 nsec. Subnanosecond pulses can be achieved for limited periods of time but only at the expense of meticulous electrode care; electrode cleaning and shaping becomes a tedious quotidian laboratory task. In addition, maximizing the repetition rate or minimizing the pulse width, both result in a decrease in pulse peak power. Other disadvantages include poor spectral output per unit bandwidth, unpolarized emission, poor focusing characteristics, and when gated by a thyratron they generate large amounts of RF noise which is difficult to suppress and may distort the measured decay curves.

The advent of pulsed lasers which are free of these shortcomings has rendered gas discharge flash lamps obsolete. Nanosecond flash lamps, however, by virtue of their simplicity and low cost will retain a niche in biochemical research for some time to come, especially when maximum sensitivity and resolution are not of paramount importance and when high data collection rates are not essential. The operation and construction of nanosecond flash lamps have been reviewed previously (Yguerabide, 1972).

Lasers may be pulsed by a variety of methods including pulsed excitation, Q-switching, mode-locking, cavity dumping, and a combination of cavity dumping and mode-locking. The choice of the best type of laser for a given application would undoubtedly entail a trade-off between cost and various other requirements such as time resolution, pulse power, repetition rates, etc. It is beyond the scope of this chapter to delve into the theory and operation of pulsed lasers (Yariv, 1975). Instead, the following discussion will be limited to those practical aspects of pulsed lasers that are relevant to the TCSP technique. For a more detailed treatment of pulsed lasers and particularly of mode-locking the reader is referred to Chapter 10 of this volume.

Pulsed excitation and Q-switching generate high pulse power but at relatively low repetition rates which makes them ill-suited for the signal averaging approach of the TCSP technique. Cavity dumping, on the other hand, produces relatively narrow pulses (~ 10 nsec) of moderate peak power ($\sim 50 - 100$ W) at variable repetition rates ranging from single shot to MHz. The cavity-dumping principle is simply to periodically diffract out of the cavity a fraction of the intracavity energy (Johnson, 1973). This is accomplished by focusing the beam onto a Bragg cell located inside of the laser cavity. When the cavity is to be dumped a pulse generator triggers a fast RF switch which allows an RF signal ($\sim 500\,\text{MHz}$) to transiently form an acoustic grating in the quartz crystal; while the grating is present a portion of the beam is deflected and steered out of the cavity through appropriate optics. The width of the resulting pulse depends on the time that the acoustic grating is sustained and on the coupling efficiency, and is of the order of the cavity round-trip time. Cavity-dumped ion lasers have been employed in time-resolved fluorimetry (Lytle and Kelsey, 1974); their major limitation is a moderately long pulse width.

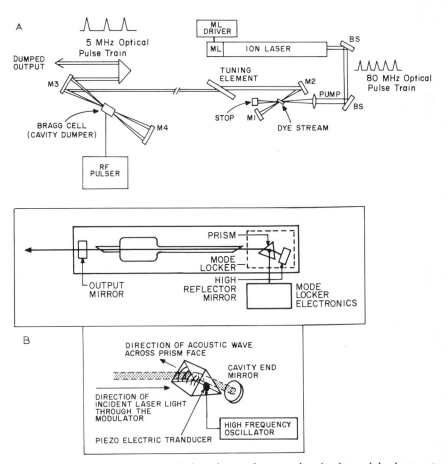

Fig. 4. (A) Schematic of mode-locked synchronously pumped cavity-dumped dye laser system. ML, mode locker; BS, beam-steering mirrors; M1–M4, cavity mirrors. (B) Mode-locked ion laser configuration. The lower panel shows a schematic representation of an acousto-optic modulator used in ion laser mode-locking.

Much shorter pulse widths (~ 200 psec) can be obtained by the process of mode-locking (Kennedy, 1975). In an inhomogeneously broadened laser a large number of longitudinal cavity modes will normally oscillate with random phase and amplitudes. If the phase and amplitude of these longitudinal modes are fixed with respect to each other the laser is said to be mode-locked and the laser output consists of a train of pulses separated by the cavity round-trip time. The duration of the pulses depends on the number of modes that can be locked together: the larger the bandwidth for gain the shorter the pulses. Mode-locking is normally achieved by modulating the loss (or gain) of the laser at a frequency whose period is equal to the cavity round-trip transit time, $2L/c$, where L is the cavity length and c is the speed of light. One way to achieve this is by inserting an acousto-optic modulator inside of the laser cavity (Fig. 4B). See the following chapter for further details on mode-locking.

Mode-locked lasers have been employed in TRF measurements by a number of investigators (Wild et al., 1977; Spears et al., 1978; Kinosita et al., 1981b). The combination of short pulse width (< 200 psec), stable shape, and pulse peak powers of the order of 50 W make mode-locked lasers an attractive light source for the TCSP technique. By changing optics, moreover, the system can be discretely tuned over the emission lines of the ion laser. Krypton and argon ion lasers are commonly used for this purpose. Table 1 shows the various lines available for mode-locking from these ion lasers. A drawback of mode-locked lasers as excitation sources for TCSP detection, however, is the very high fixed repetition rate, which results in interpulse separations (~ 10 nsec) which may be short relative to the decays being investigated and which are much shorter than the conversion time of the TAC. This requires additional external electronics to scale down the start rate of the TAC.

An improvement which combines the short pulse widths of mode-locking with the variable repetition rates of cavity dumping can be attained by cavity dumping a

TABLE 1
Emission lines of argon and krypton ion lasers

Argon wavelength (nm)	Krypton wavelength (nm)
350	337.4–356.4
454.5	413.1
457.9	468.0
465.8	476.2
472.7	482.5
476.5	520.8
488.0[a]	530.9
496.5	568.2
501.7	647.1
514.5	676.4
528.7	752.5
	799.3

[a] Mode-locked operation of this line excepted.

mode-locked laser. This configuration has also been employed with TCSP detection (Spears et al., 1978). In a mode-locked ion laser the energy is contained within a short optical pulse circulating back and forth within the resonator cavity. Under these conditions the acousto-optic modulator of the cavity dumper acts essentially as a gate which periodically couples a mode-locked pulse out of the laser cavity without affecting its shape. Thus, the cavity dumper when used in this configuration transforms the fixed high repetition rate of the mode-locked laser to a variable rate from single shot up to MHz.

A further increase in the versatility and resolution of the laser source can be attained by employing a mode-locked dye laser instead of an ion laser. One way to mode-lock the dye laser is by synchronous pumping (Chan and Sari, 1974). In this process the output pulses of a mode-locked ion laser are used to excite a dye laser whose cavity length, L, has been extended such that the intermode frequency spacing ($c/2L$) is an integral multiple of the mode-locker frequency. In this configuration, as each ion laser pulse enters the dye stream, the leading edge of the pulse brings the population of excited dye molecules up past the threshold to sustain lasing. The dye laser pulse circulating within the cavity is timed to arrive at the dye jet stream just as the dye laser medium reaches threshold. Because of the large stimulated emission cross-section of the dye medium, the inverted population is rapidly depleted by the dye laser pulse. Following this depletion, the remaining part of the pump pulse has insufficient energy to bring the dye laser medium back above threshold, and the resulting dye laser pulse is much shorter in duration than the ion laser pulse. A schematic diagram of a mode-locked synchronously pumped cavity-dumped dye laser system is shown in Fig. 4A.

Synchronous pumping of the dye laser thus results in a train of ultrashort pulses (~ 10 psec) with peak power of the order of a few hundred watts, separated by the cavity round-trip time. Of course, just as described above, a mode-locked synchronously pumped dye laser can also be cavity dumped. This result in a train of ultrashort pulses (~ 10 psec) with high peak powers (~ 1 kW), variable repetition rate (single shot up to MHz) and tunable over the gain curve of the dye. Mode-locked synchronously pumped cavity-dumped dye lasers have been employed in TRF spectroscopy with TCSP detection (Koester and Dowben, 1978).

Operating characteristics of the various laser configurations described which are of interest in TCSP detection are compared in Table 2. Table 3 additionally presents some of the dyes which have been successfully employed in synchronously pumped systems with their approximate tuning ranges.

An altogether different type of light source which also provides narrow pulses at high repetition rates is synchrotron radiation from an electron storage ring (Bisby and Munro, 1980; Munro and Sabersky, 1980). Synchrotron radiation consists of a series of equally spaced, nearly Gaussian pulses, of constant amplitude and width, and shape that is wavelength independent (Gratton and Lopez Delgado, 1979). For further details on the characteristics of synchrotron radiation see Chapter 7 of this volume. It appears that synchrotron radiation may be used to measure fluorescence lifetimes with subpicosecond resolution. Widespread use of this mode of excitation, however, is hampered by the restricted accessibility of such facilities.

All factors considered, it appears that at present mode-locked synchronously pumped cavity-dumped dye lasers constitute the most powerful and versatile excitation sources for TRF measurements. These laser systems generate highly monochromatic picosecond light pulses of high peak power, stable shape, well defined polarization, and variable repetition rate at an output wavelength which is continu-

TABLE 2

Comparison of several pulsed laser systems and air-gap discharge flash lamps as excitation sources for TRF spectroscopy[a]

	Pulse width (nsec)	Repetition rate (kHz)	Peak power (W)	Average power (W)	Wavelength range (nm)
Air-gap discharge	1–3	5–40	10^{-6}	10^{-9}	Discrete: 296, 316, 337, 358, 381
Cavity-dumped ion laser	15	Variable single shot to MHz	$10\text{--}100^{b}$	1^{b}	Discrete[c]
Mode-locked ion laser	0.2	Fixed ~ 100 MHz	$30\text{--}90^{d}$	2^{d}	Discrete[c]
Cavity-dumped mode-locked ion laser	0.2	Variable single shot to MHz	150^{b}	0.1^{b}	Discrete[c]
Mode-locked synchronously pumped cavity-dumped dye laser	0.01–0.001	Variable single shot to MHz	10^{3e}	0.02	Continuously[f] tunable

[a] Figures presented are only approximate. They should be interpreted as generalized typical values.
[b] Depends on the particular laser line employed and on the cavity dumper frequency.
[c] Discretely tunable over the laser lines available. Requires changing cavity mirrors. See Table 1 for a listing or argon and krypton ion laser lines.
[d] Depends on the particular laser line employed.
[e] Depends on the gain of the dye employed, pump line power, and cavity dumper frequency.
[f] Tunable over the gain curve of the dye employed. See Table 3 for the tunable range of some commonly used dyes.

TABLE 3

Approximate tuning range of several dyes employed in mode-locked synchronously pumped dye lasers

Dye	Tuning range (nm)	Excitation range (nm)	Pump line (nm)
Stilbene	420–460	UV	Ar 350
Coumarin 102	465–515	400–420	Ar UV Kr 413.1
Coumarin 30	490–550	400–420	Kr 413.1
Rhomadine 6G	540–650	458–514	Ar 514.5 Kr 530.9
Oxazine	695–800	647–672	Kr 647.1
DEOTC	760–870	647–672	Kr 647.1/676.4

ously tunable over the gain curve of the dye. By changing dye and optics, the tuning range of the laser may be extended from the deep blue to the infrared, and with frequency doubling techniques (Chapter 10), into the ultraviolet. Finally, because of the short pulse duration the average output power of these sources is of the order of milliwatts in spite of kilowatt peak powers; this minimizes potential thermal and radiation damage to the sample.

3.1.1.4. Photomultiplier tube response. The central role of the PMT in the TCSP instrumentation can be readily appreciated, as it is in this device that the photon detection proper takes place. The proces begins at the photocathode where fluorescence photons produce photoelectrons. The photoelectrons are accelerated through a series of dynodes in a vacuum where they generate cascades of secondary electrons which ultimately form a small current pulse at the anode. Because of the statistical nature of the photoelectric effect and of the secondary emission process the output current pulses vary from one detection event to another, even if they always arise from detection of a single photon (Leskovar and Lo, 1975). PMT output pulses, therefore, possess a distribution of amplitudes. Furthermore, the fact that photoelectrons may have different trajectories through the PMT leads to random fluctuations in the time of flight of electrons traveling from photocathode to anode. This transit time dispersion limits the resolution with which the detection of events can be timed. In fact, PMT transit time spread is at present the limiting factor in the overall timing resolution of the TCSP technique. Transit time dispersion can be optimized by judiciously adjusting the size of the illuminated photocathode area (Leskovar et al., 1976). Best results are obtained with tubes that have a small number of dynodes and a high-gain first dynode (cesium-activated gallium phosphide). RCA 8850, 8852, and C31034 PMTs have been commonly employed. The transit time spread of these tubes ranges from 0.3 to 0.8 nsec (full width at half-maximum) (Leskovar and Lo, 1975). A substantial improvement in timing resolution has been achieved with a static crossed-field PMT (Varian VPM 154) which has a transit time dispersion of less than 30 psec (Koester, 1979). Recently, a side-on type PMT (Hamamatsu TV model R928) with a response function width of 160 psec (FWHM) was employed in a TCSP system (Kinosita and Kushida, 1982).

The dependence of the PMT response on the size of the illuminated photocathode area dictates that the excitation response function $E(t)$ be determined with a detector geometry as similar as possible to that existing during the measurement of the fluorescent response $F(t)$; failure to take this phenomenon into account can lead to inaccurate results in the deconvolution computations. PMTs are prone to one other type of systematic error which may not be so easily circumvented; namely, their time response may be wavelength dependent (Lewis et al., 1973).

The difficulty arises because the wavelength λ_{em} at which the fluorescence signal $F(t)$ is collected differs from the excitation wavelength λ_{ex} due to the fluorescence Stokes shift ($\lambda_{em} < \lambda_{ex}$). Recall that the experimentally determined decay $F(t)$ is a convolution of an effective apparatus function $E(t)$ and the true fluorescence decay $f(t)$ (Eqn. 3). This apparatus function is typically obtained by replacing the fluorescent sample with a scattering medium and observing the time profile of the scattered excitation light pulse at wavelength λ_{ex}. Therefore, the $f(t, \lambda_{em})$ that is calculated from the raw data is obtained by deconvolution of

$$F(t, \lambda_{em}) = E(t, \lambda_{ex}) * f(t, \lambda_{em}) \tag{12}$$

whereas the $f(t, \lambda_{em})$ that is actually measured is given by

$$F(t, \lambda_{em}) = E(t, \lambda_{em}) * f(t, \lambda_{em}) \tag{13}$$

The inequality of $E(t, \lambda_{em})$ and $E(t, \lambda_{ex})$ thus leads to a systematic error in the determination of $f(t)$ which may be severe when fast decays and large Stokes shifts are involved and which render analysis for multiexponential decays virtually meaningless.

A number of methods have been developed to correct for this source of error. The simplest of these (Wahl et al., 1974) employs a reference compound with a well characterized single exponential decay at λ_{em}. The true fluorescence decay of the reference compound is then given by

$$f_R(t) = F_0 e^{-t/\tau} \tag{14}$$

and Eqn. 3 becomes

$$F_R(t) = E(t, \lambda_{em}) * f_R(t) \tag{15}$$

It is easy to show (Koechlin and Raviart, 1964) that Eqn. 15 leads to the relation

$$E(t, \lambda_{em}) = F_R(t) + \frac{dF_R}{dt} \tag{16}$$

from which $E(t, \lambda_{em})$ can be readily calculated numerically.

This method to correct PMT wavelength dependence requires the availability of a suitable reference compound whose emission spectrum overlaps that of the sample. Other approaches which overcome this limitation have been described (Rayner et al., 1977; Ricka, 1981). This PMT effect could be completely obviated if the excitation profile $E(t)$ were measured at λ_{em} rather than at λ_{ex}. This is not always possible with gas discharge flash lamps since the intensity profile $L(t)$ of these sources is also wavelength dependent. On the other hand, this approach may be feasible with mode-locked synchronously pumped tunable dye lasers provided that both λ_{ex} and λ_{em} fall within the tuning range of the dye.

3.1.2. Deconvolution of fluorescence decay curves
As previously indicated the experimentally determined fluorescence decay curve $E(t)$ is given by the convolution of an excitation function $E(t)$ and the true fluores-

cence response $f(t)$ (Eqn. 3). For the purposes of this discussion we now rewrite Eqn. 3 in integral form.

$$F(t) = E(t) * f(t) = \int_0^t E(t - u)f(u)du \qquad (17)$$

The central problem in the analysis of fluorescence decay data is to extract the function $f(t)$ from the convolution of Eqn. 17, given the experimentally determined $F(t)$ and $E(t)$. Mathematically, this type of problem is well defined; and, indeed, if one were dealing with ideal error-free data the recovery of $f(t)$ would be straightforward. In practice, the problem is fraught with difficulties, the sources of which have been discussed by a number of authors (Knight and Selinger, 1971; Isenberg, 1973a, b). The gist of the problem is that the solution of Eqn. 17 is markedly sensitive to numerical errors and the information available about $F(t)$ and $E(t)$ is incomplete and contains errors. In fact, knowledge about these two functions is limited to a finite number of discrete, inherently noisy, observations over a limited period of time; namely, the counts in the channels of the MCA. To further exacerbate the problem, systematic errors may distort the data. We have already encountered some of these, such as the wavelength dependence of the PMT response, which at least can be corrected prior to undertaking the deconvolution of Eqn. 17. Other sources of distortion may be more subtle and not so easily recognizable; these might include timing walk and jitter in the electronics, the presence of small amounts of stray scattered light in the fluorescence signal, polarization effects, different distribution of the incident light on the photocathode during collection of $E(t)$ and $F(t)$, RF noise, etc.

With these prefatory words of caution about the hazards of deconvolution procedures, we now examine a number of methods for the analysis of fluorescence decay data that have been widely used and extensively tested. These fall into two categories; those that assume a specific functional model for $f(t)$ and those that make no such assumption. The former methods, in general, provide more satisfactory results although they are not devoid of problems. For example, on the basis of physicochemical grounds it is often assumed that the decay law, $f(t)$, is given by a sum of exponentials. Resolution of multiexponential decays, however, results in a correlation between the various parameters so that a given error in the data leads to a whole family of solutions which are equally acceptable (Knight and Selinger, 1971). Criteria by which to estimate confidence on a given mode will be presented below.

3.1.2.1.   Method of moments.   In the method of moments (Isenberg and Dyson, 1969) it is assumed at the outset that $f(t)$ can be represented by a sum of exponentials; i.e.,

$$f(t) = \sum_{i=1}^{N} \alpha_i e^{-t/\tau_i} \qquad (18)$$

To analyze the data one calculates moments of $E(t)$ and $F(t)$ defined as

$$\mu_k = \int_0^\infty t^k F(t) dt \tag{19}$$

$$m_k = \int_0^\infty t^k E(t) dt \tag{20}$$

These moments are now considered the experimental information. From Eqns. 17 and 19 it follows that

$$\mu_k = \int_0^\infty t^k E(t - u) f(u) du dt \tag{21}$$

which upon integration leads to a set of linear equations given by the recursion relationship

$$\frac{\mu_k}{k!} = \sum_{s=1}^{k+1} G(s) \frac{M_{k+1-s}}{(k+1-s)!} \tag{22}$$

where the G_s are defined as

$$G_s = \sum_{n=1}^N \alpha_n \tau_n^s \tag{23}$$

Eqn. 23 can thus be solved for the decay parameters α_n and τ_n given the moments μ_i and m_i. A set of $2N$ G_s completely characterizes a given decay.

The paramaters obtained from Eqn. 23 are only approximate since the moments are defined from time zero to infinity, wherease the experimental decay curves are truncated at a finite time, T, and do not as a rule vanish at the last data channel. To improve the estimates of the parameters a cut-off correction is performed with an iterative procedure. The cut-off error in the moments of $E(t)$ are assumed to be negligible since $E(t)$ rapidly approaches zero at early values of t. The cut-off correction, $\delta\mu_k$, for the approximate moments of $F(t)$, μ_k^T, are given by

$$\delta\mu_k = \int_T^\infty t^k F(t) dt \tag{24}$$

$$= \sum_{i=1}^N \beta_i \int_T^\infty t^k \exp(-t/\tau_i) dt \tag{25}$$

with

$$\beta_i = \alpha_i \int_0^T E(u) \exp(u/\tau_i) du \qquad (26)$$

These corrections are initially calculated employing the approximate values of α_i and τ_i obtained from Eqn. 23. The new estimates of the moments

$$\mu_k = \mu_k^T + \delta\mu_k \qquad (27)$$

are then used to find another set of α_i and τ_i, and from them the next cut-off correction. The iterations are continued until additional looping does not change the parameters by more than some arbitrarily small amount.

The original method has now been improved to minimize the convergence time of the iterative algorithm (Isenberg et al., 1973a) and to correct for the presence of certain systematic errors in the data (Isenberg, 1973b). More recently, the method has been extended to treat other than multiexponential decays (Solie et al., 1980).

3.1.2.2. Laplace transformation. The Laplace transform of a function $X(t)$ is defined as

$$X(s) = L[X(t)] = \int_0^\infty X(t)e^{-st}dt \qquad (28)$$

The Laplace transform method (Gafni et al., 1975) for obtaining fluorescence decay parameters from Eqn. 17 is based on the convolution theorem which states

$$F(s) = L[\int_0^t E(t-u)f(u)du] = L[E(t)] \cdot L[f(t)] = E(s) \cdot f(s) \qquad (29)$$

In words, the convolution existing between $E(t)$ and $F(t)$ in the time domain is converted to a simple product in the s domain. Thus, by evaluating the Laplace transforms of $E(t)$ and $F(t)$, the Laplace transform of $f(t)$ can be calculated from

$$f(s) = \frac{F(s)}{E(s)} \qquad (30)$$

In this method it is also usually assumed that the decay law $f(t)$ is given by a sum of exponentials (Eqn. 18), the Laplace transform of which is given by

$$f(s) = L[f(t)] = \sum_{i=1}^N \frac{\alpha_i}{s + 1/\tau_i} \qquad (31)$$

Thence, from Eqns. 29 and 30

$$\frac{F(s)}{E(s)} = \frac{\alpha_i}{s + 1/\tau_i} \tag{32}$$

By evaluating this expression for $2N$ different values of s, a set of $2N$ equations is obtained the solution of which yields the desired N amplitudes and N decay constants. Care must be exercised in choosing the values of s. Multiples of $\Delta t/\tau$, where Δt is the time width of the MCA data channels and τ is the average decay time expected, offer a reasonable starting point, but it may be necessary to vary s until a range of s values is found over which the fitted parameters are independent of s.

As in the method of moments, this method also requires a cut-off correction because the transforms are defined from time zero to infinity (Eqn. 27) whereas the data extend only to a finite time. Details of this correction will not be presented here. Suffice it to say that an iterative procedure similar to that described for the method of moments is employed in which approximate parameters from Eqn. 31 are used to calculate a cut-off correction, which, in turn, is employed to further improve the parameters. Looping is continued until some convergence criterion is satisfied. Systematic errors due to the presence of a scattering component or to zero-time shifts may be corrected with this method.

3.1.2.3. Iterative convolution. As in the previous two methods, in the iterative convolution approach (Grinvald and Steinberg, 1974; Grinvald, 1976) a functional form with adjustable parameters is assumed for $f(t)$. Theoretical considerations often suggest a sum of exponentials for $f(t)$ although this method is not restricted to such forms.

In this method, the convolution, $F_c(t)$, of the model function $f(t)$ and $E(t)$ is calculated according to Eqn. 17. If we assume that $f(t)$ is again given by Eqn. 18, then

$$F_c(t) = \int_0^t E(u) \sum_{i=1}^N \alpha_i \exp\left[-(t-u)/\tau_i\right] du \tag{33}$$

This calculated $F_c(t)$ is then compared to the experimentally determined $F(t)$ and the quality of the fit is judged by the magnitude of the sum of the weighted square of the residuals, φ

$$\varphi = \sum_{j=1}^n W_j[F_c(t_j,\alpha_i,\tau_i) - F(t_j)]^2 \tag{34}$$

where the subscript j refers to the jth time interval (data channel) in the MCA, and the weighting factor W_j is related to the variance σ_j^2 of $F(t_j)$ by

$$W_j = \frac{1/\sigma_j^2}{1/n \sum_{j=1}^n 1/\sigma_j^2} \tag{35}$$

Clearly, prior knowledge about the variance σ_j^2 is needed to evaluate W_j. Fortunately, in the TCSP technique this information is available, since the counting error is Poisson-distributed and approaches a Gaussian distribution for a large number of counts (neglecting possible systematic errors). The variance σ_j^2 of the jth channel is then simply given by the number of counts in that channel; thence,

$$W_j = \frac{1/F(t_j)}{1/n \sum_{i=1}^{n} 1/F(t_j)} \tag{36}$$

A non-linear least squares iterative search procedure (Marquardt, 1963) may now be employed to find the best fit to the data; i.e., those parameters of the trial function $f(t, \alpha_i, \tau_i)$ which minimize φ, the sum of the weighted square of the residuals. At each iteration step, the minimization criteria

$$\frac{\delta\varphi}{\delta\alpha_i} = 0, \qquad \frac{\delta\varphi}{\delta\tau_i} = 0 \tag{37}$$

lead to improved values of the parameters, which are then used to generate a new $F_c(t)$ and a new φ. The iterations continue until successive values of φ differ by less than some arbitrarily chosen convergence criterion.

3.1.2.4.   Other methods.   In addition to the three methods that we have just considered, a number of others have been employed. These include the method of modulating functions (Valeur, 1978), the exponential series method (Ware et al., 1973), Fourier transform methods (Hunt, 1971), and a systems-theory approach (Eisenfeld and Ford, 1979). Simple rapid methods which are applicable only to single exponential decays have also been described (Demas and Adamson, 1971; Rockley, 1980). Space limitations do not permit detailed treatment of all these methods here, some general comments, however, would seem pertinent.

The Fourier transform method is akin to the Laplace transform one in that in both cases the convolution that exists between $f(t)$ and $E(t)$ in real space becomes a simple multiplication in transform space. While in theory the Fourier transform may provide a facile way to obtain $f(t)$, its application to photon counting data has been plagued with problems due to magnification of the statistical counting error inherent to the data. Recently, a refinement of this method less sensitive to noise was developed (Wild et al., 1977) in which deconvolution is carried out in Fourier space using the fast Fourier transform (Brigham, 1974; Makinen, 1982).

The exponential series method (Ware et al., 1973) requires, in principle, no prior assumptions about the functional form of $f(t)$. The goal is simply to obtain an accurate representation of the decay. $f(t)$ can be represented by an expression of the form

$$f(t) = \sum_{i=1}^{n} a_i e^{-t/\gamma_i} \tag{38}$$

in which the a_i may be either positive or negative and no physical significance is attached to either the a_i or γ_i. Typically, the γ_i are fixed and by a least squares procedure the a_i are varied until the best fit is obtained between the experimentel $F(t)$ and the calculated convolution of $f(t)$ with $E(t)$. For best results n, γ_i and a_i must be judiciously selected; a set of 10 evenly spaced γ_i have been generally used (O'Connor et al., 1979).

3.1.2.5. Criteria for goodness of fit. Once a set of "best" parameters has been obtained for the assumed model function $f(t)$, a decision has to be made as to whether the model chosen adequately represents the decay kinetics. This can be a difficult problem for it is not always clear whether a poor fit results from defects in the model, failure of the instrument (systematic errors) or both.

While ultimately the decision whether to accept or reject a model is a subjective one, a number of statistical parameters can be used to make the basis for such a decision as objective as possible.

The most inadequate criterion for deciding whether a given model fits the decay is a visual comparison of the calculated and experimental curves by superposition. This tactic, while widely used in other fields, is simply not sufficiently sensitive for the analysis of fluorescence decay data; examples abound in the literature in which simulated data fitted with the wrong parameters appear deceptively indistinguishable from the calculated curve by simple inspection (see Fig. 5).

Lumped statistical parameters such as the weighted sum of the square of the residuals or the root mean square deviation, by themselves are not necessarily adequate because they may fail to detect significant correlation among residuals. More informative is to plot the weighted residuals which should be randomly distributed about zero. Perhaps a more suitable diagnostic device for the randomness of scattering of the experimental points about the model function is the autocorrelation function for the residuals (Grinvald and Steinberg, 1974). The autocorrelation function $C(t)$, of the weighted residuals is given by

$$C(t_j) = 1/m \sum_{i=1}^{m} (W_i)^{1/2} \Delta_i (W_{i+j})^{1/2} \Delta_{i+j} / 1/n \sum_{i=1}^{n} W_i \Delta_i^2 \tag{39}$$

where $\Delta_i = F_c(t_i) - F(t_i)$; n is the number of data channels and m is the number of terms in the numerator. The index j can assume values 1, 2, ..., $(n-m)$. For a good fit, $C(t_j)$ shows high-frequency fluctuations of low aplitude about the zero baseline when plotted as a function of t_j. If the scatter of the experimental points about a proposed decay function is not random, $C(t_j)$ will exhibit conspicuous low-frequency oscillations around the zero line. Such behavior is symptomatic of a poor choice of the model function and is illustrated in Fig. 5.

A discussion of conditions under which the autocorrelation of the residuals may not provide an adequate test for goodness of fit has been presented recently (Irvin et al., 1981). These authors have devised a generalized chi-square statistic for such cases; unfortunately, implementation of this test requires an inordinate amount of computational effort.

3.1.2.6. Comparative performance of various methods. The diversity of deconvolution methods available for the analysis of fluorescence decay may be somewhat disconcerting to the neophyte; indeed, selecting the optimal method for a given situation may be troublesome even for the cognoscenti. Some discussion of the performance of the various methods, therefore, may be helpful at this point. Fortunately, our present task is greatly facilitated by two recent studies in which deconvolution methods were extensively and comparatively tested employing both simulated and real data (McKinnon et al., 1977; O'Connor et al., 1979). Without going into details about the relative strengths and weaknesses of each method, we shall present here some observations about which there appears to be a consensus.

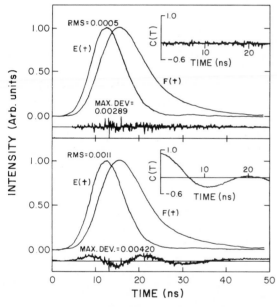

Fig. 5. Analysis of computer-simulated fluorescence decay data. $E(t)$ is the excitation light pulse and $F(t)$ the convolved fluorescence decay calculated from a sum of three exponentials with lifetimes of 1, 3 and 7 nsec. Simulated noise was assumed to be due to counting error corresponding to 10^6 counts at the peak. The smooth $F(t)$ curve was obtained by fitting the noisy curve by the least squares technique. (A) Analysis assuming the decay to be triexponential. (B) Analysis wrongly assuming the decay to be biexponential. Noisy curves below $E(t)$ and $F(t)$ are plots of the residuals, and the insets represent the autocorrelation function of the weighted residuals (Eqn. 39). Although the calculated $F(t)$ appear to give equally good fits to the "experimental" (simulated) $F(t)$ by visual inspection in both cases, the autocorrelation of the residuals reveals significant systematic deviations when the wrong model is employed. (Redrawn from Grinvald and Steinberg, 1974.)

First of all, it appears that for reasonably good data the method of moments, the Laplace transform, the method of modulating functions and the iterative convolution approach all give satisfactory results for the analysis of single exponential decays, especially if the decay time is long compared to the width of the excitation function $E(t)$.. For biexponential decays, the method of modulating functions performs better than the method of moments and the Laplace transform. This is not surprising since the latter relies on an extrapolation which may be inaccurate unless the cut-off error is insignificant, and the method of moments gives greater weight to data points at later times where the signal-to-noise ratio is poorest. Fourier methods yield generally unacceptable results in most cases, and are not recommended. Finally, there is general agreement that the iterative convolution approach provides the best results in all cases. It appears to be superior to the other methods in its ability to find fast decays, in its tolerance of noise in the data, and in its power to resolve closely spaced time constants in biexponential decays. Therefore, the iterative convolution approach is recommended as the method of choice.

3.2. Phase shift method

The phase shift method has been employed for the measurement of fluorescence lifetimes since long before the advent of the TCSP technique (Bailey and Rollefson, 1953). The main advantages of the latter are twofold: first, as a pulse method it generates a complete time spectrum of the fluorescence intensity decay; second, compared to analog signal processing, single-photon counting results in a substantial enhancement in signal-to-noise ratio. If the phase shift can be measured as a function of modulation frequency, however, both methods are, at least in principle, equivalent since the amplitude-time data and the phase-frequency information are related through a Fourier transform. Recent advances in the technology of high-frequency optical modulation now make it feasible to construct phase fluorometers with continuously variable modulation frequency up to about 100 MHz (Hauser and Heidt, 1975), and single-photon counting techniques have been incorporated into the phase method (Schlag et al., 1974). Thus, at present, the phase shift technique remains a powerful, versatile and experimentally convenient approach to TRF measurement. To be sure, both methods have advantages and disadvantages, and intrinsic differences between the two may render one or the other better suited for a certain type of application. The decision as to which method to employ, however, is often based on the investigator's convenience or bias rather than on some inherent superiority of either approach.

3.2.1. Instrumentation
Fig. 6 shows a schematic diagram of a typical experimental arrangement for phase fluorometry. The principal components of the system are the light source, an optical modulator, and the phase-sensitive detector. The system works as follows: a portion of the modulated exciting light is deflected by a beam splitter onto a PMT whose output forms the reference input to the phase-measuring device where it is

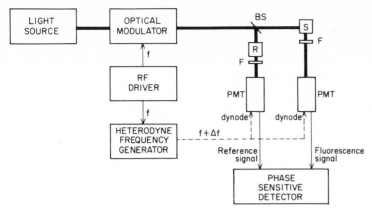

Fig. 6. Schematic of phase modulation apparatus. The light source may be a broad band high-pressure arc discharge or a cw laser; the light modulator is typically a Debye–Sears or an electro-optic device. The gain of the PMTs may be optionally (dashed line) modulated at a heterodyne frequency, and a reference fluorophore (R), instead of a scattering solution may be used to generate the reference signal in order to correct for the PMT wavelength-dependent response. BS, beam splitter; S, sample; F, optical filter or monochromator.

compared to the signal from a second PMT which views the fluorescence. The absolute phase angle between the modulated excitation and the fluorescence is obtained by measuring the difference in phase shift which results when the sample is replaced by a scattering medium. Optionally, the gain of the PMTs may be modulated at a heterodyne frequency which differs only slightly (Δf) from the modulation frequency, so that the phase measurement is performed at the low frequency.

The light source should be stable, of high intensity, and possess a broad optical bandwidth, Xenon arcs have been commonly employed. The modulator should produce a pure sinusoidal signal at a wide range of frequencies, and the transit time response of the PMTs should be independent of wavelength, photocurrent and dynode voltage. Finally, the amplifiers and associated electronics in the phase measurement system should be free of spurious phase shifts.

3.2.1.1. Modulations of the exciting light. The main instrumental problem in the phase shift method is the modulation of the exciting light. Ideally, the modulation depth should be as large as possible to maximize the signal-to-noise ratio and the modulation frequency should be variable over a large range. Measurement of lifetimes between 0.05 and 10 nsec require frequencies ranging from 100 to 20 MHz.

In the past, modulation has been accomplished with acousto-optic devices such as Debye–Sears type modulators in which an ultrasonic standing wave in a liquid causes it to behave as an intermittent diffraction grating (Bailey and Rollefson, 1953; Spencer and Weber, 1970). These devices require modest voltages ($\sim 10^{-2}$ V) for operation but their design for frequencies above 20 MHz is problematic and the

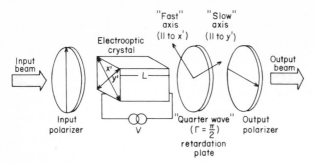

Fig. 7. Schematic diagram of a longitudinal electro-optic modulator. An electro-optic crystal is placed between two crossed polarizers oriented at 45° relative to the electrically induced birefringent axes X' and Y'. The total retardation is the sum of the constant retardation bias ($\Gamma_B = \pi/2$) introduced by a quarter-wave plate and that caused by the crystal. (Redrawn from Yariv, 1975.)

modulation frequency can only be varied by replacing the transducer or by driving it at harmonics of the fundamental. The latter tactic results in a decrease in acoustic power and, therefore, causes a reduction in modulation depth (Salmeen and Rimai, 1977).

Alternately, electro-optic modulators such as Pockels cells have been employed in phase fluorometry (Hauser and Heidt, 1975; Muller et al., 1965). These devices are based on the linear electro-optic effect (Yariv, 1975); i.e., birefringence is induced in a crystal which is proportional to an externally applied electric field. Only crystals which do not possess inversion symmetry exhibit this effect; uniaxial crystals of class $\overline{4}2m$ group symmetry such as potassium dihydrogen phosphate (KH_2PO_4) known as KDP are commonly used for this purpose. Fig. 7 depicts a typical configuration for a longitudinal electro-optic modulator employing a KDP crystal. In this arrangement the electric field is applied along the optic axis of the crystal which is parallel to the direction of propagation of the light. The electrically induced birefringence causes vertically polarized light entering the crystal to acquire a horizontal component which grows with distance along the crystal at the expense of the vertical component. Choosing the length of the crystal to be the distance required to produce a retardation $\Gamma = \pi$ will cause the exciting light to be completely polarized in the horizontal direction. At this point a polarizer is inserted at right angles to the input polarization so that with the field on, the optical beam is transmitted unattenuated whereas with the field off ($\Gamma = 0$), the output beam is blocked off completely. This control of the optical energy flow serves as the basis of electro-optic amplitude modulation of light as shown in Fig. 8.

In principle, the modulation frequency of this type of device can be varied continuously up to 100 MHz when crystal heating becomes significant. A drawback of longitudinal modulators, however, is that they require applied voltages of the order of 10 kV to achieve experimentally useful modulation. This results in RF leakage which can interfere with the detection equipment and is difficult to suppress.

This problem can be circumvented by employing a transverse electro-optic mod-

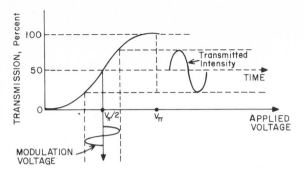

Fig. 8.   Electro-optic modulation of light is illustrated in this figure which shows the transmission of a modulator such as the one shown in Fig. 7. The modulator is biased with a fixed retardation $\Gamma_B = \pi/2$ to the 50% transmission point. An applied sinusoidal voltage modulates the transmitted intensity about the bias point. (Drawn after Yariv, 1975.)

ulator. The principle of operation is similar to that of the longitudinal device, but in this case the crystal orientation is changed and the electric field is applied normal to the direction of light propagation. The advantage of this arrangement is that the retardation in this case is proportional to VL/d, where V is the applied voltage, L is the crystal length and d its thickness; thus, by utilizing long thin crystals the driving voltage can be reduced. The crystal dimensions that are needed to fully exploit this phenomenon, however, require a light beam of such directionality and low divergence, that a laser must be employed as a light source. A phase fluorometer based on this technique which requires driving voltages of the order of 100 V has been described recently (Salmeen and Rimai, 1977). Use of a tunable dye laser would considerably enhance the versatility of this apporach.

3.2.1.2.   Phase-sensitive detection.   The measurement of the phase difference between two periodic signals can be performed in diverse ways (Toffler and Winters, 1975). Comparison methods using a calibrated phase shifter are rather simple; the two signals, one of which is put through a calibrated phase shifter, are applied to the detector. One signal is then shifted until it is in phase with the other and the phase angle is read on the phase shifter dial. Time delay methods and pictorial vector display techniques can also be used (Haggai, 1964). More sophisticated phase-sensitive detectors can take a variety of forms; the basic principle, however, is always the same: the detector provides an output voltage which is proportional to the component of the input signal which is in phase with a chosen reference signal. By the use of two such detectors with reference signals displaced by 90°, it is possible to derive dc output voltages which are proportional to the in-phase and quadrature components of an input signal of arbitrary phase.

Instruments known as lock-in amplifiers used for measuring the amplitude of periodic signals buried in noise incorporate a phase-sensitive detector and are commonly used in phase fluorometry (Lakowicz and Cherek, 1981). In a typical lock-in amplifier the signal channel of the phase-sensitive detector contains a wide-band preamplifier, a bandpass filter that can be centered around the reference frequency, and a main amplifier. The reference channel contains wave-shaping circuitry so that signals of any shape are acceptable. It also contains a calibrated phase shifter so that components of the signal at the reference frequency but at a different phase can be found. A low-pass filter with a variable time constant is included in the output so that a suitable integrating time constant can be selected (A de Sa, 1981). Commercial lock-in amplifiers are available that provide direct reading of phase angles and of the quadrature components, as well as direct readout of the vector sum of the in-phase and quadrature components.

The reliability of phase shift measurements can be improved if a heterodyning procedure is employed in which the photocurrent produced by the PMTs is mixed with a voltage of fixed frequency near but not equal to the modulation, and the phase shift is measured in the amplified difference signal (Δf). This can be accomplished by modulating the gain of the PMTs (Spencer and Weber, 1970). This has the advantage that the sensivity of the PMT is improved and that all frequencies above Δf may be filtered off with an attendant enhancement in signal-to-noise ratio and signal isolation.

3.2.1.3. Photomultiplier tube response. Measurements of fluorescence lifetimes by the phase shift method are subject to the same type of systematic errors that were discussed in connection with the TCSP technique (Muller et al., 1965). In the present case, any dependence of the PMT transit time on wavelength and photocathode area may introduce spurious phase shifts that can cause significant errors, especially when measuring very short lifetimes. It has been reported that for lifetimes less than 0.5 nsec these errors result in apparent phase lifetimes which are zero or negative (Lakowicz et al., 1981). Furthermore, the magnitude of these effects is modulation-frequency dependent; being more pronounced at smaller frequencies.

A simple method to minimize these sources of systematic error analogous to that described for the pulse technique has been described recently (Lakowicz et al., 1981). Instead of employing a scattering solution to determine the absolute phase angle of the modulated excitation, a reference fluorophore of known lifetime whose emission overlaps that of the sample is used. The observed phase difference, ϕ_{obs}, with the reference solution is smaller than that obtained with scattered light because of the phase shift due to the lifetime of the reference fluorophore. The phase angle of the reference solution, ϕ_R, calculated from the known lifetime (Eqn. 6) is used to calculate the actual phase angle of the sample:

$$\varphi = \varphi_{obs} + \varphi_R \tag{40}$$

The corrected phase angle is then used to calculate the lifetime of the sample.

4. Fluorescence lifetimes

By virtue of its exquisite sensitivity to environmental perturbations, the fluorescence lifetime — or preferably decay — of a fluorophore strategically located in a biological system can reveal subtle details about the structure of such a system, and more important about its dynamic behavior on a nanosecond time scale.

Although fluorescence lifetimes can be quite informative by themselves. as we will see below, they also find utility as ancillary measurements in certain steady-state fluorescence studies. In particular, fluorescence lifetimes are needed to calculate mean rotational correlation times of molecules from steady-state fluorescence anisotropy measurements employing the well known Perrin equation (Perrin, 1929),

$$\frac{r_0}{r} = 1 + \frac{\tau}{\varphi} \tag{41}$$

where r_0 is the limiting anisotropy, r is the anisotropy under the conditions of interest, τ is the fluorescence lifetime, and ϕ is the rotational correlation time.

Fluorescence lifetimes are also needed in the study of quenching reactions. Collisional quenching of fluorescence is described by the Stern–Volmer equation (Pesce et al., 1971)

$$\frac{I_0}{I} = 1 + K\tau_0[Q] \tag{42}$$

where I_0 and I are the fluorescence intensities in the absence and presence of quencher, respectively, τ_0 is the lifetime in the absence of quencher, [Q] is the quencher concentration and K is the bimolecular quenching constant. From knowledge of the latter obtained from Eqn. 42 the diffusion constant of the quencher can be calculated from the Einstein–Smoluchowski relationship (see, for example, Schurr, 1970). Similar considerations apply to excimer formation.

Energy transfer efficiencies, E, and thus determination of intermolecular distances, can also be readily calculated from fluorescence lifetimes measurements (Schiller, 1975) according to:

$$E = 1 - \frac{\tau_D}{\tau_D^0} \tag{43}$$

where τ_D and τ_D^0 represent the fluorescence lifetime of the donor fluorophore in the presence and absence of acceptor, respectively.

4.1. Technical considerations

In order to avoid potential artifacts a few precautions must be observed when performing fluorescence lifetime measurements by either the phase or the pulse method. One possible source of error is the presence of scattered exciting light in the fluorescence signal. In the pulse technique, this may be particularly troublesome if a multiexponential analysis of the decay is to be performed or if the lifetime is short relative to the exciting pulse width. When the presence of scattered light cannot be avoided experimentally it may be possible to correct for it by choosing a suitable model function in the deconvolution procedure. Another potential source or error arises if the emission is polarized; i.e., if the rate of depolarization is of the same order of magnitude as the rate of emission (Shinitzsky, 1972); the source of the depolarization may be rotational diffusion of the fluorophore or energy transfer (Spencer and Weber, 1970). When such polarization effects are present, a second exponential decay with a lifetime of about half of the correct one may appear. This effect is present when unpolarized excitation light is used and becomes more pronounced when employing vertically polarized light as it is often the case with laser excitation. These ratotional transport artifacts can be obviated by exciting the sample with light polarized at an angle of 35.3° from the vertical (Spencer and Weber, 1970). This can be accomplished with a polarizer or a polarization rotator depending on the nature of the source. Alternatively, vertical excitation is used and the fluorescence is viewed through a polarizer oriented at 54.7° from the vertical position.

4.2. Representative biological applications

4.2.1. Photophysics of aqueous tryptophan
Tryptophan, the most photochemically active of the amino acids, plays an important role in the photolysis and inactivation of proteins and enzymes. A great deal of interest, therefore, exists in the photophysics of this molecule. A recent TRF spectroscopy study in which the fluorescence decay of aqueous tryptophan was investigates as a function of temperature and pH (Fleming et al., 1978) illustrates the kind of information that fluorescence lifetimes can provide about photophysical processes.

For sample excitation these workers employed single pulses selected from the output of a mode-locked Nd^+: phosphate glass laser, amplified, and then converted to the fourth harmonic by two frequency-doubling stages ($\lambda \sim 264$ nm, pulse width ~ 6 psec, pulse energy ~ 0.3 mJ). Detection was carried out with a streak camera coupled to an optical multichannel analyzer. A non-linear least squares optimization routine was used to fit the decay curves.

Fluorescence decay curves were collected at various temperatures in the range 19.5 – 78°C at pH 7 and pH 11. At the lower pH, tryptophan exists in the zwitterion form whereas at the higher pH the animo group is unprotonated. It was found that at room temperature at pH 7 the decay was biexponential with lifetimes $\tau_1 = 2.1$

nsec, $\tau_2 = 5.4$ nsec, and a relative weight $f = 0.77$ for the faster component. As the temperature was raised the lifetime of both components decreased markedly, and the fraction f of the slow component decreased until at temperatures above 60°C it became negligible. At pH 11 the decay curves were single exponentials at all temperatures and the lifetimes decreased from 8.2 nsec at 19.5°C to 0.60 nsec at 78°C.

A homogeneous ensemble of molecules would be expected to yield a single exponential decay irrespective of how many competing non-radiative processes are present. Even the presence of two or more separate emitting states would lead to a single exponential provided that the communication between the states was rapid on the time scale of the emission event. The data, therefore, imply the presence of two separate emitting states slowly interconverting on the time scale of the fluorescence lifetimes. On the basis of background information about the photophysics of indole derivatives these authors concluded that the slow component of the decay was mainly correlated with indole-like radiationless processes: intersystem crossing plus ionization; whereas the fast component arises from three processes in competition: intersystem crossing, photoionization plus intramolecular electron transfer; the latter leading to a non-radiative product. The single long-lived component at pH 11 is presumably due to the fact that the electron-withdrawing power of the carbonyl has been decreased sufficiently by deprotonation of the $-NH_3^+$ so that electron transfer is no longer effective.

4.2.2. Dynamics of protein conformation revealed by fluorescence of tryptophan residues

There is a great deal of evidence which suggests that the functional properties of proteins may be dependent on their dynamic behavior on a time scale ranging from picoseconds to nanoseconds. As an illustration of the phase shift method to measure fluorescence lifetimes we now consider a study of dipolar relaxation in proteins on the nanosecond time scale (Lakowicz and Cherek, 1980). In this study the ability of the protein matrix to reorient around the increased dipole moment of tryptophan in the excited state is examined. The reorientation of adjacent dipoles around an excited fluorophore results in shifts of the fluorescence emission to longer wavelengths. If the dipolar relaxation time is comparable to the fluorescence lifetime, emission occurs from a distribution of fluorophores which have relaxed to varying degrees. This is manifested by an increase in fluorescence lifetime across the emission spectrum. Intuitively, this can be understood as follows: observation of the short wavelength (blue) edge of the emission selects for unrelaxed fluorophores which are decaying by both fluorescence emission and by transport to longer wavelengths. In contrast, observation of the red edge of the emission selects for already relaxed fluorophores, which therefore will have longer apparant lifetimes.

These investigators measured the fluorescence lifetime of tryptophan in human serum albumin (HSA) and melittin which possess a single tryptophan residue, and of liver alcohol dehydrogenase. Interference filters were used to isolate the blue (317 nm), central (344 nm) and red (400 nm) regions of the fluorescence emission spectra. The phase method was employed with a modulation frequency of 30 MHz. Rotational diffusion effects on the measurement of lifetimes were avoided by exciting with vertically polarized light and observing the emission through a polarizer oriented at 55° from the vertical position. Artifacts due to the wavelength dependence of the PMT response were eliminated by employing p-terphenyl in ethanol as a reference compound with a known lifetime of 0.9 nsec. It was verified that the contribution of scattered light to the signal was negligible employing a glycogen scattering suspension which scattered an equivalent amount of exciting light as did the protein solutions.

It was found that the lifetimes of all three proteins increased significantly with increasing wavelength (317–400 nm); the increase being 0.7 nsec for melittin and about 1.5 nsec for HSA and alcohol dehydrogenase. The smaller increase for melittin is consistent with the known exposure of its single tryptophan residue to the aqueous phase. These data were interpreted as resulting from dipolar relaxation of the protein matrix on the nanosecond time scale.

Pulse methods have been employed for a similar purpose. Grinvald and Steinberg (1976) investigated the fluorescence decay kinetics of a number of proteins at different ranges of the emission spectrum and under a variety of conditions. Only a few pertinent findings from this study will be presented here.

The fluorescence decays of proteins possessing more than one tryptophan residue were always multiexponential reflecting the existence of different microenvironments for the individual residues. Under favorable conditions, the various residues could in principle be differentiated in a single decay measurement. Perhaps more interestingly, most proteins that contained a single tryptophan residue also displayed multiexponential decays, indicating that variability in conformation may exist for most proteins. It was suggested by these authors that the observed heterogeneity in the decay kinetics is due to variability in the quenching of the tryptophan residues resulting from differences in their interaction with neighboring groups that originate from structural fluctuations on the nanosecond time scale. Different decay kinetics were observed for the blue and the red spectral ranges of the emission as well, the general trend being that the decay is relatively slower for light emitted at the long wavelength end of the spectrum. A particularly interesting result was obtained with chicken pepsinogen. In this protein, light emitted at the red edge of the spectrum decayed biexponentially with one of the components having a negative amplitude. This begavior indicates that the fluorescence builds up before it decays, suggesting that the electronically excited species involved has been created during the fluorescence lifetime. This subnanosecond relaxation process was attributed to relaxation of the protein matrix around the excited chromophore or to the formation of a more specific excited state complex; e.g., an exciplex.

5. Time- and phase-resolved emission spectra

The chemical and physical properties of excited molecules can differ significantly from those in the ground state. As a result of this, fluorophores may undergo a variety of interactions with surrounding molecules during their excited state lifetime. Often, the dynamics of such interactions can be followed directly by observation of the fluorescence spectra as a function of decay time.

5.1. Time-resolved emission spectra (TRES)

TRES are fluorescence spectra obtained at discrete times during the fluorescence decay (Ware et al., 1971). There are two methods in common use for obtaining TRES using the TCSP technique.

In the first method (Ware et al., 1971) the conversion output of the TAC is processed through a single channel analyzer (SCA) before being sent to the MCA. The upper and lower level discriminators in the SCA are set at voltages V_1 and V_2 corresponding to times t_1 and t_2 in the TAC. This sets a time window of width $\delta t = t_2 - t_1$ outside of which events are not accepted. This window mat be centered around any time Δt after an arbitrarily chosen time zero which usually is the rising edge or the peak of the excitation pulse $E(t)$. Photons arriving within the time window are spectrally analyzed by scanning the monochromator over the desired range. The MCA is operated in the multiscaling mode with the channel advance synchronized to the stepping of the monochromator, thus a time-resolved spectrum is constructed in the memory of the MCA. In this approach, the spectral resolution is set by the monochromator slit width and the time resolution is determined by the width δt.

In essence, the method just described analyzes the spectral content of narrow "time slices" of the experimental decay $F(t)$, which as we have already seen is the convolution of the true decay $f(t)$ with the excitation function $E(t)$. Depending on the magnitude of the fluorescence lifetime and on the excitation pulse width, TRES obtained by the above procedure may be distorted, particularly at short times after excitation. When accurate kinetic data are desired these distortions can be avoided by reconstructing the TRES from deconvolved decay curves obtained at various wavelengths (Easter et al., 1976). For this type of application it is best to recover $f(\lambda, t)$ empirically without assuming a physical decay law in the deconvolution procedure. A sum of exponentials in which the amplitude and time constants are treated as free parameters and the number of components is arbitrary is a suitable function (see Eqn. 38). Since decay curves are collected at several wavelengths, it is essential to take proper account of the wavelength dependence of the PMT response. Furthermore, since the decay curves at the various wavelengths are collected for different periods of time, the deconvolved decay curves $f(\lambda, t)$ must be normalized to the steady-state emission spectrum $F(\lambda)$ of the sample which is obtained at the same spectral bandwidth as $F(\lambda, t)$. The properly normalized TRES is then given by

$$I(\lambda, t) = K(\lambda)f(\lambda, t) \qquad (44)$$

where the normalization factor $K(\lambda)$ is defined as

$$K(\lambda) = \frac{F(\lambda)}{\int_0^\infty f(\lambda, t)\mathrm{d}t} \tag{45}$$

5.1.1. Representative applications

TRES have been employed to investigate a variety of biologically interesting processes such as excited state proton transfer in organic acids (Loken et al., 1972) and dynamic solvation interactions of naphthalene derivatives (DeToma et al., 1976). Here we examine an application in which TRES provide information about the dynamic relaxation of lipid bilayers on the nanosecond time scale (Easter et al., 1976). These authors employed 2-*p*-toluidinonaphtalene-6-sulfonate (2,6 *p*-TNS) as a fluorescent probe adsorbed to *L*-α-egg lecithin single bilayer vesicles. TRES ware reconstructed from deconvolved decay curves taken at different wavelengths, a process which is illustrated in Fig. 9 and 10. These data show that the mean decay

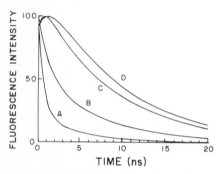

Fig. 9. Peak normalized fluorescence decay curves at several wavelengths generated from the deconvolution parameters of 2,6*p*-TNS (11 μM) adsorbed to L-α-egg lecithin vesicles (0.86 mM) at 7°C. A, 390 nm; B, 423 nm; C, 485 nm; D, 530 nm. (Redrawn from Easter et al., 1976.)

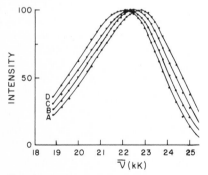

Fig. 10. Peak normalized time-resolved emission spectra of 2,6*p*-TNS adsorbed to L-α-egg lecithin vesicles. Spectra were reconstructed from deconvolved decay curves such as those shown in Fig. 9. A, 1.02 nsec; B, 2.04 nsec; C, 4.08 nsec; D, 12.24 nsec. Kilo Kayser (kK) = 10^3 cm^{-1}. (Redrawn from Easter et al., 1976.)

time increases with increasing wavelength and that a component with a negative amplitude appears toward the red edge of the spectrum. The latter phenomenon demonstrates the presence of a reaction occurring in the excited state, and the overall behavior is characteristic of excited state solvation. These results were interpreted in terms of a general relaxation of polar residues about the excited fluorophore. Similar time-dependent red shifts are observed with 2,6 p-TNS dissolved in glycerol but not with the probe dissolved in non-viscous solvents. These findings, therefore, imply not only that the probe is surrounded by polar residues but that these groups have restricted motion compared to that expected in a liquid solution.

More recently, Ghiggino et al. (1981) employed a cavity-dumped laser system to obtain convolved TRES from the dansyl fluorescent probe (5-dimethyl amino-1-naphthalene sulfonate) in a variety of systems. In lipid bilayers, TRES exhibited a marked time dependence, with red shifts of up to 40 nm over a 40 nsec period. These workers indicated that this behavior could be at least partly explained by solvent reorientation, but they also suggested that heterogeneity of the binding site could contribute to the complex kinetics observed.

5.2. Phase-resolved spectra

The phase method cannot be employed to obtain TRES when a gradual shift of the fluorescence emission with time is observed and a continuum of species appear to be involved, as in the situations considered above. On the other hand, the phase method does provide a means to directly resolve individual spectra from a mixture of two fluorophores which decay with different lifetimes. These so-called phase-resolved spectra are, in fact, lifetime-resolved spectra.

The principle of this technique is based on the following considerations (Lakowicz and Cherek, 1981). In the phase method the fluorescence emission is demodulated and delayed by an angle δ relative to the exciting light (Eqn. 5). The fluorescence signal measured with a phase-sensitive detector is porportional to the modulated intensity and to the cosine of the difference between the phase of the sample and that of the detector, φ_D:

$$F(\varphi_D) = K \cos (\varphi_D - \delta) \tag{46}$$

where K is a constant. If the emission arises from a heterogeneous two-component mixture (A and B), the modulated emission is described by

$$F_M(t) = K_A \sin (\omega t - \delta_A) + K_B \sin (\omega t - \delta_B) \tag{47}$$

where the δ_i are the phase angles of the individual components. If measured with a phase-sensitive detector, the resulting signal is

$$F_M(\varphi_D) = K_A \cos (\varphi_D - \delta_A) + K_B \cos (\varphi_D - \delta_B) \tag{48}$$

By changing the detector phase φ_D, the contributions of A and B to the total signal $F_M(\varphi_D)$ can be varied. In particular, if φ_D is chosen to be 90° out of phase with either component ($\varphi_D - \delta_i = 90°$), then the phase-sensitive signal contains contributions only from the other component, and its emission spectrum can then be recorded directly. Note that the intensity of the unsuppressed component is decreased to $\sin(\delta_A - \delta_B)$ of the original intensity, which results in a progressive loss of signal-to-noise ratio as the lifetimes of the components become similar. Under favorable conditions spectra from species differing in lifetime by a few hundred picoseconds can be resolved.

This method offers a valuable tool for the analysis of two-state excited state reactions such as proton transfer, excimer formation and energy transfer. It may be useful in studies of fluorophore–biopolymer binding via suppression of the signal from the unbound fluorophore.

6. Time-resolved fluorescence anisotropy measurements

In a TRF emission anisotropy experiment the sample is excited by a pulse of vertically polarized light and the fluorescence decay is alternately measured through a polarizer oriented parallel $[I_\parallel(t)]$ or perpendicular $[I_\perp(t)]$ to the polarization of the exciting light.

Fluorophores whose absorption dipole moments are near parallel to the electric field vector of the exciting light will be preferentially excited; thus, excitation with polarized light introduces an anisotropy into the system. After pulsed excitation this anisotropy decays with time due to the orientational relaxation of the system brought about by rotational diffusion. If fluorophore motion occurs on the same time scale as the fluorescence emission, then this motion can be monitored by measuring the fluorescence emission anisotropy, $r(t)$, as a function of time, where $r(t)$ is defined as (Wahl, 1975)

$$r(t) = \frac{d(t)}{s(t)} \tag{49}$$

and

$$d(t) = i_\parallel(t) - i_\perp(t) \tag{50}$$

$$s(t) = i_\parallel(t) + 2i_\perp(t) \tag{51}$$

$s(t)$ is the total fluorescence decay; $i_\parallel(t)$ and $i_\perp(t)$ are related to the experimentally determined $I_\parallel(t)$ and $I_\perp(t)$ through the usual convolutions with the exciting function $E(t)$ (Eqn. 3).

The time-dependent emission anisotropy is an experimentally convenient function for monitoring molecular rotational diffusion in macroscopically isotropic samples. In such cases, if the angle between the emission and absorption dipoles is known, $r(t)$ can provide information not only about the dynamics of fluorophore motion but also about its orientation within an anisotropic microenvironment. Orientational information can be obtained from steady-state fluorescence polarization measurements only from macroscopically oriented systems (Johansson and Lindblom, 1980).

6.1. Technical considerations

The time-dependent anisotropy, $r(t)$, is determined as follows. First, $D(t)$ and $S(t)$ curves are obtained from the experimental polarized decays through a point-by-point calculation.,

$$D(t) = I_{\parallel}(t) - I_{\perp}(t) \tag{52}$$

and

$$S(t) = I_{\parallel}(t) + 2I_{\perp}(t) \tag{53}$$

These functions are related to $d(t)$ and $s(t)$ by the usual convolutions

$$S(t) = s(t) * E(t) \tag{54}$$

$$D(t) = d(t) * E(t) \tag{55}$$

$s(t)$ is obtained by deconvolution of Eqn. 54 employing one of the methods previously discussed (section 3.1.2). Having obtained $s(t)$, a model function with adjustable parameters is chosen for $r(t)$ from which $d(t)$ is calculated according to Eqn. 49; i.e.,

$$d(t) = r(t) \cdot s(t) \tag{56}$$

This $d(t)$ is convolved with $E(t)$ to produce a calculated difference curve $D(t)_{\text{calc}}$, and a least squares procedure is used to find values of the parameters for $r(t)$ which generate the best fit between $D(t)_{\text{calc}}$ and the experimental $D(t)$.

Certain precautions must be observed when collecting the $I_{\parallel}(t)$ and $I_{\perp}(t)$ decay curves. The types of error introduced by the presence of scattered light in the fluorescence signal have already been discussed in connection with lifetime measurements (section 4.1). Since scattered light is strongly polarized, the scattering component in an anisotropy experiment will be larger in $I_{\parallel}(t)$ than in $I_{\perp}(t)$. This will result in an artifactual rapid initial decay in $r(t)$ which may lead to erroneous interpretation of the data. In addition, in an anisotropy experiment information about the difference in intensity between two signals must be preserved. This is in contrast to a lifetime measurement in which only the shape of the decay is of interest. Therefore, $I_{\parallel}(t)$ and $I_{\perp}(t)$ must be collected under similar conditions; manipulations which alter the relationship between signal intensity and TAC conversion rate while collecting the two polarized decays must be avoided. Intensity and frequency fluctuations of the exciting source can introduce similar errors. These can usually be eliminated by correcting one of the polarized components so that the ratio of the integrated intensities of the decay curves $I_\parallel(t)$ and $I_\perp(t)$ is equal to the ratio of the sums of the TAC outputs $I_\parallel$ and $I_\perp$ alternately measured for short periods of time (short relative to the fluctuations of the source). For example, $I_\perp(t)$ is correctd by a point-by point multiplication by a factor K,

$$K = \frac{\int_0^T I_\parallel(t)\mathrm{d}t}{\int_0^T I_\perp(t)\mathrm{d}t} \times \frac{I_\perp}{I_\parallel} \tag{57}$$

where T is the measuring time defined by the TAC range.

6.2. Information provided by TRF anisotropy measurements

In general, the time-dependent emission anisotropy $r(t)$ can be expressed as a sum of exponentials,

$$r(t) = \sum_{i=1}^{N} a_j e^{-t/\varphi} \tag{58}$$

where the number of components, N, may range from one to infinity depending on the complexity of the situation. The simplest case arises from rotational diffusion of a spherically symmetric molecule in an isotropic solvent, whose fluorescence decays as a single exponential. In this instance the anisotropy decays exponentially to zero with time,

$$r(t) = r_0 e^{-t/\varphi} \tag{59}$$

where φ, the rotational correlation time is related to the rotational diffusion coefficient ($D_R = 1/6\,\varphi$), and is given by

$$\varphi = V\eta/KT = M(\bar{v} + h)\eta/KT \tag{60}$$

in which V is the molecular volume, η is the viscosity of the solvent, K is the Boltzmann constant, T is the absolute temperature. M is the molecular weight, $\bar{v}$ is the partial specific volume, and h is the degree of hydration. Thus, by analysis of $r(t)$ in terms of Eqn. 59 the shape, flexibility and conformation of biopolymers can be investigated.

The number of components required to describe $r(t)$ increases as the symmetry of the fluorophore decreases. For a molecule of arbitrary symmetry in an isotropic solvent, $r(t)$ decays to zero as a sum of 5 exponentials (Tao, 1969) with the time constants given by different linear combinations of the rotational diffusion coefficients for motions about the various molecular axes. When the fluorophore is not spherically symmetric and in addition is located in an anisotropic environment, the problem of obtaining an analytical expression for $r(t)$ becomes a formidable one. Kinosita et al. (1977) have treated the motion of several special classes of fluorophores in anisotropic environments of uniaxial symmetry. These cases are of current interest because of their applicability to the study of membranes. It is beyond the scope of this work to discuss theoretical aspects of anisotropic rotational diffusion which have been investigated recently (Lipari aand Szabo, 1980; Johansson and Lindblom, 1980). Instead, a phenomenological approach will be adopted and several approximations which have been derived recently will be intuitively presented.

It has been observed that for fluorophores embedded in lipid bilayers or biomembranes the anisotropy does not decay to zero at long times. This has been interpreted as indicating that these molecules undergo librational motions in membranes rather than free rotation. In such cases it has been found that $r(t)$ can be represented to a good approximation by

$$r(t) = (r_0 - r_\infty) \exp(-t/\varphi) + r_\infty \tag{61}$$

where r_0 is the initial anisotropy at $t = 0$, r_∞ is the limiting anisotropy at $t = \infty$ and φ is some kind of relaxation time characterizing the anisotropic rotational motion.

For a fluorophore of cylindrical symmetry with either the emission dipole μ_e or the absorption dipole μ_a parallel to its symmetry axis, it can be shown that (Lipari and Szabo, 1980)

$$r_0 = 0.4\, P_2(\cos \psi) \tag{62}$$

and

$$r_\infty = 0.4\, P_2(\cos \psi)\, \langle P_2(\cos \theta) \rangle^2 \tag{63}$$

where ψ is the angle between μ_a and μ_e and $P_2(x) = (3x^2 - 1)/2$ is the second Legendre polynomial. θ is the angle between the symmetry axes of the fluorophore (wobbling axis) and the membrane, and the brackets denote an equilibrium average.

It has been recognized (Heyn, 1979) that

$$\frac{r_\infty}{r_0} = \langle P_2(\cos \theta) \rangle^2 = S^2 \tag{64}$$

where $S = P_2(\cos \theta)$ is the order parameter, which has been frequently employed in NMR and ESR spectroscopy.

The order parameter is the first non-trivial term in a Legendre polynomial expansion of the equation that describes the orientational distribution of the probe; thus, it provides model-independent information about the equilibrium orientational distribution of the fluorophore.

Recently, the decay of anisotropy from membrane-bound fluorophores has been frequently interpreted in terms of wobbling diffusion of the probe within a cone of half-angle θ_{max}. Kinosita et al. (1977) have derived expressions which apply to several special cases of practical interest. In the case of a rod-shaped molecule with both absorption and emission dipoles parallel to the long axis of the molecule, then

$$\frac{r_\infty}{r_0} = [\tfrac{1}{2} \cos \theta_{max} (1 + \cos \theta_{max})]^2 \tag{65}$$

Thus, it is seen that TRF anisotropy measurements provide both dynamic (φ) and static ($r_\perp/r_0$) information about the structure of membranes.

An analogous situation arises when a fluorophore attached to a biopolymer is able to undergo some local restricted motion. Eqn. 61 has been generalized to encompass the situation when the diffusional motion of the biopolymer cannot be neglected on the time scale of the experiment (Lipari and Szabo, 1980)

$$r(t) = (r_0 - r_\infty) \exp \left(-t(\varphi_m^{-1} + \varphi_{eff}^{-1})\right) + r_\infty \exp \left(-t/\varphi_m\right) \tag{66}$$

where ϕ_m is the rotational correlation time of the macromolecule and ϕ_{eff} is related to the wobbling coefficient.

If the local relaxation of the probe is rapid compared to the macromolecular motion ($\phi_m \gg \phi_{eff}$) the term ϕ_m^{-1} in the argument of the first exponential in Eqn. 66 may be neglected. An example will be presented below which demonstrates the utility of this model for the analysis of protein structural dynamics.

6.3. Representative applications

6.3.1. Structural dynamics of biological membranes

It is now widely recognized that the spatial organization of membrane proteins, and the structure and dynamics of the lipid matrix play essential roles in the many chemical processes that take place at or within membranes. Information about molecular orientation and mobility of membrane components is therefore necessary for unraveling the functional mechanisms of biological membranes.

Among the various techniques which have been utilized to study membrane lipid dynamics fluorescence polarization methods have played a central role because of their high sensitivity. 1,6-Diphenyl-1,3,5-hexatriene (DPH) is one fluorescent probe that has been widely used for this purpose. The popularity of DPH stems from its favorable spectral and physicochemical characteristics. First of all, it is a rod-shaped molecule whose emission and absorption dipole moments coincide with the long symmetry axis of the molecule; this makes TRF anisotropy measurements interpretable in terms of the model described in section 6.2 above. In addition, it has a high extinction coefficient and a quantum yield which approaches unity when dissolved in hydrocarbon solvents; it is practically insoluble in aqueous solutions; it has a lifetime of about 11 nsec in hydrocarbon solvents, and its large Stokes shift results in little spectral overlap between its absorption and emission bands which minimizes depolarization of the fluorescence by energy transfer.

TRF anisotropy techniques employing DPH as a fluorescent probe have been employed to investigate the dynamic structure of lipid bilayers (Chen et al., 1977; Kawato et al., 1977; Stubbs et al., 1981) and how it is affected by cholesterol (Kawato et al., 1978). More recently, the method has been applied to the study of biological membranes (Kinosita et al., 1981a; Gallay et al., 1982). For illustration purposes some findings from the studies of Kawato et al. are briefly described.

Dipalmitoyl-phosphatidylcholine (DPPC) bilayers in aqueous suspension undergo a gel-to-liquid-crystalline phase transition at about 40°C. $r(t)$ measurements of DPH in DPPC bilayers closely follow the predictions of Eqn. 61 both below and above the phase transition temperature as shown in Fig. 11. The motion of DPH was analyzed in terms of the wobbling-in-cone model outlined in the previous section. The cone angle θ_{max} showed a sigmoidal dependence on temperature: θ_{max} was about 20° at low temperatures and abruptly increased to about 70° at the phase transition. In contrast, the wobbling diffusion constant, D_w, did not exhibit a dis-

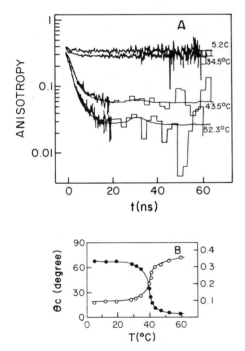

Fig. 11. (A) Time dependence of the emission anisotropy $r(t)$ of DPH (3 mg ml^{-1}) in DPPC vesicles at different temperatures. Noisy curves are the experimentally determined $r(t)$ and the solid curves represent the best fit according to Eqn. 61. (B) Temperature dependence of r (o) and of the cone angle (•).

continuity at the transition. Since the dimensions of DPH are similar to those of the lipid acyl chains, the motion of the probe is assumed to directly reflect the thermal motion of the lipids. Addition of cholesterol decreases the cone angle in the liquid-crystalline and increases it in the gel phase. The relaxation time φ, on the other hand, remains almost constant as the cholesterol content is varied. Thus, the effect of cholesterol is to increase the molecular ordering (S) of the hydrocarbon chains while only minimally affecting the molecular dynamics.

Recently, TRF anisotropy studies of membrane lipid dynamics employing parinaric acid (Wolber and Hudson, 1981) and n-(9-anthroyloxy) fatty acids (Vincent et al., 1982) as fluorescent probes have been reported. The latter probes are interesting in that they are rod-shaped but possess the emission moment perpendicular to the long axis; thus, "in plane" and "out of plane" rotations can be detected by varying the excitation wavelength (Weber, 1971).

6.3.2. Dynamics of biopolymers

The use of fluorescence lifetime measurements to study the dynamics of protein conformations was discussed in section 4.2.2. Now, we investigate how TRF anisotropy measurements can be applied for a similar purpose.

As an example of this approach a recent study of tryptophan fluorescence in a series of proteins containing a single tryptophan residue is examined (Munro et al., 1979). This study employed a synchrotron radiation source in conjunction with the TCSP technique and provided information about the angular range and kinetics of internal motions of this fluorophore in the subnanosecond time range.

The proteins investigated exhibited different degrees of freedom of their tryptophan residues ranging from almost no mobility to nearly complete freedom in the subnanosecond scale. These authors found that the tryptophan residue in nuclease B (20 000 dalton) has a single rotational correlation time (φ) of 9.9 nsec at 20°C. This can be compared with the value of 7.6 nsec calculated from Eqn. 60 assuming $v = 0.73 \text{ cm}^3 \text{ g}^{-1}$. Thus, the tryptophan residue in nuclease B has little rotational freedom. By contrast, myelin basic protein (18 000 dalton) exhibits rotational correlation times of 0.09 and 1.26 nsec showing that the tryptophan residue of this protein is highly flexible. Two distinct rotational motions were also observed in holoazurin (14 000 dalton). It was calculated that in this protein the tryptophan residue rotates within a cone of semiangle of 34°. In apozurin, on the other hand, this angular range increased to 44° and the longer of the correlation times shortened, suggesting that removal of the active site copper ion increases the flexibility of the molecule.

It may be noted that the correlation times in this study were obtained from an equation equivalent to Eqn. 66 when $\phi_m \gg \phi_{eff}$. The values of the cone half-angles were calculated from Eqn. 65. This equation, however, is valid only for rod-shaped fluorophores having both emission and absorption dipoles parallel to the long axis of the molecule; its application to tryptophan is, therefore, questionable.

In contrast to this study which relies on the availability of an intrinsic fluorophore (tryptophan), in many cases an extrinsic chromophore can be introduced into the system. An example of this type of probe is ethidium bromide which has been recently employed to study torsional dynamics in linear molecules such as DNA (Genest et al., 1981; Thomas et al., 1980).

7. *Differential polarized phase fluorometry*

The time dependence of the emission anisotropy cannot be obtained by the phase method. However, in combination with steady-state measurements the phase technique can supply information about anisotropic motion. In particular, for the case of an isotropic hindered rotator, information analogous to that obtained by the pulse technique can be extracted; namely, a rotational rate R (rad sec^{-1} and the degree of orientational constraint (r_∞). The technique employed for this purpose is called differential polarized phase fluorometry (DPPF). It entails measurement of the difference in lifetime between the parallel and perpendicular components of the fluorescence emission when the sample is excited by sinusoidally modulated vertically polarized light.

The fluorometer has a T configuration in which two PMTs oriented at 90° relative to the excitation beam view the fluorescence through polarizers. To measure the differential phase between the polarized components of the fluorescence, one of the two emission polarizers is first rotated to render it parallel to the other and the phase difference is nulled. Then one of the polarizers is rotated and the phase difference between the signals is measured.

7.1. *Information provided by the differential polarized phase method*

The theoretical basis of this method was developed by Weber (1977) and elaborated further by Lakowicz et al. (1979).

Following the latter author, the differential phase angle is given as the tangent of the phase difference between the polarized components and is related to the difference in lifetimes ($\Delta\tau$) by

$$\tan \Delta = \omega\Delta\tau \tag{67}$$

The tangent of the differential phase angle is a function of T, r_0, τ, and ω. For an unhindered isotropic rotator

$$\tan \Delta = \frac{(2R\tau)\omega\tau r_0}{\dfrac{1}{9} m_0(1 + \omega^2\tau^2) + \dfrac{(2R\tau)}{3}(2 + r_0) + (2R\tau)^2} \tag{68}$$

where

$$m_0 = (1 + 2r_0)(1 - r_0) \tag{69}$$

R may be obtained from Eqn. 68 from measurements of $\tan \Delta$ and τ. For isotropic rotations, the maximum value of $\tan \Delta$ is a function of only r_0, ω, and τ.

$$\tan \Delta_{max} = \frac{3\omega\tau r_0}{(2 + r_0) + 2[m_0(1 + \omega^2\tau^2)]^{\frac{1}{2}}} \tag{70}$$

Anisotropic rotations can be detected by $\tan \Delta_{max}$ values which are smaller than predicted by Eqn. 70.

For an isotropic hindered rotor

$$\tan \Delta = \frac{(r_0 - r)(2R\tau)}{\dfrac{1}{9} m_0(1 + \omega^2\tau^2) + \dfrac{1}{3} s(2R\tau) + m_\infty(2R\tau)^2} \tag{71}$$

where

$$m_\infty = (1 + 2r_\infty)(1 - r_\infty) \tag{72}$$

$$s = 2 + r_0 - r_\infty(4r_0 - 1) \tag{73}$$

and the maximum value of $\tan \Delta$ is given by

$$\tan \Delta_{max} = \frac{3\omega\tau(r_0 - r_\infty)}{s + 2[m_0 m_\infty(1 + \omega^2\tau^2)]^{\frac{1}{2}}} \tag{74}$$

An ambiguity exists in Eqn. 71 in that a change in $\tan \Delta$ may be caused by a change in either R or r_∞, or both. As a result, this equation cannot be used to obtain either parameter. Lakowicz et al. (1979) have recognized that by combining DPPF measurements with steady-state ones, unique solutions for R and r_∞ can be obtained. Combining the steady-state expression (Perrin's equation modified for hindered motion)

$$r_\infty = r + (r - r_0)/6R\tau \tag{75}$$

where τ is the steady-state anisotropy, with Eqn. 71 yields

$$(m \tan \Delta)(2R\tau)^2 + (C \tan \Delta - A)(2R\tau) + (D \tan \Delta - B) = 0 \tag{76}$$

where

$$\begin{aligned}
A &= 3B = \omega\tau(r_0 - r) \\
C &= (\tfrac{1}{3})(2r - 4r^2 + 2) \\
D &= (\tfrac{1}{9})(m + m_0\omega^2\tau^2) \\
m &= (1 + 2r)(1 - r)
\end{aligned} \tag{77}$$

Therefore, measurement of r, $\tan \Delta$ and τ allows R to be obtained from Eqn. 76. The calculated value of R is then substituted into Eqn. 75 along with the steady-state anisotropy to obtain r_∞.

7.2. Applications

This technique has been used to study anisotropic motion of aromatic fluorophores in homogeneous solutions (Mantulin and Weber, 1977). More recently, it has been employed with the fluorescent probe DPH to examine the motion of this probe in lipid bilayers (Lakowicz et al., 1979). The findings of this latter study are in excellent agreement with those obtained by pulse methods, as described in section 6.3.1. More specifically, DPPF measurements reveal that the degree of orientational constraint of DPH (r_∞) decreases sharply at the phase transition temperature but that the rate of rotation R is not as dramatically affected.

8. TRF spectroscopy in the study of resonance energy transfer

Electronic excitation energy can be transferred non-radiatively from a fluorescent energy donor to a suitable acceptor over distances of the order of 10–100Å. The various factors that contribute to the efficiency of this resonant dipole–dipole energy transfer process can be evaluated from Forster's (1965) quantum mechanical treatment of the problem. The efficiency, E, of energy transfer is given by

$$E = \frac{R_0^6}{R_0^6 + r^6} \tag{78}$$

where r is the distance between donor (D) and acceptor (A) fluorophores and R_0 is a characteristic distance at which transfer is 50% efficient. R_0 is given by

$$R_0 = \frac{9000 \ln 10 \varkappa^2 q_D}{128 \, \Pi^5 \, N \, n^4} \int_0^\infty f(v)\varepsilon(v) \, \frac{dv}{v^4} \tag{79}$$

where $\varkappa^2$ is a dimensionless geometric factor determined by the relative orientation of the emission dipole moments of D and A; q_D is the quantum yield of the donor in the absence of acceptor; N is Avogrado's number, and n is the index of refraction of the medium. $f(\bar{v})$ and $\varepsilon(\bar{v})$ denote the fluorescence intensity of the donor and the molar extinction coefficient of the acceptor, respectively, at wave number v. The integral in Eqn. 79 thus represents the spectral overlap between the emission of the donor and the absorption of the acceptor.

Energy transfer has long been recognized as a useful phenomenon for determining intermolecular separations and has now been widely used in studies of biological macromolecules. Many of these applications have been recently reviewed (Stryer, 1978; Fairclough and Cantor, 1978). The present discussion will be restricted to how TRF measurements can be employed in the study of energy transfer processes.

From Eqn. 79 it follows that the energy transfer efficiency depends on the relative orientation between D and A through the x^2 term defined as

$$x^2 = (\cos \alpha - 3 \cos \beta \cos \gamma)^2 \tag{80}$$

where α is the angle between the emission dipole moments of D and A, β is the angle between the donor emission dipole and the vector $\mathbf{R}$ joining the centers of D and A, and γ is the angle between the emission dipole of A and $\mathbf{R}$. If both donor and acceptor rotate freely in a time that is short relative to the fluorescence lifetime of the donor, then a dynamic average can be used and $x^2 = 2/3$. On the other hand, when the motion of the fluorophores is restricted x^2 can assume values ranging from 0 to 4. In many applications of biochemical interest, energy transfer takes place between D-A pairs in which both D and A have some degree of orientational freedom with respect to a rigid macromolecule that defines their fixed separation, thus it becomes important to have some means of quantitating the value of x^2 (Dale et al., 1979). It was shown previously (section 6.2) that TRF anisotropy measurements can supply information on the rotational diffusion and the rotational equilibrium distribution of fluorophore. Thus, it can be readily appreciated how TRF anisotropy measurements can be valuable in estimating the possible range of values for x^2. A detailed treatment of this problem has been formulated by Dale and Eisinger (1975).

Fluorescence lifetime measurements are also useful in studies of energy transfer. Efficiencies, for example, can be calculated from the lifetime of the donor in the presence and absence of acceptor (see Eqn. 43). The utility of lifetime measurements in energy transfer studies, however, goes beyond that of calculating the efficiency of the process. For instance, many cases of biological interest arise when the D–A pairs are not uniformly separated; i.e., their relative distance follows a certain distribution function. Such situations may occur when the D–A pairs are situated on biopolymers that do not assume a unique conformation of when the donor and acceptor separately reside on different macromolecules which are statistically distributed in space.

Under such circumstances the efficiency of energy transfer is given by (Cantor and Pechukas, 1971)

$$E = \int_0^\infty F(r) \frac{R_0^6}{R_0^6 + r^6} \, dr \tag{81}$$

where $F(r)$ represents the normalized distribution of distances between D–A pairs and r is assumed no to vary during the donor lifetime. Grinvald et al. (1972) have shown that the distribution $F(r)$ can be obtained from energy transfer studies by measuring the fluorescence decay of the donor and acceptor. If the distance between D–A pairs is large so that transfer occurs only within each pair, then the fluorescence decay $f(t)$ is given by

$$f(t) = K \int\limits_0^\infty F(r) \exp\left[-t/\tau - \frac{t}{\tau}(R_0/r)^6\right] dr \qquad (82)$$

where K is a constant. The first and second terms in the exponential argument represent the decay due to fluorescence emission and to energy transfer respectively. The distance distribution function can be evaluated from the fluorescence decay by assuming a plausible model $F(r)$ with adjustable parameters and employing a least squares method to optimize the fit between the calculated (Eqn. 82) and the experimentally determined decays. More recently, the problem of obtaining the surface density of membrane-bound fluorophores from energy transfer measurements has been treated theoretically (Kwok-Keung Fung and Stryer, 1978; Koppel et al., 1979).

Finally, when the acceptor is fluorescent, time-resolved emission spectra can be useful in following the kinetics of energy transfer processes. If the donor and acceptor emission bands are well separated, however, the kinetics can be determined by measuring the fluorescence decay at only two wavelengths.

8.1. Biochemical applications

The method outlined above to obtain distance distributions from the fluorescence decay of the donor in energy transfer experiments has been implemented by Haas et al. (1975) to investigate the distribution of end-to-end distances of oligopeptides in solution. For this purpose, these authors synthesized a homologous series of oligopeptides made up of 4–9 amino acid residues each containing donor (naphtalene) and acceptor (dansyl) fluorophores at its carboxyl and amino termini respectively. Experiments were performed in glycerol solution in order that the relative translational diffusion of donor and acceptor during the fluorescence lifetime of the donor could be neglected.

It was found that the donor fluorescence decay from oligomers labeled only with donor was monoexponential whereas the decays from oligomers labeled with donor and acceptor were markedly non-exponential. By choosing analytical expressions for $F(r)$ independently derived by other workers on the basis of theoretical considerations, Haas et al. found that the average end-to-end distance and the spread of the distance distribution increases systematically with chain length which suggested that the oligomers adopt a large number of conformations in solution.

In a subsequent study, Haas et al. (1978) extended these energy transfer studies to solvent mixtures of different viscosities so that diffusion of the fluorophores could now occur during the excited state lifetime of the donor. It was found that the decay rate of the donor increases when the solvent viscosity decreases reflecting diffusion of the molecular ends toward one another. Assuming a modified Fick equation to describe this diffusional motion, the fluorescence decay data were analyzed in terms of a diffusion coefficient describing the Brownian motion of the molecular ends. The analysis revealed that the diffusion coefficients are about an order of magnitude lower than those expected for the unattached chromophores which shows that the polypeptide backbone possesses appreciable rigidity, the magnitude of the internal friction being higher for the shorter chain molecules.

9. Application of TRF spectroscopy to cell biology

It is evident, from the various biological applications examined thus far, that TRF spectroscopy is a powerful and versatile analytical tool for the physical biochemist. Clearly, an extension of the technique to the study of intact cells would constitute a valuable addition to the repertoire of biophysical methods available to the cell biologist. While time-resolved fluorescence methods have been extensively employed for the study of isolated membranes and other purified biochemical systems, their application to the study of intact cells has been hindered in the past by the inherent biochemical and spatial (compartmental) heterogeneity of a cell which may render the fluorescent signal difficult to interpret unambiguously, and by the complexities which arise from studying a dynamic system such as a living cell. At present, the application of TRF spectroscopy to living cells is just beginning to be exploited.

Difficulties associated with biochemical heterogeneity can be minimized by employing extrinsic fluorescent probes of high specificity and possessing spectral characteristics separable from those of intrinsic cellular fluorescence; the extent to which this goal can be achieved varies with the type of application. Recent advances have resulted in the current availability of a large selection of fluorescent labeling reagents with a diversity of functional groups and spectral properties. Often, a biological ligand of known specificity can be covalently tagged with a fluorescent probe and the resulting conjugate used to label a particular cellular component. Progress has also been made in the development of methods for incorporating exogenous components into living cells, which include microinjection and fusion techniques; in the latter the fluorescent molecules are trapped inside carriers such as liposomes which are subsequently fused with target cells. For a recent review on methods to fluorescently label living cells see Taylor and Wang (1980). The state of the art is such that, at present, much meaningful information can be extracted from TRF measurements from intact cells.

Uncertainties in interpretation due to spatial cellular hererogeneity can be made less significant by employing a microscopic approach which permits examination of small regions of single cells; a feat which is feasible by virtue of the high sensitivity of fluorescence spectroscopy, especially when employing photon-counting techniques. A promising new approach which would permit study of membrane and cell surface phenomena without interference from cytoplasmic signals, even in the case of a probe wich partitions into internal cellular compartments, is total internal reflection spectroscopy. In this method, fluorescence in labeled cells attached to a quartz slide is excited by evanescent radiation only to within a depth of a fraction of the wavelength employed (Axelrod, 1981). Although to date time-resolved fluorescence methods have not been used in conjunction with total internal reflection techniques, the combination of these two approaches would constitute a significant advance for the study of molecular dynamics in living cells. Finally, the dynamic nature of a living system requires that measurements in cells be performed in relatively short times. In certain cases fixing the cells prior to measurement circumvents this problem while still providing the type of information sought.

In spite of these potential difficulties, the application of TRF spectroscopy to intact cells possesses an obvious advantage: It makes possible the study of fast (sub-nanosecond) physicochemical processes and thus the gathering of intelligence that can be derived thereof, under conditions which preserve the dynamic state of regulation found in a living cell. A microscopic approach further enhances the usefulness of the method by combining the spatial resolution of the optical microscope ($\sim$ 1 μm) with subnanosecond time resolution: knowledge about the spatial organization of the cell with respect to a particular dynamic property becomes experimentally attainable. When used in conjunction with resonance energy transfer, this approach effectively endows the fluorescence microscope with a resolution of the order of 10–100Å while maintaining the dynamic conditions of an intact cell.

A microscopic approach to TRF spectroscopy employing the TCSP technique has been developed by Fernandez and coworkers who have applied the method to the study of cell surface topography and membrane dynamics in single living cells as detailed below. It should be noted that in addition to the usual sources of error inherent to TRF methods, measurements through the microscope are subject to additional systematic errors. In particular, the high numerical aperture of the optical system results in underestimation of fluorescence anisotropy values which, therefore, must be corrected (Von Sengbush and Thaer, 1973). Theoretical discussions of the effect of high numerical aperture on the measurement of fluorescent polarization have been presented recently (Lindmo and Steen, 1977; Axelrod, 1979; Eisert and Beisker, 1980).

Time-resolved fluorescence anisotropy measurements from single living cells measured through the microscope were first reported by Feinstein et al. (1975), who showed that the rotational correlation time of the hydrophobic fluorescent probe perylene incorporated into Ehrlich ascites cells decreased from 8.7 to 6.5 nsec upon incubation of the cells with the local anesthetic tetracaine.

More recently, in a study of membrane dynamics during myoblast fusion (Herman and Fernandez, 1978, 1979), fluorescence lifetimes and steady-state fluorescence anisotropy measurements were made on single intact cultured skeletal muscle cells labeled with the amphipathic dye ANS (8-anilino-1-naphthalenesulfonate). Perrin's equation (Eqn. 41) was employed to calculate apparent rotational correlation times for the probe at different times during development. Small regions ($\sim$ 9 $\mu\mu$) of single cells were examined and regions of cell contact or fusion were compared with other areas of the cell surface. It was found that myoblast fusion is preceded by a generalized transient increase in the rotational diffusion of ANS and that the mobility of the probe was higher *locally* at areas of fusion. These results were interpreted in terms of membrane lipid fluidity although an order parameter could not be obtained. The ability to collect fluorescence decay curves from small

274

regions of a single cell illustrate the high sensitivity of fluorescence spectroscopy and demonstrate its utility for investigating the spatial organization of membrane and cell surface dynamic properties.

The phenomenon of resonance energy transfer has also been utilized in the study of single cells. Fernandez and Berlin (1976) introduced this technique to investigate the cell surface distribution of lectin receptors on normal and transformed cultured fibroblasts. Dansyl and rhodamine concanavalin A (Con A) conjugates were employed as donor and acceptor fluorophores respectively. In this study, time-resolved emission spectra were collected from single cells from which the kinetics of the energy transfer process could be followed, and energy transfer efficiencies were calculated from either spectral or fluorescence lifetime data — both approaches giving qualitatively similar results. The results revealed that Con A-receptor complexes are more clustered on transformed cells and that the kinetics of ligand-induced receptor aggregation are different in the normal and transformed cells. The data also suggested that movement of Con A-receptor complexes was not the result of simple diffusion but rather it appeared to be limited by other mechanisms.

More recently, energy transfer between pyrene-labeled Con A (Herman and Fernandez, 1982a) and fluorescein-Con A was used to examine the topography and mobility of cell surface glycoproteins in single cultured skeletal muscle cells at various times during differentiation (Herman and Fernandez, 1982b; Fernandez and Herman, 1982). Energy transfer efficiencies were calculated from donor fluorescence lifetimes and average receptor separations were obtained at different times during development. Distances derived from lifetime measurement qualitatively agreed with those obtained from spectral data. Time-resolved spectra collected from single fixed cells supplied information on the kinetics of the energy transfer process. Evidence was obtained which suggested that Con A receptors undergo a marked alteration in topographical distribution during myoblast fusion: Before the onset of fusion these receptors exist predominantly in microclusters, they disperse during the period of fusion activity, and then become clustered again after the fusion process is completed. In addition, prior to fusion, receptors appear to be more clustered at regions of cell contact than at other areas of the cells, and the translational mobility of Con A receptors appeared to be higher during the period of myoblast fusion.

The time-resolved fluorescence studies of single cells cited above were performed with a thyratron-gated flash lamp as the pulsed excitation source. More recently, a time-resolved microspectrofluorimeter has been developed (Fernandez, 1982) which employs a mode-locked synchronously pumped cavity-dumped tunable dye laser as the light source. The advent of such systems will significantly enhance the sensitivity, speed, and resolution with which measurements can be performed in single cells. Although the utilization of TRF spectroscopy in cellular studies is only beginning to be exploited, it seems reasonable to predict that this technique will find increased applicability in cell biology in the years to come.

References

A de Sa (1981) Principles of Electronic Instrumentation, John Wiley, New York, p. 264.

Axelrod, D. (1979) Carbocyanine dye orientation in red cell membrane studied by microscopic fluorescence polarization, Biophys. J., 26, 557–574.

Axelrod, D. (1981) Cell-substrate contacts illuminated by total internal reflection fluorescence, J. Cell Biol., 89, 141–145.

Badea, M.G. and Georghiou, S. (1976) Versatile nanosecond fluorometer employing a boxcar averager, Rev. Sci. Instrum., 47, 314–317.

Bailey, E.A. and Rollefson, G.K. (1953) The determination of the fluorescence lifetimes of dissolved substances by a phase shift method, J. Chem. Phys., 21, 1315–1322.

Bisby, R.H. and Munro, I.H. (1980) Synchrotron radiation as a source for time-resolved fluorescence studies, UV Spectrum, Group Bull., 8, 60–67.

Bollinger, L.M. and Thomas, G.E. (1961) Measurement of the time dependence of scintillation intensity by a delayed-coincidence method, Rev. Sci. Instrum., 32, 1011–1050.

Bradley, D.J. (1977) Methods of generation, in: S.L. Shapiro (Ed.), Topics of Applied Physics, Vol. 18, Springer, New York, pp. 17–81.

Brigham, E.O. (1974) The Fast Fourier Transform, Prentice Hall, Englewood Cliffs, NJ.

Cantor, C.R. and Pechukas, P. (1971) Determination of distance distribution functions by singlet–singlet energy transfer, Proc. Natl. Acad. Sci. (U.S.A.), 78, 2099–2101.

Chan, C.K. and Sari, S.O. (1974) Tunable dye laser pulse converter for production of picosecond pulses, Appl. Phys. Lett., 25, 403–406.

Chen, L.A., Dale, R.E., Roth, S. and Brand, L. (1977) Nanosecond time-dependent fluorescence depolarization of diphenylhexatriene in dimyristoyl-lecithin vesicles and the determination of "microviscosity", J. Biol. Chem., 252, 2163–2169.

Coates, P.B. (1968) The correction of photon "pile-up" in the measurement of radiative lifetimes, J. Sci. Instr. (J. Phys. E.), 1, 878–879.

Crosby, D.A. and MacAdam, K.B. (1981) Simple fast transient digitizer, Rev. Sci. Instrum., 52, 297–300.

Dale, R.E. and Eisinger, J. (1975) Polarized excitation energy transfer, in: R.F. Chen and H. Edelhoch (Eds.), Biochemical Fluorescence, Marcel Dekker, New York, pp. 115–284.

Dale, R.E., Eisinger, J. and Blumberg, W.E. (1979) The orientational freedom of molecular probes: The orientation factor in intramolecular energy transfer, Biophys. J., 26, 161–194.

D'Alessio, J.T., Ludwig, P.K. and Burton, M. (1964) Ultraviolet lamp for the generation of intense, constant-shape pulses in the subnanosecond region, Rev. Sci. Instrum., 35, 1015–1017.

Davis, C.C. and King, T.A. (1970) Single photon counting pile-up corrections for time-varying light sources, Rev. Sci. Instrum., 41, 407–408.

Demas, J.N. and Adamson, A.W. (1971) Evaluation of photoluminescence lifetimes, J. Phys. Chem., 75, 2463–2466.

DeToma, R.P., Easter, J.H. and Brand, L. (1976) Dynamic interactions of fluorescence probes with the solvent environment, J. Am. Chem. Soc., 98, 5001–5007.

Docchio, F., Longoni, A. and Zaraga, F. (1981) Subnanosecond fluorescence waveforms measurements with a dual time-scale microprocessor-controlled averager, Rev. Sci. Instrum., 52, 1671–1675.

Dorsey, C.C., Pelletier, M.J. and Harris, J.M. (1979) Time correlation method for measuring fluorescence decays with a cw laser, Rev. Sci. Instrum., 50, 333–336.

Easter, J.H., DeToma, R.P. and Brand, L. (1976) Nanosecond time-resolved emission spectroscopy of a fluorescence probe adsorbed to egg lecithin vesicles, Biophys. J., 16, 571–583.

Eisenfeld, J. and Ford, C.C. (1979) A systems-theory approach to the analysis of multiexponential fluorescence decay, Biophys. J., 26, 73–84.

Eisert, W.G. and Beisker, W. (1980) Epi-illumination optical design for fluorescence polarization measurements in flow systems, Biophys. J., 31, 97–112.

Fairclough, R.H. and Cantor, C.R. (1978) The use of singlet–singlet energy transfer to study macromolecular assemblies, Methods Enzymol., 48, 347–379.

Feinstein, M.B., Fernandez, S.M. and Sha'afi, R.I. (1975) Fluidity of natural membranes and phosphatidylserine and ganglioside dispersions: Effects of local anesthetics, cholesterol, and protein, Biochim. Biophys. Acta, 413, 354–370.

Fernandez, S.M. (1982) Time resolved microspectrofluorimeter, Biophys. J., 37, 73a.

Fernandez, S.M. and Berlin, R.D. (1976) Cell surface distribution of lectin receptors determined by resonance energy transfer, Nature (London), 264, 411–415.

Fernandez, S.M. and Herman, B. (1982) Topography and mobility of Concanavalin A receptors during myoblast fusion, in: M.L. Pearson and H.F. Epstein (Eds.), Muscle Development: Molecular and Cellular Control, Cold Spring Harbor Laboratory, New York.

Fleming, G.R., Morris, J.M., Robbins, R.J., Wolfe, G.J., Thistlethwaite, P.J. and Robinson, G.W. (1978) Nonexponential fluorescence decay of aqueous tryptophan and two related peptides by picosecond spectroscopy, Proc. Natl. Acad. Sci. (U.S.A.), 75, 4652–4656.

Forster, T.H. (1975) Delocalized excitation and excitation transfer, in: O. Sinanoglu (Ed.), Modern Quantum Chemistry, Part III, Academic Press, New York, pp. 93–137.

Gafni, A., Modlin, R.L. and Brand. L. (1975) Analysis of fluorescence decay curves by means of the Laplace transformation, Biophys. J., 15, 263–280.

Gallay, J., Vincent, M. and Alfsen, A. (1982) Dynamic structure of bovine adrenal cortex microsomal membranes studied by time-resolved fluorescence anisotropy of all-*trans*-1,6-diphenyl-1,3,5-hexatriene, J. Biol. Chem., 257, 4038–4041.

Gedcke, D.A. and McDonald, W.J. (1968) Design of the constant fraction of pulse trigger for optimum time resolution, Nucl. Instrum. Methods, 58, 253–260.

Genest, D., Sabeur, G., Wahl, P. and Auchet, J. (1981) Fluorescence anisotropy decay of ethidium bound to chromatin, Biophys. Chem., 13, 77–87.

Ghiggino, K.P., Lee, A.G., Meech, S.R., O'Connor, D.V. and Phillips, D. (1981) Time-resolved emission spectroscopy of the dansyl fluorescence probe, Biochemistry, 20, 5381–5389.

Gratton, E. and Lopez Delgado, R. (1979) Use of synchrotron radiation for the measurement of fluorescence lifetimes with subpicosecond resolution, Rev. Sci. Instrum., 50, 789–790.

Grinvald, A. (1976) The use of standards in the analysis of fluorescence decay data, Anal. Biochem., 75, 260–280.

Grinvald, A. and Steinberg, I.Z. (1974) On the analysis of fluorescence decay kinetics by the method of least-squares, Anal. Biochem., 59, 583–598.

Grinvald, A. and Steinberg, I.Z. (1976) The fluorescence decay of tryptophan residues in native and denatured proteins, Biochim. Biophys. Acta, 427, 663–678.

Grinvald, A., Haas, E. and Steinberg, I.Z. (1972) Evaluation of the distribution of distances between energy donors and acceptors by fluorescence decay, Proc. Natl. Acad. Sci. (U.S.A.), 69, 2273–2277.

Haar, H.P. and Hauser, M. (1978) Phase fluorometer for measurement of picosecond processes, Rev. Sci. Instrum., 49, 632–633.

Haas, E., Wilcheck, M., Katchalski-Katzir, E. and Steinberg, I.Z. (1975) Distribution of end-to-end distances of oligopeptides in solution as estimated by energy transfer, Proc. Natl. Acad. Sci. (U.S.A.), 73, 1807–1811.

Haas, E., Katchalski-Katzir, E. and Steinberg, I.Z. (1978) Brownian motion of the ends of oligopeptide chains in solution as estimated by energy transfer between the chain ends, Biopolymers, 17, 11–31.

Haggai, T.F. (1964) In: E. Bleuler and R.O. Maxby (Eds.), Phase Measurements in Methods of Experimental Physics, Vol. 2, Academic Press, New York, pp. 549–557.

Haugen, G.R., Wallin, B.W. and Lytle, F.E. (1979) Optimization of data-acquisition rates in time-correlated single-photon fluorimetry, Rev. Sci. Instrum., 50, 64–72.

Hauser, M. and Heidt, G. (1975) Phase fluorometer with a continuously variable frequency, Rev. Sci. Instrum., 46, 470–471.

Herman, B.A. and Fernandez, S.M. (1978) Changes in membrane dynamics associated with myogenic cell fusion, J. Cell. Physiol., 94, 253–263.

Herman, B.A. and Fernandez, S.M. (1979) A microfluorimetric study of membrane dynamics during development of dystrophic muscle in vitro, Arch. Biochem. Biophys., 196, 430–435.

Herman, B.A. and Fernandez, S.M. (1982a) A fluorescent pyrene derivative of concanvalin A: Preparation and spectroscopic characterization, Biochemistry, 21, 3271–3275.

Herman, B.A. and Fernandez, S.M. (1982b) Dynamics and topographical distribution of surface glycoproteins during myoblast fusion: A resonance energy transfer study, Biochemistry, 21, 3275–3283.

Heyn, M.P. (1979) Determination of lipid order parameters and rotational correlation times from fluorescence depolarization experiments, FEBS Lett., 108, 359–364.

Hieftje, G.M. and Haugen, G.R. (1981) Correlation based approaches to time-resolved fluorimetry, Anal. Chem., 53, 755A–765A.

Hunt, B.R. (1971) Biased estimation for nonparametric identification of linear systems, Math. Biosci., 10, 215–237.

Irvin, J.A., Quickenden, T.I. and Sangster, D.F. (1981) Criterion of goodness of fit for deconvolution calculations, Rev. Sci. Instrum., 52, 191–194.

Isenberg, I. (1973a) On the theory of fluorescence decay experiments, I. Nonrandom distortions, J. Chem. Phys., 59, 5695–5713.

Isenberg, I. (1973b) On the theory of fluorescence decay experiments, II. Statistics, J. Chem. Phys., 59, 5708–5713.

Isenberg, I. (1975) Time decay fluorometry by photon counting, in: R.F. Chen and H. Edelhoch (Eds.), Biochemical Fluorescence, Marcel Dekker, New York, pp. 43–77.

Isenberg, I. and Dyson, R.D. (1969) The analysis of fluorescence decay by a method of moments, Biophys. J., 9, 1337–1350.

Isenberg, I., Dyson, R.D. and Hanson, R. (1973) Studies on the analysis of fluorescence decay data by the method of moments, Biophys. J., 13, 1090–1115.

Johansson, L.B. and Lindblom, G. (1980) Orientation and mobility of molecules in membranes studied by polarized light spectroscopy, Quart. Rev. Biophys., 13, 63–118.

Johnson, R.H. (1973) Characteristics of acousto-optic cavity dumping in a mode-locked laser, IEEE J. Quant. Electr., QE-9, 255–257.

Kawato, S., Kinosita Jr., K. and Ikegami, A. (1977) Dynamic structure of lipid bilayers studied by nanosecond fluorescence techniques, Biochemistry, 16, 2319–2324.

Kawato, S., Kinosita Jr., K. and Ikegami, A. (1978) Effect of cholesterol on the molecular motion in the hydrocarbon region of lecithin bilayers studied by nanosecond fluorescence techniques, Biochemistry, 17, 5026–5031.

Kennedy, C.J. (1975) Frequency-domain analysis of the mode-locked and frequency doubled homogeneous laser, IEEE J. Quant. Electr., QE-11, 857–862.

Kinosita, S. and Kushida, T. (1982) High-performance, time-correlated single photon counting apparatus using a side-on type photomultiplier, Rev. Sci. Instrum., 53, 469–474.

Kinosita Jr., K., Kawato, S. and Ikegami, A. (1977) Theory of fluorescence polarization decay in membranes, Biophys. J., 20, 289–305.

Kinosita Jr., K., Kataoka, R., Kimura, Y., Gotoh, O. and Ikegami, A. (1981a) Dynamic structure of biological membranes as probed by 1,6-diphenyl-1,3,5-hexatriene: a nanosecond fluorescence depolarization study, Biochemistry, 20, 4270–4277.

Kinosita, S., Ohta, H. and Kushida, T. (1981b) Subnanosecond fluorescence-lifetime measuring system using single photon counting method with mode-locked laser excitation, Rev. Sci. Instrum., 52, 572–575.

Knight, A.E.W. and Selinger, B.K. (1971) The deconvolution of fluorescence decay curves: A non-method for real data, Spectrochim. Acta, 27A, 1223–1234.

Koechlin, Y. and Raviart, A. (1964) Analyse par échantillonage sur photons individuels des liquides fluorescents dans le domaine de la sub-nanoseconde, Nucl. Instrum. Methods, 29, 45–53.

Koester, V.J. (1979) Improved timing resolution in time-correlated photon counting spectrometry with a static crossed-field photomultiplier, Anal. Chem., 51, 458–459.

Koester, V.J. and Dowben, R.M. (1978) Subnanosecond single photon counting fluorescence spectroscopy using synchronously pumped tunable dye laser excitation, Rev. Sci. Instrum., 49, 1186–1191.

Koppel, D.E., Fleming, P.J. and Strittmatter, P. (1979) Intramembrane positions of membrane-bound chromophores determined by excitation energy transfer, Biochemistry, 18, 5450–5457.

278

Kwok-Keung Fung, B. and Stryer, L. (1978) Surface density determination in membranes by fluorescence energy transfer, Biochemistry, 17, 5421–5248.

Lakowicz, J.R. (1980) Fluorescence spectroscopic investigations of the dynamic properties of proteins, membranes and nucleic acids, J. Biochem. Biophys. Methods, 2, 91–119.

Lakowicz, J.R. and Cherek, H. (1980) Dipolar relaxation in proteins on the nanosecond timescale observed by wavelength-resolved phase fluorometry of tryptophan fluorescence, J. Biol. Chem., 255, 831–834.

Lakowicz, J.R. and Cherek, H. (1981) Phase-sensitive fluorescence spectroscopy: A new method to resolve fluorescence lifetimes or emission spectra of components in a mixture of fluorophores, J. Biochem. Biophys. Methods, 5, 19–35.

Lakowicz, J.R., Prendergast, F.G. and Hogen, D. (1979) Differential polarized phase fluorometric investigations of diphenylhexatriene in lipid bilayers, Quantitation of hindered depolarizing rotations, Biochemistry, 18, 508–519.

Lakowicz, J.R., Cherek, J. and Balter, A. (1981) Correction of timing errors in photomultiplier tubes used in phase-modulation fluorometry, J. Biochem. Biophys. Methods, 5, 131–146.

Leskovar, B. and Lo, C.C. (1975) Single photo-electron time spread measurement of fast photomultipliers, Nucl. Instrum. Methods, 123, 145–160.

Leskovar, B., Lo, C.C., Hartig, P.R. and Sauer, K. (1976) Photon counting system for subnanosecond fluorescence lifetime measurements, Rev. Sci. Instrum., 47, 1113–1121.

Lewis, C., Ware, W.R., Doemeny, L.J. and Nemzek, T.L. (1973) The measurement of short-lived fluorescence decay using the single photon counting method, Rev. Sci. Instrum., 44, 107–114.

Lindmo, T. and Steen, H.B. (1977) Flow cytometric measurement of the polarization of fluorescence from intracellular fluorescein in mammalian cells, Biophys. J., 18, 173–187.

Lipari, G. and Szabo, A. (1980) Effect of librational motion on fluorescence depolarization and nuclear magnetic resonance relaxation in macromolecules and membranes, Biophys. J., 489–506.

Loken, M.R., Hayes, J.W., Gohlke, J.R. and Brand, L. (1972) Excited-state proton transfer as a biological probe, determination of rate constants by means of nanosecond fluorometry, Biochemistry, 11, 4779–4786.

Lytle, F.E. and Kelsey, M.S. (1974) Cavity-dumped argon-ion laser as an excitation source in time-resolved fluorimetry, Anal. Chem., 46, 855–860.

Makinen, S. (1982) New algorithm for the calculation of the Fourier transform of discrete signals, Rev. Sci. Instrum., 53, 627–630.

Mantulin, W.W. and Weber, G. (1977) Rotational anisotropy and solvent-fluorophore bonds: An investigation by differential polarized phase fluorometry, J. Chem. Phys., 66, 4092–4099.

Marquardt, D.W. (1963) An algorithm for least-squares estimation of nonlinear parameters, J. Soc. Industr. Appl. Math., 11, 431–441.

McKinnon, A.E., Szabo, A.G. and Miller, D.R. (1977) The deconvolution of photoluminescence data, J. Phys. Chem., 81, 1564–1570.

Menzel, E.R. and Popovic, Z.D. (1978) Picosecond resolution fluorescence lifetime measuring system with a cw laser and a radio, Rev. Sci. Instrum., 49, 39–44.

Muller, A., Lumry, R. and Kokubun, H. (1965) High performance phase fluorometer constructed from commercial subunits, Rev. Sci. Instrum., 36, 1214–1226.

Munro, I.H. and Sabersky, A.P. (1980) Synchrotron radiation as a modulated source for fluorescence lifetime measurements and for time-resolved spectroscopy, Synchrotron Radiat. Res., 323–352.

Munro, I., Pecht, I. and Stryer, L. (1979) Subnanosecond motions of tryptophan residues in proteins, Proc. Natl. Acad. Sci. (U.S.A.), 76, 56–60.

O'Connor, D.V., Ware, W.R. and Andre, J.C. (1979) Deconvolution of fluorescence decay curves, A critical comparison of techniques, J. Phys. Chem., 83, 1333–1343.

Perrin, F. (1929) La flouorescence des solutions: Induction moléculaire — polarisation et durée d'émission — Photochimie. Ann. Phys., XII, 169–249.

Pesce, A.J., Rosen, C. and Pasby, T. (1971) Fluorescence Spectroscopy, Marcel Dekker, New. York.

Rayner, D.M., McKinnon, A.E. and Szabo, A.G. (1977) Correction of instrumental time response vari-

279

ation with wavelength in fluorescence lifetime determinations in the ultraviolet region, Rev. Sci. Instrum., 48, 1050–1054.

Ricka, J. (1981) Evaluation of nanosecond pulse-fluorometry measurements — no need for the excitation function, Rev. Sci. Instrum., 52, 195–199.

Rigler, R. and Ehrenberg, M. (1973) Molecular interactions and structure as analysed by fluorescence relaxation spectroscopy, Quart. Rev. Biophys., 6, 139–199.

Rockley, M.G. (1980) A nomogram for deconvolution of single exponential fluorescence decays, Biophys. J., 30, 193–197.

Salmeen, I. and Rimai, L. (1977) A phase-shift fluorometer using a laser and a transverse electrooptic modulator for subnanosecond lifetime measurements, Biophys. J., 20, 335–342.

Schiller, P.W. (1975) The measurement of intramolecular distances by energy transfer, in: R.F. Chen and H. Edelhoch (Eds.), Biochemical Fluorescence, Marcel Dekker, New York, pp. 285–303.

Schlag, E.W., Selzle, H.L., Schneider, S. and Larsen, J.G. (1974) Single photon phase fluorimetry with nanosecond time resolution, Rev. Sci. Instrum., 45, 364–367.

Schurr, J.M. (1970) The role of diffusion in biomolecular solution kinetics, Biophys. J., 10. 700–716.

Schuyler, R. and Isenberg, I. (1971) A monophoton fluorometer with energy discrimination, Rev. Sci. Instrum., 42, 813–817.

Shinitzsky, M. (1972) Effect of fluorescence polarization on fluorescence intensity and decay measurements, J. Chem. Phys., 56, 5979–5981.

Solie, T.N., Small, E.W. and Isenberg, I. (1980) Analysis of nonexponential fluorescence decay data by a method of moments, Biophys. J., 29, 367–378.

Spears, K.G., Cramer, L.E. and Hoffland, L.D. (1978) Subnanosecond time-correlated photon counting with tunable lasers, Rev. Sci. Instrum., 49, 255–262.

Specht, L.T. (1980) Signal averager interface between a Biomation 6500 transient recorder and an LSI-11 microcomputer, Rev. Sci. Instrum., 51, 1704–1709.

Spencer, R.D. and Weber, G. (1970) Influence of Brownian rotations and energy transfer upon the measurements of fluorescence lifetime, J. Chem. Phys., 52, 1654–1663.

Steingraber, O.J. and Berlman, I.B. (1963) Versatile technique for measuring fluorescence decay times in the nanosecond region, Rev. Sci. Instrum., 34, 524–529.

Stryer, L. (1978) Fluorescence energy transfer as a spectroscopic ruler, Annu. Rev. Biochem., 47, 819–846.

Stubbs, C.D., Kouyama, T., Kinosita Jr., K. and Ikegami, A. (1981) Effect of double bonds on the dynamic properties of the hydrocarbon region of lecithin bilayers, Biochemistry, 20, 4257–4262.

Tao, T. (1969) Time-dependent fluorescence depolarization and Brownian rotational diffusion coefficients of macromolecules, Biopolymers, 8, 609–632.

Taylor, I.J. and Becker, T.H. (1972) A high speed time-to-amplitude converter using integrated circuits, Nucl. Instrum. Methods, 99, 387–395.

Taylor, D.L. and Wang, Y.L. (1980) Fluorescently labelled molecules as probes of the structure and function of living cells, Nature (London), 284, 405–410.

Thomas, J.C., Allison, S.A., Appelhof, C.J. and Schurr, J.M. (1980) Torsion dynamics and depolarization of fluorescence of linear macromolecules, II. Fluorescence polarization anisotropy measurements on a clean viral 29 DNA, Biophys. Chem., 12, 177–188.

Toffler, J.E. and Winters, P.N. (1975) In: E. Bleuler and R.O. Maxby (Eds.), Phase Measurements in Methods in Experimental Physics, Vol. 2, Part B, Academic Press, New York, pp. 38–47.

Valeur, B. (1978) Analysis of time-dependent fluorescence experiments by the method of modulating functions with special attention to pulse fluorometry, Chem. Phys., 30, 85–93.

Vincent, M., de Foresta, B., Gallay, J. and Alfsen, A. (1982) Nanosecond fluorescence anisotropy decays of n-(9-anthroloxy) fatty acids in dipalmitoylphosphatidylcholine vesicles with regard to isotropic solvents, Biochemistry, 21, 708–716.

Von Sengbush, G. and Thaer, A. (1973) Some aspects of instrumentation and methods as applied to fluorometry at the microscale, in: A.A. Thaer and M. Sernetz (Eds.), Proceedings of the Conference on Quantitative Fluorescence Techniques as Applied to Cell Biology, Springer, New York, pp. 31–40.

Wahl, P. (1975) Decay of fluorescence anisotropy, in: R.F. Chen and H. Edelhoch (Eds.), Biochemical Fluorescence, Marcel Dekker, New York, pp. 2–41.

Wahl, P., Auchet, J.C. and Donzel, B. (1974) The wavelength dependence of the response of a pulse fluorometer using the single photoelectron counting method, Rev. Sci. Instrum., 45, 28–32.

Walden, G.L. and Winefordner, J.D. (1980) A streak camera system for time resolved fluorimetry, Spectrosc. Lett., 13, 793–801.

Ware, W.R., Lee, S.K., Brant, G.J. and Chow, P.P. (1971) Nanosecond time-resolved emission spectroscopy: Spectral shifts due to solvent-excited solute relaxation, J. Chem. Phys., 54, 4729–4737.

Ware, W.R., Doemeny, L.J. and Menzek, T.L. (1973) Deconvolution of fluorescence and phosphorescence decay curves, A least square method, J. Phys. Chem., 77, 2038–2048.

Weber, G. (1971) Theory of fluorescence depolarization by anisotropic Brownian rotations, Discontinuous distribution approach, J. Chem. Phys., 55, 2399–2407.

Weber, G. (1977) Theory of differential phase fluorometry: detection of anisotropic molecular rotations, J. Chem. Phys., 66, 4081–4091.

Wild, U.P., Holzwarth, A.R. and Good, H.P. (1977) Measurement and analysis of fluorescence decay curves, Rev. Sci. Instrum., 48, 1621–1627.

Wolber, P.K. and Hudson, B.S. (1981) Fluorescence lifetime and time-resolved polarization anisotropy studies of acyl chain order and dynamics in lipid bilayers, Biochemistry, 20, 2800–2810.

Yariv, A. (1975) Quantum Electronics, John Wiley, New York.

Yguerabide, J. (1965) Generation and detection of subnanosecond light pulses: Application to luminescence studies, Rev. Sci. Instrum., 36, 1734–1742.

Yguerabide, J. (1972) Nanosecond fluorescence spectroscopy of macromolecules, Methods Enzymol., 26, 498–579.

CHAPTER 10

Picosecond spectroscopy

DEWEY HOLTEN

Department of Chemistry, Washington University, St. Louis, MO 63130, U.S.A.

Contents

1. Introduction 282
2. Instrumentation 282
   2.1. The laser 282
   2.2. Mode-locked lasers 285
   2.3. Repetition rates, energies, and single-pulse selection 289
   2.4. Wavelength shifting 291
   2.5. Experimental configurations 296
        2.5.1. Transient absorption studies 296
        2.5.2. Emission measurements 300
3. Photosynthesis 302
   3.1. Overview 302
   3.2. Bacterial photosynthesis 303
        3.2.1. Intermediary electron acceptors 303
        3.2.2. Multiphoton absorption 306
        3.2.3. Picosecond studies with pre-reduced electron acceptors 306
        3.2.4. Subpicosecond investigations 307
        3.2.5. Temperature-dependent electron transfer 309
   3.3. Green-plant photosynthesis 310
        3.3.1. Photosystem I 311
        3.3.2. Photosystem II 315
   3.4. Emission studies 316
   3.5. Model systems 319
4. Visual pigments and the purple membrane 322
   4.1. Rhodopsin 322
   4.2. Bacteriorhodopsin 324
5. Hemoglobin 326
6. Additional studies and concluding remarks 327
Acknowledgements 328
References 328

R.I. Sha'afi and S.M. Fernandez (Eds.), Fast Methods in Physical Biochemistry and Cell Biology
© *1983 Elsevier Science Publishers*

1. Introduction

Investigations of biologically interesting molecules, complexes, and intact functional units with ultrashort light pulses is currently an active and exciting area of research. During the past decade, and particularly over the last 5 or 6 years, picosecond spectroscopy has made a dramatic impact on our understanding of basic photochemical behavior in such areas as energy transfer and charge separation in photosynthesis, energy transduction in visual pigments, photodissociation of hemoglobin complexes of oxygen and carbon monoxide, and in model reactions for several of these processes.

In this article, studies with laser pulses having durations ranging from about 50 psec down to several tenths of a picosecond will be reviewed. The shorter end of this range is, of course, the regime of subpicosecond spectroscopy. Currently available short-pulse laser equipment is surveyed. Because of the centrality to picosecond work, several instrumental topics are considered in some detail. Among these are mode-locking and wavelength shifting via non-linear optical techniques. In keeping with the scope of this volume, an attempt has been made to present both the instrumental methods and the application of picosecond techniques to systems of biological interest in a way so as to be a useful technical review for the expert and at the same time beneficial on a fundamental level to the novice in the field.

2. Instrumentation

2.1. The laser

A typical laser consists of an active gain medium placed between two mirrors. One of the mirrors is 100% reflective over the lasing bandwidth, while the second is usually partially transmissive, allowing escape of radiation from the cavity. Lasers can now be designed around a large variety of active media (Yariv, 1975; Tang, 1979). However, the short-pulse lasers used to date for studies on molecular systems of biological interest have been of two main types: solid-state lasers such as Nd (neodymium):glass, Nd:YAG (yttrium aluminum garnet), or ruby and organic dye lasers. Lasing occurs when the gain in the active medium is greater than the losses of all components in the cavity. The minimum gain for lasing is called the threshold value. Losses in the laser cavity arise mainly from mirror transmission, partial reflection from optical surfaces, and scattering in the lasing material. Gain is achieved by first creating a population inversion in the active medium. In most solid-state lasers and in some organic dye lasers capable of producing short pulses, this condition is reached by pumping optical radiation into the gain medium with a helical flashlamp or several linear flashlamps. The flashlamp(s) and lasing medium are surrounding by a cylinder of highly reflective material so that the maximum electrical energy discharged through the flashlamp can be delivered to the laser rod or dye solution. Population inversion can also be attained by pumping the active medium with a second laser; that most common example being an organic dye laser pumped by an argon ion or nitrogen laser.

The arrangement of an active laser medium between two mirrors is called a laser oscillator (Yariv, 1971; Beesley, 1971). A laser medium without mirrors can act as an amplifier. The two mirrors of the laser cavity define a Fabry–Perot interferometer or resonator. The electromagnetic radiation that can be supported by the resonator forms standing wave patterns with nodes at the mirrors. The stationary waves satisfy the relation $n\,\lambda/2 = L$, or $v = nc/2L$ or $\omega = \pi nc/L$, where n is an integer, c is the velocity of light, and L is the length of the cavity (distance between the mirrors). The different values of n define modes of oscillation in the cavity called axial or longitudinal modes. Each axial mode is labeled as TEM_{pqn} (transverse electromagnetic). The subscripts p and q define the transverse modes of the cavity, which manifest themselves as intensity distributions in directions transverse or perpendicular to the laser axis (Yariv, 1975). For each transverse mode there are a large number of axial modes. The most stable operation of the laser is usually attained when an aperture is placed in the cavity in order to supress all but the lowest-order transverse mode, called the TEM_{00n} or simply the TEM_{00} mode. There are a number of advantages of the TEM_{00} mode, including high-energy density and low beam divergence. One disadvantage of running a laser in this mode is that the output energy is much lower than when many transverse modes are allowed to receive gain. Some lasers, particularly the short-pulse Nd:glass laser, are often operated in the multitransverse mode configuration (no apertures in the cavity).

At optical and near-infrared wavelengths, on the order of million different axial modes can be supported by a cavity having a length of 1 or 2 m, a typical mirror separation for short-pulse solid-state lasers. The frequency separation between adjacent axial modes is $c/2L$, independent of the number of modes. Thus, increasing the length of the cavity increases the number of possible modes that can be supported.

Actually, only those modes for which the laser gain is greater than the threshold value will be active in the laser oscillator. This further restricts the axial mode frequencies to lie within the emission bandwidth of the lasing medium. In fact the relationship between the minimum pulsewidth, Δt, obtainable from a lasing material having a spectral bandwidth, $\Delta\omega$, is given by the uncertainty principle, $\Delta t\,\Delta\omega \geq K$, where K is a constant that depends on the intensity profile and is on the order of unity. For a Gaussian profile, $K = 0.441$. The larger the emission bandwidth, the greater the number of modes than can be active in the laser oscillator, and the shorter the minimum possible pulsewidth. Furthermore, the instantaneous intensity increases with decreasing pulsewidth. The shortest duration pulse for a given laser material set by uncertainty principle is called a "bandwidth-limited" or "transform-limited" pulse. The latter term arises because the description of the laser output in the time and frequency domains are related to each other by a Fourier transform relationship. Knowledge of the phase and amplitude of each mode is, therefore, sufficient to describe the laser output in both the time and frequency domains (Bradley, 1977; Kaufmann, 1979).

In a conventional laser the amplitudes and phases of the axial modes are uncorre-

284

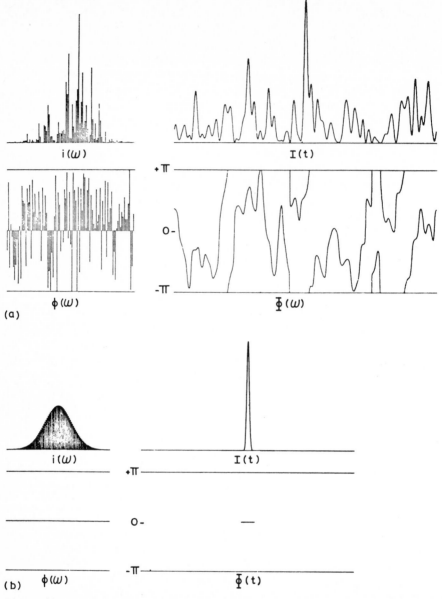

Fig. 1. (a) Simulation of the signal structure of a non-mode-locked laser. In the frequency domain (left) the intensities $i(\omega)$ of the 101 discrete longitudinal modes have a Rayleigh distribution about a Gaussian mean, and the phases $\phi(\omega)$ are randomly distributed in the range $-\pi$ to $+\pi$. In the time domain (right), the phases $\phi(t)$ fluctuate randomly, while the intensity $\phi(t)$ has the characteristics of thermal noise. (b) The signal structure of an ideally mode-locked laser. The 101 spectral intensities have a Gaussian distribution, while the spectral phases are identically zero. In the time domain, the signal is a transform-limited Gaussian pulse. The temporal intensity should be scaled up by about +20 to correspond quantitatively with (a) (Bradley, 1977).

lated due to the random oscillation of the various axial frequency modes. As shown in Fig. 1a, the temporal pattern has the same characteristics as the thermal noise (Bradley and New, 1974; New, 1980). If the modes are made to oscillate together by locking together the spectral amplitudes and phases, a transform-limited pulse of high intensity is produced (Fig. 1b). The process by which this condition is achieved is called "mode-locking" the laser.

2.2. Mode-locked lasers

Several techniques have been developed for locking together the longitudinal modes (DeMaria et al., 1969; Bradley, 1977). Active mode-locking is accomplished with an externally driven electro-optic or acousto-optic device that modulates the Q (quality) of the cavity at the intermode frequency spacing $c/2L$. This method is most frequently used to mode-lock cw (continuous wave) lasers, such as argon ion or krypton ion, where pulses on the order of 100–200 psec duration are generated (Adams et al., 1980). Flashlamp-pumped dye lasers have also been actively mode-locked (Schneider, 1980). The stability of the RF driving circuit is critical to produce the shortest pulses reliably with this technique.

Passive mode-locking is the preferred method for production of picosecond pulses from flashlamp-pumped lasers. Rather than utilizing an externally driven device, a cell containing a saturable absorber is placed in the laser cavity. The cell is placed either at Brewster's angle to the laser axis, or is optically contacted with one of the mirrors. Non-contacted cells typically have pathlengths on the order of 1 cm, while contacted cells have optical paths of 1 mm or less. There are advantages and disadvantages to both arrangements (Kaufman, 1979; Robinson et al., 1978). The saturable absorber is an organic dye that provides a non-linear loss in the resonator. It absorbs low-intensity light emitted by the active medium, but bleaches at high intensities and becomes more transmitting.

Spontaneous emission begins to occur in the active medium with the firing of the flashlamp. A random multimode radiation pattern develops in the cavity exhibiting temporal fluctuations due to the independence of the excited states giving rise to the emission. Stimulated emission begins to dominate when gain in the active medium overcomes the cavity losses and threshold is reached. The temporal intensity fluctuations in the resonator resemble random noise, as already mentioned (Fig. 1a). During this and subsequent periods of the mode-locking process, the saturable absorber acts as an intensity-dependent shutter, selecting one of the largest intensity fluctuations for growth at the expense of the weaker ones through a "survival of the fittest" mechanism. The largest fluctuations are on the order of the inverse bandwidth of the gain medium. The mode-locking dye absorbs low-intensity radiation and allows a large noise spike to pass essentially unattenuated. After a number of transits through the resonator, an intense optical pulse develops in the cavity out of the initial random "noise" (Fig. 1b). The weaker front edge of the pulse is absorbed by the saturable absorber while the trailing edge is chopped-off as the dye relaxes. Additional low-level fluctuations that develop in the gain medium are also suppressed. In this way, low-intensity radiation is attenuated, allowing the strong pulse to become stronger as it predominates in receiving additional gain as it passes through the active medium. Each time this single pulse reaches the partially reflecting mirror, a fraction of the radiation is transmitted from the resonator, while the remaining portion returns for additional passes through the active medium where it is further amplified, and through the saturable absorber where the short-pulse duration is maintained. This process continues until the gain falls below the threshold value as the flashlamp pumping cycle (pulse) comes to an end. More detailed discussions of this general time-domain description of mode-locking and of frequency-domain approaches can be found in reviews by DeMaria et al. (1969), Bradley (1977), and New (1980).

Thus, the output of the mode-locked laser is a train of intense pulses that are separated from each other by the time, $2L/c$, it took the single pulse in the cavity to undergo one "round-trip". The number of pulses in the train is determined by the concentration of the mode-locking dye and the stimulated emission cross-section of the active medium. The output from the mode-locked Nd:YAG and Nd:glass lasers have on the order of 10 and 100 pulses per train, respectively, and the pulse durations are approximately equal to the relaxation time of the saturable absorbers employed. For example, pulses from the Nd:glass laser mode-locked by Eastman dye 9860 are usually in the range of 5–10 psec while those for Nd:YAG laser mode-locked by Eastman dye 9740 are found to be in the range of 25–35 psec. These two saturable absorbers are in the carbocyanine family of dyes. Under well controlled conditions, pulses of about 3 psec duration can be obtained from the Nd:glass laser. (The theoretical minimum pulsewidth for this laser is 0.3 psec, but factors such as inhomogeneous broadening which limits the effective bandwidth, and frequency chirp in the carrier lead to the production of longer duration pulses.) A common mode-locking dye for the ruby laser is cryptocyanine; pulses of about 20 psec are obtained. The ruby laser is usually more difficult to operate in the mode-locked configuration, so that to date most experimentalists requiring high-intensity short pulses for studies of molecular phenomena have chosen the mode-locked Nd:glass or Nd:YAG laser. The fundamental wavelengths of these lasers are 694, 1054, 1060 and 1064 nm for ruby, Nd:phosphate glass, Nd:silicate glass, and Nd:YAG, respectively. Weber (1979) has tabulated many of the important properties of solid-state laser media.

In addition to the saturable absorber, gain depletion (loss of the population inversion via stimulated emission) of the active medium plays an important role in mode-locking. In the giant-pulse solid-state lasers just described, the saturable absorber is almost entirely responsible for selection and compression of the pulse in the cavity and, therefore, the time duration is determined primarily by the absorber relaxation time (New, 1980). However, gain depletion assists the absorber by slowing the growth of competing fluctuations in the early stages of the mode-locking process. In dye lasers, on the other hand, these roles are reversed and gain depletion plays the dominant role in pulse selection and compression (New, 1980; Ippen and Shank, 1979). This difference in behavior is a result of the short (nanosecond) relaxation times of the lasing dyes, compared with the emission lifetimes of several hundred microseconds or greater exhibited by solid-state gain media. Therefore, the pulsewidths obtainable from mode-locked dye lasers are routinely shorter than the relaxation times of the saturable absorbers employed. The mode-locking dye absorbs the front edge of the pulse while gain depletion effectively attenuates its trailing edge. Polymethine dyes such as DODCI (3,3′-diethyloxadicarbocyanine iodide) are often used as saturable absorbers for mode-locking dye lasers.

Pulses of 3–10 psec duration over the wavelength range 450–800 nm have been obtained from passively mode-locked flashlamp-pumped dye lasers (Schneider, 1980; Bradley, 1977). Since organic laser dyes fluoresce over broad spectral bands, tunable outputs can be obtained by replacing the totally reflective mirror in the laser cavity with a diffraction grating. The tuning range is usually limited to about 25 nm because of the necessary overlap of the emission spectrum of the lasing dye and the absorption spectrum of the mode-locking dye used in a particular configuration. Of course, the dyes can be changed to extend the tuning range. Output pulse energies are generally in the range 10–100 μJ. Although flashlamp-pumped dye lasers were one of the first sources of tunable picosecond pulses, they have not found widespread use in spectroscopic studies of molecular processes. One reason is the relatively high output divergence, which makes it difficult to achieve efficient amplification and to use non-linear optical processes for extending the wavelength range (section 2.4).

Saturable absorbers have also been used to passively mode-lock cw dye lasers. Folded-cavity arrangements similar to that shown in Fig. 2 are typically used (Ippen and Shank, 1975, 1977, 1979; Ruddock and Bradley, 1976). Rather than using a flashlamp, the laser dye (commonly Rhodamine 6G) is pumped by one or more lines from a (non-mode-locked) cw argon ion laser. The active medium is in the form of a thin jet stream placed near the center of the resonator. The jet stream is made by dissolving the laser dye in a viscous solvent such as ethylene glycol and passing it through a pair of nozzles. A second jet stream containing the saturable absorber is placed near one end of the cavity. Optically thin cells containing the saturable absorber contacted to one of the mirrors have also been used. The output of these lasers is tunable over a small range with an intracavity prism or Lyot filter. For pulses in the 1–10 psec range, output wavelengths in the 580–630 nm range have been obtained, limited by currently available saturable absorbers (Bradley, 1977). Subpicosecond transform-limited pulses as short as 0.3 psec duration have been generated by further compressing the pulses with a grating pair, which compensates for frequency chirp in the pulse. Added stability can be obtained by using a mixture of saturable absorbers to mode-lock the resonator. For example, with Rhodamine 6G as the laser dye and the mixture of saturable absorbers DODCI and malachite green, reproducible subpicosecond pulses tunable over the range 598–615 nm have been obtained with the set-up shown in Fig. 2 (Ippen and Shank, 1975, 1977, 1979; Ippen et al., 1980). Single pulses can be extracted from the resonator (Fig. 2) with an acousto-optic deflector (cavity dumper) at a repetition rate adjustable up to 10^6/sec. Pulse energies obtained from these passively mode-locked dye lasers are on the order of a few nJ, a factor of about 10^6 lower than those obtained from the giant-pulse solid-state lasers described above.

The problem of low energies can be overcome by sending the output pulses through a multistage pulsed amplifier (Fig. 3), consisting of dye cells pumped by the 532 nm second harmonic output of a Q-switched Nd:YAG laser (Ippen et al., 1980; Shank et al., 1979). The stages are isolated from each other with spatial filters or saturable absorbers, such as malachite green. Total amplification on the order of 10^6 can be achieved. Thus subpicosecond pulses having energies of a few mJ can be generated. The Q switch of the pump laser is synchronized with the cavity dumper in the dye laser resonator, so that the entire system can be operated at a repetition rate of 10 Hz.

The problem of tunability of the passively mode-locked dye laser can be overcome by using forced mode-locking. This technique involves pumping of a dye laser with a mode-locked oscillator such as ruby, Nd:glass, Nd:YAG, He:Ne, argon ion, or krypton ion (Kaufmann, 1979; Adams et al., 1980). The most common example for the production of picosecond pulses is that of pumping a dye laser resonator with the output of an actively mode-locked argon ion or krypton ion laser. The

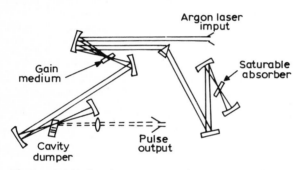

Fig. 2. A folded-cavity arrangement for a continuously operated, passively mode-locked dye laser. Under the proper conditions, subpicosecond pulses can be generated (Ippen and Shank, 1979).

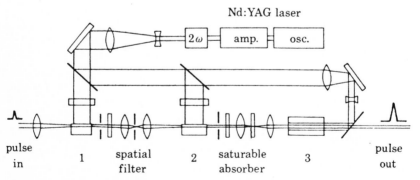

Fig. 3. Three-stage dye amplifier for amplification of subpicosecond pulses to peak powers of 5 GW at a repetition rate of 10 Hz (Ippen et al., 1980).

length of the dye laser cavity is finely adjusted to be equal to or a submultiple of the pump laser cavity length. In this way, the dye laser is mode-locked in synchrony with the ion laser; the configuration is called a synchronously pumped or sync-pumped dye laser system. As described above, the ion laser is actively mode-locked with an acousto-optic modulator to produce pulses of about 100–200 psec duration. Pulses obtained from the dye laser are generally in the range of about 1–10 psec. Subpicosecond pulses were first obtained with this technique by using the output of one dye laser to synchronously pump a second dye laser in tandem (Heritage and Jain, 1978). This technique is useful for extending the output to longer wavelengths, by choosing the dye in the second dye laser to absorb radiation emitted by the first, but which lases further to the red. Two dye lasers can be synchronously pumped in parallel by the same mode-locked ion laser (Jain and Heritage, 1978). This configuration is useful for studies of molecular phenomena, since two different wavelengths are obtained simultaneously by using different dyes in the two resonators. Subpicosecond pulses have also been obtained by incorporating Fabry–Perot etalons or wedge-shaped filters in the dye laser cavity (Kuhl et al., 1979).

Many of the strongest lines of the argon ion (514.5 nm) and the krypton ion (407 and 647 nm) lasers have been actively mode-locked and used to synchronously pump dye lasers. The krypton ion laser is better for pumping the near-infrared emitting laser dyes. The available output wavelengths at present extend from about 420 to 835 nm (Adams et al., 1980). Pulse energies are in the nJ range, similar to those of passively mode-locked dye systems. Therefore, multistage amplifiers must be employed if higher energy pulses are required (Fig. 3). Higher output energies and lower repetition rates are obtained by incorporating a cavity dumper in the dye laser cavity, but pulsewidths generally increase to about 25 psec.

2.3. Repetition rates, energies, and single-pulse selection

High repetition rates (short time intervals between pulses) are highly advantageous for a number of reasons. The most notable of these is that powerful signal-averaging techniques can be employed. Extremely weak signals can be resolved and shorter time periods are required to obtain information on samples from biological systems that would otherwise decompose under prolonged illumination. Of course, samples can be flowed and this should be done when at all possible, but often biological samples are precious and the large quantities necessary for most flow systems are frequently difficult to acquire. As we shall see (sections 3–5), many of these systems are not totally reversible and products build up in the sample cell (bacteriorhodopsin) in which case flowing is necessary. Or the transient reactions have components that decay on time scales as slow as 1 sec (photosynthetic reaction centers) so that even at repetition rates as low at 10 Hz (one pulse every 100 msec) there is the probability for re-excitation of the unrelaxed components in the sample. These cumulative effects can cause misinterpretation of the true course of events and lead to side reactions that do not normally take place under physiological conditions. In these instances, flowing the sample is absolutely essential.

The low (nJ) energies produced by passively mode-locked or synchronously pumped dye lasers cause small percentage changes per pulse in the sample, so that cumulative effects are minimized, and the high (up to the MHz range) repetition rates along with signal averaging permits weak signals often present in biological systems to be extracted. The low intensities and high repetition rates of synchronously pumped dye lasers are excellent for measuring fluorescence lifetimes by the time-correlated single photon-counting method (Chapter 9). Cavity dumpers can be used to reduce the pulse repetition rate where necessary. With low repetition rates (10 Hz or less), it is advantageous to have higher energy pulses in order to obtain larger signals per pulse. Less signal-averaging, and thus shorter total data accumulation periods, are required. A further advantage to the higher energy pulses is that non-linear optical techniques for generating new wavelengths (section 2.4) can be efficiently utilized.

Mode-locked solid-state oscillators produce higher energies (μJ–mJ) per pulse than do mode-locked dye resonators (nJ), but must be operated at lower repetition rates due to thermal instabilities and other factors. Most mode-locked Nd:silicate glass lasers (5–10 psec pulses at 1060 nm) are fired only once every minute or so to give reliable operation. The Nd:YAG oscillators can be operated at higher repetition rates (typically up to 10 Hz), are more easily mode-locked reliably in the TEM$_{00}$ mode than Nd:glass, but produce longer pulses (25–30 psec) due to the narrower lasing bandwidth. The new mode-locked Nd:phosphate glass systems offer a good compromise, shorter pulses (5–10 psec at 1054 nm) and higher repetition rate (1–10 Hz).

When the flashlamp is fired in a solid-state laser oscillator a train of pulses is output by the resonator, separated from one another by $2L/c \simeq 10$ nsec (a 1.5 m mirror separation). Although there are some studies in which excitation of a sample with entire pulse trains is useful, it is generally preferable to employ a single pulse out of the train so as to avoid problems due to the accumulation and/or re-excitation of transient states having relaxation times longer than $2L/c$. A typical single-pulse selector consists of a Pockels cell between crossed polarizers (Glan–Thompson or Glan–Foucalt prisms). The Pockels cell is an electro-optic device that rotates the polarization of light in response to an applied electric field. The voltage at which the polarization of a specific wavelength of light is rotated by 90° is called the half-wave voltage. The electric field vector of the radiation output by the solid-state laser is normally horizontally polarized, due to the orientation of the Brewster–Brewster rod and the (non-contacted) saturable absorber cell. The first polarizing prism is set to transmit horizontally polarized light and reject any spurious vertically polarized radiation. With zero potential difference across the two electrodes of the Pockels cell, it is oriented to also pass these horizontally polarized pulses. The second polarizing prism is rotated 90° from the first so that it deflects the horizontally polarized pulses onto a photodiode or into the nitrogen gas between the electrodes of a laser-triggered spark gap. Upon receipt of a pulse of sufficient intensity, the photodiode circuity is activated or the spark gap breaks down. The half-wave voltage is applied across the Pockels cell for a time long enough that only one later pulse in the train has its polarization rotated by 90°. This vertically polarized pulse is then transmitted by the second Glan–Focault prism. The remaining pulses in the train are rejected since their polarization remains horizontal. Newer devices employ double Pockels cells, each rotating the pulse by one-quarter wave. Lower voltages are required and the usable lifetime of the Pockels cell is increased.

The single pulse selected out of the train can be boosted in energy to 10 mJ or more by passing it through one or more flashlamp-pumped amplifier rods. Solid-state laser amplifiers are normally run to give single-pass amplification factors of between 5 and 10. The output energies are suitable for efficient generation of new wavelengths with non-linear optical techniques.

2.4. Wavelength shifting

One of the more demanding prospects facing experimentalists using picosecond lasers to study systems of biological interest is that of configuring the instrument to provide flashes at wavelengths necessary to excite the sample being investigated and to monitor the transient states produced. This means that the apparatus must be capable of exciting at least one band of one of the chromaphores present. More appropriately, the sample should be pumped at a wavelength corresponding to the excitation energy that would normally be put into the biological system under physiological conditions. Studies on photosynthetic reaction centers are a prime example (section 3). The best situation would be if the instrument could generate pulses continuously tunable, or at least tunable in discrete steps, over a broad spectral range encompassing the absorption spectrum of the sample.

There are a number of techniques for wavelength-shifting short-duration pulses (and of longer ones also), and in the past several years many of the necessary devices have been incorporated into picosecond laser instruments being used for studies of biological systems. Anisotropic crystals are employed for producing harmonics of the laser fundamental, for mixing pulses at two wavelengths to give pulses at the sum or difference frequencies, and in two-crystal optical parametric amplifiers that can be used with the appropriate pump wavelengths to generate tunable picosecond pulses from the near UV to the mid IR (Yarborough and Massey, 1971; Kung, 1974; Seilmeier et al., 1978; Laubereau et al., 1978; Kryukov et al., 1978; Kranitzky et al., 1980; Campillo et al., 1979a, b). Crystals have also been utilized in a number of detection schemes and in pulsewidth measurements (Ippen and Shank, 1977; Ippen et al., 1980). Another technique is to focus radiation at one frequency into a suitable liquid to produce flashes at the Stokes and anti-Stokes frequencies by stimulated Raman scattering (Colles, 1969; Yariv, 1975). For certain liquids such as CCl_4, water, and phosphoric acid a weak "white-light" or "continuum" pulse is produced (Alfano and Shapiro, 1970, 1971; Magde and Windsor, 1974; Nakashima and Mataga, 1975), which is useful as a broad-band monitoring flash in picosecond transient absorption measurements (section 2.5.1). The versatility of these techniques lies in the fact that the devices are "easy" additions to an apparatus in that they operate in a single-pass mode; pulses at one or two wavelengths are put in, and fraction of the energy comes out as radiation at a new wavelength(s). Major changes in the instrument are not necessary.

Much of the original literature on these topics can be found in the references cited above and in Ippen et al. (1980). Examples of all of these devices and how they are incorporated into instruments set up for specific spectroscopic applications will be discussed below. In view of their importance and ever increasing use in picosecond (and slower time scale) studies on biological systems, it will be useful to first review the terminology, and the basic principles underlying the field of wavelength shifting in anisotropic cystals. It is hoped that this discussion will be helpful to those not familiar with this area and to those who might need short pulses in regions of the spectrum currently not available on their instruments.

When a plane-polarized electromagnetic wave is incident upon a transparent dielectric medium such as a crystal, the electric field, $\bar{E}$ associated with the wave induces dipoles in the material. For a low-intensity wave, the induced dipole moment per unit volume is $\bar{P} = \varepsilon_0 \chi \bar{E}$, where χ is the electric susceptibility of the medium. (For the non-linear anisotropic substances discussed below χ is a second-rank tensor.) Thus, as the input wave propagates through the material, it sets up an array of induced dipoles (a charge polarization wave) that in turn reradiates a free electromagnetic wave having the same frequency as the incident wave, but with a different phase (Zernike and Midwinter, 1973). As the reradiated wave propagates through the material, it can interfere with the original wave. The total field in the medium is the sum of the incident and induced fields and is proportional to the electric displacement $\bar{D} = \varepsilon_0 \bar{E} + \bar{P}$. Because of the phase difference between $\bar{P}$ and $\bar{E}$, the velocity of the total field is reduced to c/n, where c is the velocity in vacuum and n is the refractive index of the material. The refractive index changes with wavelength, generally increasing with decreasing wavelength. This is called dispersion. In regions of the spectrum near natural frequencies of the crystal, the radiation is absorbed and is not useful for optical processes.

At the high applied electric fields associated with laser radiation, the induced polarization no longer varies linearly with the applied electric field, but is dependent on higher powers of E. This takes us into the regime of non-linear optics (Zernike and Midwinter, 1973; Zernike, 1979; Young, 1977; Baldwin, 1974). The non-linearity is commonly introduced by expanding the induced dipole moment per unit volume in a power series, $\bar{P} = \varepsilon_0(\chi_1 \bar{E} + \chi_2 \bar{E}^2 + \chi_3 \bar{E}^3 + \ldots)$, where $\chi_2, \chi_3, \ldots$ are known as non-linear polarizabilities or non-linear susceptibilities. The second-order non-linear susceptibility, χ_2, determines the magnitude of interactions involving the product of two waves, not necessarily having the same frequency. These three-wave interactions (two inputs and one output) give rise to effects that play important roles in a number of devices (Auston, 1977) contained in a picosecond laser instrument set up for spectroscopic work. These range from the single pulse selector (Pockels effect) to crystals for wavelength shifting, detection schemes, and pulsewidth measurements.

A standard solution to Maxwells' equations containing the non-linear polarization is a plane wave of frequency ω characterized by a phase velocity c/n and a direction of propagation (Z-axis) along the wave vector $\bar{k} = (\omega s/c)\bar{s}$, where $\bar{S}$ is the wave normal. The interaction of two collinear input waves of frequencies ω_1 and ω_2 and propagation constants k_1 and k_2 in a crystal induces components of the second order non-linear polarization ($\bar{P}_2 = \chi_2 \bar{E}_A \bar{E}_B$) at the second harmonic frequencies $2\omega_A$ and $2\omega_B$, and at the sum ($\omega_A + \omega_B$) and $|\omega_A - \omega_B|$ difference frequencies. Other commonly used terms for these interactions include frequency doubling for the first process, frequency mixing or parametric upconversion for the second, and parametric downconversion or parametric amplification for the third. Which of the four induced charge polarization waves will reradiate waves that can be sustained depends critically on a number of physical parameters of the crystal in which the interaction occurs and on the polarizations, frequencies, and direction of propagation of the incident light wave(s). Typically only one of the four interactions will occur for a particular set of conditions. The most stringent criterion for a non-linear optical process to take place with useful efficiency is that the input and output waves interfere constructively over the longest possible distance in the crystal; they must maintain the proper phase relationship. In other words, the light waves must travel with the same phase velocity (c/n) through the crystal. For example, in sum frequency generation $\omega_C = \omega_A + \omega_B$, the polarization wave set up by ω_A and ω_B will have a propagation constant $k_A + k_B = n_A \omega_A/c + n_B\omega_B/c$. But this is not necessarily equal to the propagation constant of the generated wave $k_C = n_C\omega_C/c$ because of dispersion ($n_A \neq n_B \neq n_C$) in the crystal. For the velocities of the polarization wave and the new wave it reradiates to be equal, then $k_C = k_A + k_B$ or $\Delta k = k_A + k_B - k_c = 0$. The criterion $\Delta k = 0$ is called the phase velocity synchronism, momentum conservation, k-vector matching, or "phase matching" condition. In addition, the interaction must obey energy conservation ($\Delta\omega = 0$).

Phase matching is most frequently accomplished by making use of the birefringence of anisotropic crystals; the refractive index "seen" by a light wave of a particular frequency depends on its polarization and its direction of propagation through the crystal. By choosing these last two properties correctly, the effect of dispersion can be overcome and phase matching of the polarization wave set up by the input radiation and the new wave it generates can be achieved. The simplest kind of anisotropic crystal has a single axis of symmetry, called the optic axis. These are known as uniaxial crystals. A light wave whose E vector is polarized perpendicular to the optic axis is called an ordinary ray (o-ray), since the refractive index seen by this ray is independent of the direction of propagation, and it will propagate through the crytal in its original polarization. A ray whose E vector is polarized parallel to the optic axis is called an extraordinary ray; the refractive index seen by an e-ray is dependent on the direction of propagation, varying elliptically between the ordinary index, n_o, and the extraordinary index, n_e. When the k vector is in a direction θ to the optic axis, the extraordinary index is given by the Fresnel equation (Eqn. 1.26 of Zernike and Midwinter, 1973). If the birefringence ($n_e - n_o$) is greater than zero, the crystal is said to be positive uniaxial, if $n_e - n_o$ is less than zero then it is called negative uniaxial.

For negative uniaxial crystals if the two input waves, of the same wavelength in the case of frequency doubling, have the same polarization (both o-rays) the phase matching is called a type I process. The output wave will be an e-ray, polarized orthogonally to the input wave(s). If the two input waves have orthogonal polarizations (one o-ray and one e-ray) then the interaction is called type II, and the output would be an e-ray. For positive uniaxial crystals the ray specifications are reversed.

The polarization of the radiation at different wavelengths emerging from a crystal must be taken into account for use in spectroscopic work, since effects such as orientational relaxation and particularly photoselection in complex biological systems in rigid media or at low temperatures are highly dependent on the absolute and relative polarizations of the excitation and monitoring pulses. The polarizations required for subsequent wavelength-shifting stages must also be considered (type I versus type II). Quarter-wave and half-wave plates can be used to "clean-up" the beams and produce the desired polarizations.

Thus for waves propagating at some angle θ to the optic axis of the appropriate crystal, the refractive indices for the o-rays and e-rays at the specified wavelengths will be such that phase matching will occur. This angle for the desired process is obtained by solving the momentum and energy conservation relationships along with the Fresnel equation. Angle-tuning curves for a wide variety of crystals and interactions can be found in the literature and obtained from companies who grow, cut and mount crystals (see also Zernike and Midwinter, 1973). The efficiency of the interaction depends on a number of parameters including how well phase matching is achieved, the crystal length, the intensity (or intensities) of the input wave(s), the wavelengths involved, and on the appropriate elements of the non-linear susceptibility tensor of the medium. The non-linear properties of the material, taking into account the direction of propagation and a number of other factors, are usually lumped into what is called the effective d coefficient. The divergence of the beam is also important in determining the efficiency of the interaction, one reason why it is desirable to operate the laser in the TEM_{00} mode. The appropriate parameter of the crystal is called its acceptance angle, larger by a factor of 2 for type II over type I processes (Zernike and Midwinter, 1973).

The problem of walk-off in anisotropic crystals often needs to be considered, particularly in two crystal schemes. Walk-off of the o-ray and e-ray occurs because the ray direction (the direction of energy flow or intensity axis for the beam) and the k-vector direction (the direction of phase propagation) for an e-ray are parallel only when $\theta = 0°$ or $\theta = 90°$. In other words, for an e-ray $\overline{D}$ and $\overline{E}$ are not generally collinear (Born and Wolf, 1965). At intermediate angles, the o-ray(s) and e-ray(s) do not overlap along the entire interaction length and efficiency is reduced. Walk-off is more of a problem in type II than in type I processes. In two-crystal parametric amplifiers, it is often desirable to point the optic axes of the two crystals in different directions, in order to minimize the effect of walk-off in the first crystal on the second one (Seilmeier et al., 1978). The effect of walk-off of the extraordinary ray(s) is not present when the phase-matching angle is made to be 90° (Eqn. 1.29 of Zernike and Midwinter, 1973). This can be accomplished in a variety of crystals by temperature tuning, rather than by angle tuning. Extraordinary rays are generally more

295

temperature dependent than o-rays. The allowable acceptance angle is generally larger in temperature tuning, and the interaction takes place over a longer distance in the medium. For these reasons, 90° phase matching (temperature tuning) is called "non-critical" phase matching. A disadvantage of temperature tuning is that expensive temperature controlling devices are required, frequently costing more than the mounted crystal.

There are usually trade-offs in choosing the length of a crystal. Generally, long crystals are preferable, but more costly than shorter ones. For picosecond-duration pulses, the crystal length is typically kept near or slightly longer than the pulse spatial "width". For example, with 30 psec pulses, crystals having lengths of 10 mm or so are adequate. Very long crystals sometimes have disadvantages, since with very high intensity picosecond pulses and high conversion efficiency processes such as type II second harmonic generation, the output face of the crystal can become damaged by the second harmonic even though the input face is not damaged by the fundamental. Under any circumstances, the damage threshold of the particular material should be considered. Fortunately the damage threshold increases as the pulsewidth is reduced, from several hundred MW/cm^2 with nanosecond pulses to several GW/cm^2 for picosecond pulses. The possibility of damage at the crystal faces in many cases is reduced by using "dry" mounted crystals purged with nitrogen, rather than those filled with index matching fluid. The most commonly used crystals are potassium dihydrogen phosphate (KDP), ammonium dihydrogen phosphate (ADP), and cesium dihydrogen arsenate (CDA). Their deuterated analogs, denoted for example d-KDP or KD*P, in many cases give higher conversion efficiencies, higher damage thresholds, and extended infrared transmission; a disadvantage is that they cost more. Lithium niobate (LiNiO$_3$) and the iodate (LiIO$_3$) are popular because of their higher non-linear coefficients, requiring shorter crystals, but are more expensive and have lower damage thresholds than the other materials. In the latter regard, the iodate is usually a better choice than the niobate. Rhenium dihydrogen phosphate (RDP) is useful for tripling Nd lasers. Starting with a single pulse of about 10 mJ at 1060 nm (1064 nm) from a system consisting of a mode-locked TEM$_{00}$ mode Nd:glass (Nd:YAG) oscillator a single pulse selector and an amplifier, conversion efficiencies to the second, third, and fourth harmonics of about 40%, 15%, and 5% can be routinely obtained with the proper choice of crystal and interaction (type I or II).

Tunable picosecond excitation and monitoring flashes can also be generated with a parametric amplifier. This is a specific example of difference frequency generation that has developed its own terminology. The result of the overall process is that a strong pump pulse (ω_p) is split into two photons of lower frequency called the signal (ω_s) and idler (ω_i), such that $\omega_p = \omega_s + \omega_i$. But, unlike the typical case of difference frequency generation, the signal frequency is not provided externally, but rather by zero-point energy fluctuations (noise) in the crystal. The signal is amplified over the interaction length and the idler intensity grows. The output of this device is sometimes referred to as parametric fluorescence. A parametric oscillator is made by placing the crystal in a cavity. A more useful approach is the single-pass scheme employing two crystals. Angle-tuned KDP or LiNiO$_3$, or temperature-tuned ADP crystals, are common examples. In the first crystal, the pump radiation generates the signal and idler waves, whose bandwidths depend on a number of factors including the divergence of the pump. The pump, signal, and idler waves must satisfy the energy conservation requirements and the phase-matching conditions set by the type and cut of the crystal and the angle its optic axis makes with the incident beam (or on the temperature in case of temperature tuning). The second crystal is placed at a distance from the first and similarly angle (or temperature) tuned so as to amplify preferentially a narrow signal wavelength band. Depending on the crystal and pump wavelength, essentially tunable picosecond pulses covering the near UV to the mid IR can be generated (Kung, 1974; Seilmeier et al., 1978; Laubereau et al., 1978; Kryukov et al., 1978; Kranitzky et al., 1980; Campillo et al., 1979a, b). Parametric amplifiers have also been used to shorten the time duration of picosecond pulses (Laubereau et al., 1978; Kranitzky et al., 1980). In fact, all of the non-linear effects discussed above shorten the pulse duration to some extent.

Specific examples of all these wavelength-shifting techniques will be mentioned along with the applications of picosecond instrumentation in sections 3–5.

2.5. Experimental configurations

Two main classes of picosecond measurements on biological systems to date can be defined, the transient absorption experiment and the emission experiment. The instruments used for either application are built up from the basic components discussed in the preceding sections, but differ in how they are configured together and what detection system or scheme is employed. The capabilities and limitations of a particular set-up are thus defined. Other picosecond measurement techniques employing the transient grating method, inverse Raman scattering, or picosecond CARS, will not be discussed here, since their application to biological systems is minimal or non-existent to date.

2.5.1. Transient absorption studies

Picosecond transient absorption measurements are based on the pump-probe technique. A strong pump or excitation pulse creates excited states in the sample, initiating the chemical process to be investigated. A weak probe or monitoring pulse arrives at the sample at a specified time after the excitation pulse and monitors the change in transmission (absorption) of the sample due to the transient state(s) present. Most instruments are dual beam in nature, in that a portion of the probe pulse also monitors unexcited sample, similar to the operation of a conventional absorption spectrometer. In the transient experiment the measured quantity is the difference in absorption (ΔA) between the ground and transient state(s) at the selected wavelength and delay time. The monitoring flash is normally of the same time duration as the excitation pulse, but is either at a different wavelength or contains a broad band of wavelengths (the picosecond continuum mentioned in section 2.4).

Fig. 4 shows schematically the apparatus set up in our laboratory, illustrating the general configuration of most picosecond instruments based on solid-state lasers used for transient absorption studies (cf. Magde and Windsor, 1974). The important parameters pertaining to instruments used in specific applications will be discussed in sections 3–5. The apparatus is based on a mode-locked Nd:YAG laser system that delivers single, 30 psec, 1064 nm, 10 mJ pulses at a repetition rate adjustable up to 10 Hz. The pulses are in the TEM_{00} mode, resulting in low divergence and good beam quality necessary for efficient wavelength shifting via non-linear optical techniques. The high repetition rate of the laser is useful for increasing signal to noise with signal-averaging techniques, while lower repetition rates are used for samples with long-lived transient state when flowing the sample is not possible.

Radiation at 1064 nm is split into two parts with a 50% beam splitter to drive the excitation and monitoring (pump and probe) legs of the apparatus. In the excitation leg, radiation at 1064 nm can be converted to the harmonics at 532, 355, and 266 nm in crystals such as KDP (potassium dihydrogen phosphate). For a particular configuration of crystals, the unwanted residual radiation is removed from the desired harmonic(s) with a dichroic filter. Any of the harmonics can be used directly to excite the sample by focusing to a 1–2 mm diameter spot in the center of a 1–5 mm pathlength sample cuvette, or can be sent first through an additional wavelength-shifting device (Fig. 4). For example, when 532 nm pulses are focused into a cell containing cyclohexane or its fully deuterated analog, 630 or 600 nm excitation

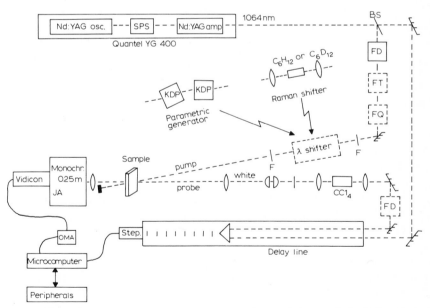

Fig. 4. Picosecond transient absorption apparatus. FD, FT, FQ, crystals for frequency doubling, tripling, or quadrupling; F, colored glass filters and/or dichroic mirrors; SPS, single-pulse selector; BS, 50% beam splitter.

pulses, respectively are produced. Conversion efficiencies of about 15% are typical. Higher Stokes harmonics and anti-Stokes radiation at shorter wavelengths are also generated. The second type of wavelength-shifting device is a parametric generator consisting of two angle-tuned KDP crystals. With 532 nm pump pulses, the signal wave output of the parametric generator is tunable over the range 780–1400 nm. The efficiency is about 5%. Tunable visible pulses can be produced with the 355 nm pump.

The 1064 nm radiation in the probe leg of the apparatus traverses a stepping motor-driven optical delay line. This provides the time resolution for the instrument by making use of the fact that light travels 3.33 psec/mm (about 1 ft/nsec). Since the pulses travel up and back the delay line, a 1 mm step gives a 2 mm pathlength difference or a change of 6.7 psec between arrival of the pump and probe pulses at the sample. The delay line permits delays from a few picoseconds to greater than 10 nsec. The broad band monitoring pulses are geneated by focusing the 1064 nm pulses into a cell containing CCl_4. The probe pulses are also of about 30 psec duration and extend from about 450 nm through the visible and past 1064 nm into the IR. Stronger monitoring flashes that extend down to about 380 nm are produced when 532 nm radiation is used to generate the continuum. The 1064 or 532 nm light used to generate the monitoring flashes is removed with a dichroic beam splitter. The probe light is elongated vertically by a pair of cylindrical lenses, passed through excited and unexcited regions of the sample, and focused onto the entrance slit of a $1/4$ m monochromator. A Princeton Applied Research vidicon detector is placed at the output of the spectrograph and is coupled to an optical multichannel analyser (OMA) console.

In the 1D acquisition mode, the vidicon is oriented to give 500 active detector channels down the exit slit of the monochromator. This gives a digitized measure of the intensity profile (at the selected wavelength) of the probe light vertically down the sample, which contains information on the light transmitted by the excited and unexcited regions of the sample. In this way the difference in absorption of the two regions at the preset wavelength and delay time is measured. The 1D mode is particularly useful for measuring kinetics at a single wavelength by repetitive scanning of the probe delay line.

The 2D mode is more convenient for measuring transient absorbtion spectra (Greene et al., 1979; Holten and Windsor, 1980). An 80–300 nm wide spectrum can be obtained in a single laser shot; the spectral width is determined by the dispersion of the grating used. The OMA electronics can dissect the vidicon detector active area into a number of tracks, each of which is 500 channels wide. Two tracks are normally used. In this mode the exit slit is removed from the monochromator and the vidicon rotated 90° from its position in the 1D mode. The two tracks are set so that one is aligned with the dispersed spectrum of probe light passing through the excited sample, while the second 500 channel track is aligned with the light transmitted through the unexcited or reference region of the sample. The digitized spectral information can be temporarily stored in the two OMA memories or read directly in real time into the microcomputer. Baseline subtraction, calculation of the absorption changes, signal averaging, display, and storage are all handled by the computer, which also controls the firing of the laser and the stepping motor driving the delay line in both the 1D and 2D modes.

On some instruments the continuum is generated before traversing the delay line, or is produced by a parametric generator and then delayed. The probe is sometimes fixed and the pump traverses a delay line. Rather than using cylindrical lenses, the probe light can also be broken down into two parallel beams with a beam splitter and mirror, and then sent through excited and unexcited regions of the sample. A novel way to measure kinetics in a single shot is to pass an elongated probe beam through an echelon, a glass device that contains steps of different thicknesses (Topp et al., 1971). Portions of the monitoring light traverse the various steps and arrive at the sample with different time delays. The intensity profile of the sample probed by the different portions of the probe beam is sent through a monochromator and recorded on a vidicon or other multichannel detector. Reflection echelons are also used. This scheme can be employed in a 2D experiment to give simultaneous spectral and kinetic information (Huppert et al., 1977). The echelon is most useful for delays of a few hundred picoseconds or less.

Similar pump-probe techniques are used with passively mode-locked or synchronously pumped dye lasers producing pulses in the 1–10 psec or subpicosecond range. With high repetition rate (MHz) operation and low energy pulses (nJ), one practice is to split the beam into two pieces; one is the pump, one is the probe. One of these beams traverses a delay line before reaching the sample, providing the time delays. The pump beam is chopped. The change in transmission by the sample of the probe beam at the same wavelength as the pump is measured as a function of time delay by a photomultiplier tube coupled with a lock-in amplifier set to respond to signals at the chopper frequency (Ippen and Shank, 1977). Pumping and probing at different wavelengths can be done by frequency doubling one of the beams. This generally requires tight focusing because of the low input energy. With synchronously pumped lasers, two different wavelengths can be obtained simultaneously by pumping in parallel two dye lasers containing different dyes with the output of the same mode-locked argon or krypton ion laser (section 2.2). To obtain sufficient energy to generate a continuum or to Raman shift, the pulses must be amplified in a multistage scheme similar to that discussed in section 2.2 (Fig. 3).

The read-in–read-out appears to be another promising method for pumping and probing at different wavelengths when a continuum or tunable picosecond or subpicosecond probe pulses cannot be generated because of a low-energy fundamental (Wiesenfeld and Ippen, 1979; Ippen et al., 1980). Expensive amplifiers are not required. Convenient sources of monitoring light are continuous or pulsed (long pulse durations are satisfactory) tunable dye laser. The detection scheme makes use of upconversion by mixing (gating) in a crystal, after the sample, the monitoring light with a time-delayed portion of the original excitation pulse that has not passed through the sample. Kinetics of the transmission (absorption) changes induced in the sample by the pump are monitored with a monochromator and detector at the upconverted wavelength as a function of time delay between the excitation and gat-

300

ing pulses. Spectral data is acquired by tuning the monitoring light source to a new wavelength, which will also change the detection wavelength following upconversion. This gating technique can also be used to detect fluorescent light at near-infrared wavelengths where the appropriate detectors are not sensitive or not available (Mar et al., 1972; Hallidy and Topp, 1977; Beddard et al., 1980a).

2.5.2. Emission measurements

Fluorescence lifetimes with subnanosecond resolution are measured with several techniques. Lifetimes of 10 psec or less can be obtained with a streak camera or an optical Kerr gate. Both methods are used mainly with Nd:glass or Nd:YAG solid-state laser systems, which can induce emission signals of sufficient intensity on a single laser shot. Time-correlated single-photon counting yields emission lifetimes of 50 psec less with signal averaging and deconvolution procedures. Since low counting rates and high laser repetition rates are required, synchronously pumped dye lasers are ideal sources of short-duration excitation pulses. This technique is described in detail in Chapter 9. Subnanosecond fluorescence lifetimes are also measured with the phase-shift method, which does not require short-duration laser pulses (cf. Kaufmann, 1979).

A typical picosecond emission spectrometer employing a streak camera is shown schematically in Fig. 5. This apparatus is similar to that used by Porter et al. (1977). Streak camera technology and applications have been reviewed by several authors (Robinson et al., 1978; Kaufmann, 1979; Bradley, 1977; Bradley et al., 1980; Hus-

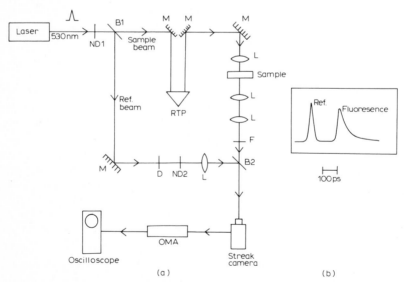

Fig. 5. (a) Schematic for picosecond emission studies with a streak camera. M, 100% mirrors; B1, B2, beam splitters; ND1, ND2, neutral density filters; RTP, rooftop prism; L, lenses; D, diffuser; OMA, optical multichannel analyzer. (b) A typical trace showing reference pulse and fluorescence pulse (Porter et al., 1977).

ton and Helbrough, 1980; Suzuki et al., 1980a). The pulse or train of pulses generated by the laser and wavelength-shifting devices is split into two parts with a beam splitter. The portion of the pulse (or train) containing most of the intensity traverses the delay line and excites fluorescence in the sample, and is then removed with a filter. The second harmonic from a Nd:glass (or YAG) laser has been the most common pump. The emission is directed to the streak camera coupled with an OMA or other multichannel detector to digitize the streak intensities. The weaker portion of the pulse (or train) acts as a reference or marker pulse to trigger the streak camera electronics. The delay line is normally set so that the reference pulse arrives at the detector a few hundred picoseconds to several nanoseconds before the fluorescence. Signal averaging is achieved by aligning the reference pulse from each shot with a specific channel in the multichannel analyzer, thus ensuring that the fluorescence decay curves will coincide and average properly. This is necessary to compensate for shot-to-shot jitter in the streak camera triggering. The excitation intensity can be varied with a neutral density filter. Relative quantum yields are obtained by integrating the fluorescence profile in the digitizer and measuring the excitation intensity by directing a known fraction to a calibrated photodiode.

Fluorescence lifetimes with about 10 psec resolution can also be measured by employing the optical Kerr shutter introduced by Duguay and Hansen (1969). Fig. 6 illustrates a typical setup, similar to that used by Yu et al. (1975). The fluorescence is excited with a pulse at 530 nm (or other wavelength) and is directed through a CS_2 light gate between crossed polarizers situated in front of the detector. Filters or a monochromator can be used to isolate the fluorescence prior to the detector, which is generally a photomultiplier or a photodiode. The 1060 nm analyzing or gating pulse opens the light gate at a preset time relative to excitation pulse by inducing a birefringence in the cell. This briefly (a few picoseconds) rotates polarization of the fluorescent light, allowing it to pass the second polarizer and reach the detector. In the conventional arrangement, the fluorescence intensity at only one time delay between excitation and analysis (gating) pulses is obtained on each laser shot. The decay curve is built up by successive measurements with different delays. However, entire short decays can be obtained on a single shot by using an oblique propagation of the gating pulse through the Kerr cell (Ho and Alfano, 1977).

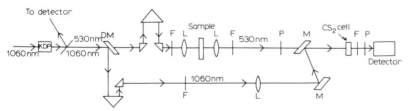

Fig. 6.   Schematic for picosecond emission studies with a fast optical gate. At the Kerr cell (CS_2 cell), the gating 1064 nm pulse is aligned collinearly with the collected fluorescence light. L, lens; F, filter; M, mirror; P, polarizer (Yu et al., 1975).

3. Photosynthesis

3.1. Overview

The photochemical reactions of plant and bacterial photosynthesis are initiated by the absorption of photons by large arrays of antenna pigments that funnel the energy rapidly to special pigment–protein complexes called reaction centers. Within the reaction center a multistep charge separation process takes place, effectively converting a portion of the incident energy into chemical potential (Govindjee, 1975; Olson and Hind, 1977; Blankenship and Parson, 1978, 1979; Clayton and Sistrom, 1978; Parson and Ke, 1981). In plants there are two such photosystems, called PSI and PSII, each containing antenna pigments and the corresponding reaction centers, which cooperate in the photosynthetic process (Williams, 1977; Parson and Ke, 1981). The total chemical potential developed together by the two photosystems is sufficient for the oxidation of water to molecular oxygen (PSII) and the reduction of NADP to NADPH (PSI), which with the aid of ATP and other components reduces carbon dioxide to carbohydrate.

Photosynthetic bacteria, on the other hand, contain only one photosystem. Although the photochemical reactions that occur in bacteria are basically similar to those that occur in plants, the primary oxidant produced in the photochemical reaction in bacteria is not high enough in potential to oxidize water. Therefore, bacteria require alternative sources of electrons, such as H_2S, succinate, or other substrates.

For several years now it has been possible with suitable detergent treatments to obtain preparations of bacterial reaction centers free of antenna pigments. To date this has not been possible for the two plant photosystems, although subchloroplast fragments have been obtained from both PSI and PSII that contain about 30–40 antenna pigments per reaction center. This fact, together with the relative simplicity of the bacterial photosynthetic process, has made bacterial reaction centers the preferred choice for early studies of the primary photochemical reactions employing picosecond transient absorption techniques (Campillo and Shapiro, 1977; Holten and Windsor, 1978; Rentzepis et al., 1978; Parson and Ke, 1981). Reaction center photochemistry in subchloroplast fragments from plant photosystems has only recently been investigated with picosecond spectroscopy (Parson and Ke, 1981).

Subnanosecond-resolved emission studies have been performed to investigate the organization of the photosynthetic units in both bacteria and plants, to measure the rate(s) of energy transfer (or exciton migration) within the antenna complexes, and to probe the mechanism by which the harvested energy is trapped by the reaction center (Campillo and Shapiro, 1977; Holten and Windsor, 1978; Pellegrino et al., 1978).

Studies of electron transfer from photoexcited chlorophylls and related porphyrins to a variety of electron acceptors have been carried out to probe one or more aspects of photosynthetic electron transfer (Seely, 1977, 1978). Measurements on model systems, particularly picosecond studies (Holten and Windsor, 1978; Blankenship and Parson, 1979; Kaufmann, 1979) have been helpful in attempting to understand the molecular factors responsible for the high efficiency of the electron transfer steps in vivo. They provide an opportunity to investigate electron transfer reactions in simple, more well-defined environments.

In the following subsections we review the impact picosecond spectroscopy has made on these various areas of photosynthesis research. The review articles cited above should be referred to for additional background material.

3.2. Bacterial photosynthesis

Reaction centers isolated from several strains of photosynthetic bacteria have been found to contain four molecules of bacteriochlorophyll (BChl), two of the Mg-free analog, bacteriopheophytin (BPh), one or two quinones (Q_I, Q_{II}), a non-heme iron atom, and three polypeptides. The energy pumped into the reaction center, either from the antenna system in chromatophores, or through direct optical excitation in the case of isolated reaction centers, prepares the lowest excited singlet state (P*) of the primary electron donor believed to be a complex involving two of the four molecules of BChl. This "dimer" is usually referred to by the position of its long-wavelength absorption band, i.e. P870 in *Rhodopseudomonas sphaeroides* or *Rhodospirillum rubrum*, P960 in *Rps. viridis*, or simply as P. *Rps. viridis* reaction centers contain BChl *b* and BPh *b*, and thus absorb at longer wavelengths than do reaction centers of *Rps. sphaeroides* and *Rhodospirillum rubrum*, which contain BChl *a* and BPh *a*. Optical and magnetic resonance studies have shown that the primary reaction involves the oxidation of P to the cation radical (P$^+$) and reduction of the "primary" acceptor (Q) (Norris et al., 1971; Feher et al., 1975; Fajer et al., 1977; Katz et al., 1978; Blankenship and Parson, 1978) and that the spin of the unpaired electron on P$^+$ is delocalized over the two BChl molecules.

3.2.1. Intermediary electron acceptors
The major incentive for time-resolved measurements on bacterial reaction centers has been to identify any electron acceptors between P* and Q. Our current understanding of the electron transfer sequence is summarized in Fig. 7. Abbreviations used in this section are included in the legend. When *Rps. sphaeroides* or *Rps. viridis* reaction centers are excited with a 7–10 psec flash, absorption changes attributable to P$^+$ appear within the time resolution ($\sim$ 10 psec) of the instrumentation (cf. Fig. 4). These include prompt bleaching in the long-wavelength absorption band of P at 870 nm in *Rps. sphaeroides* and at 960 nm in *Rps. viridis* (Rockley et al., 1975; Kaufmann et al., 1975; Holten et al., 1978a; Schenck et al., 1981a), and the appearance of a new absorption band attributable to P$^+$ at 1250 nm in *Rps. sphaeroides* and at 1310 nm in *Rps. viridis* (Dutton et al., 1975; Netzel et al., 1977; Mosckowitz and Malley, 1978).

Fig. 8A shows the absorption changes obtained 20 psec (open circles) and 3 nsec (filled circles) after excitation of *Rps. sphaeroides* R-26 reaction centers with non-saturating, 8 psec flashes at 600 nm (Schenck et al., 1981a). Similar spectra have been obtained following excitation with 7–10 psec flashes at 530, 600, or 627 nm of

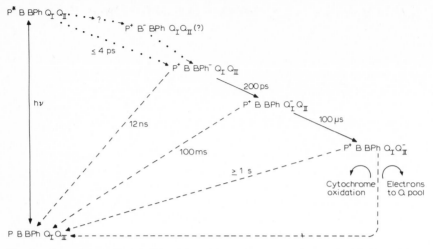

Fig. 7. Our current understanding of the primary electron transfer reactions in bacterial reaction centers with all electron acceptors in their normal states prior to excitation. BChl, bacteriochlorophyll; BPh, bacteriopheophytin; P, BChl dimer; B, a special BChl molecule; Q, quinone.

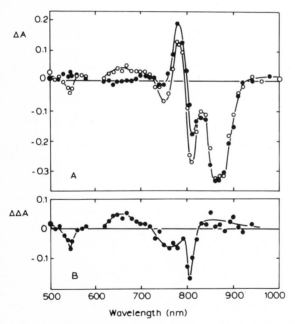

Fig. 8. Spectra of absorbance changes caused by excitation with single 8 psec, 600 nm flashes of *Rps. sphaeroides* reaction centers at moderate potential. Panel A: ○, measurements made 20 psec after the excitation pulse, assigned to P^+I^-; ●, measurements made 3 nsec after the excitation pulse, due to P^+Q^-. Panel B shows the difference between the filled and open circles of A, and represents absorbance changes due mainly to the reduction of the intermediary electron carrier, I (Schenck et al., 1981a).

Rps. sphaeroides (Rockley et al., 1975; Kaufmann et al., 1975; Dutton et al., 1975; Kaufmann et al., 1976a; Holten et al., 1980) and of *Rps. viridis* (Netzel et al., 1977; Holten et al., 1978a). The spectra in *Rps viridis* are, of course, shifted to longer wavelengths. Most of the picosecond measurements have employed excitation flashes at 530 nm, the second harmonic of the Nd:glass laser. These are not absorbed by P, but rather by BPh. Energy transfer from BPh* to P evidently occurs within 10 psec of excitation with the 530 nm flash (Netzel et al., 1973; Mosckowitz and Malley, 1978). Excitation pulses at 600 or 627 nm are generated by stimulated Raman scattering from the 530 nm pulses in liquids like perdeuterocyclohexane or cyclohexane, respectively. Flashes at these wavelengths are absorbed mainly by P, and probably also to a lesser extent by one or both of the other two BChls (B) in the reaction center that are not intimately involved in P. Similar spectra to those shown in Fig. 8A have also been obtained in studies of *Rds. rubrum* and *Rps. sphaeroides* reaction centers with 30–50 psec, 880 nm excitation flashes that excite P directly in its long-wavelength band (Shuvalov et al., 1978; Kryukov et al., 1978; Akhamanov et al., 1980). Pulses at this wavelength were produced by sending the 532 nm pulses from a mode-locked Nd:YAG laser through a single-pass parametric generator consisting of two KDP crystals (section 2.4).

The 3 nsec difference spectrum in Fig. 8A can be assigned to the state P^+Q^-, in comparison with previous slower time scale and steady-state measurements. In fact, these absorbance changes essentially reflect only the oxidation of P, as the reduction of Q causes negligible changes in the regions shown. The 20 psec spectrum is due to an earlier state referred to as P^+I^-, where I is an intermediary electron carrier. The electron then moves from I^- to Q with a time constant of 150–250 psec (Rockley et al., 1975; Kaufmann et al., 1975, 1976a; Netzel et al., 1977; Peters et al., 1978; Holten et al., 1978a, 1980a; Shuvalov et al., 1978; Pellin et al., 1978; Akhamanov et al., 1980). The 20 psec spectrum is identical to that of state P^F first observed by Parson et al. (1975) in nanosecond studies of reaction centers in which electron transfer to Q was blocked by its chemical reduction prior to excitation.

Based on these identifications, the difference between the 20 psec and 3 nsec spectra (Fig. 8B) gives the absorbance changes due mainly to the reduction of I. Bleaching at 545 and 760 nm attributable to BPh suggests that I is a BPh. However, bleaching near 800 nm, and near 600 nm observed with 530 nm excitation, indicate the involvement of BChl. The BChl molecule responsible for this bleaching, and which is not one of the BChls intimately involved in P, is referred to as B. This indicates that I is the complex (B BPh). Similar spectra and conclusions have been drawn from studies of reaction centers from a number of bacterial strains in which I^- is trapped by continuous illumination at low redox potential (Q reduced) in the presence of bound or exogenous cytochrome (cf. Blankenship and Parson, 1978, and below).

In fact, flash-induced absorbance changes resulting from excitation of *Rds. rubrum* reaction centers with 30 psec 880 nm pulses indicate that the B may serve as an additional electron carrier, between P* and BPh. Shuvalov et al. (1978) and Kryukov et al. (1978) observed initially no bleaching in the PBh bands at 540 and

760 nm, but rather absorbance increases, which then bleached with ∼ 30 psec kinetics. An extra component to the initial bleaching in the 800 nm band was also reported, which decayed with a time constant of ∼ 30 psec. The authors interpreted these observations as reflecting the transient reduction of B, which then passed the electron to BPh with a 30 psec time constant. The electron was found to move from BPh⁻ to Q with a time constant of 250 ± 50 psec, in agreement with the previous observations. The fact that bleaching at 540 and 760 nm was observed within the response time (∼ 10 psec) of previous measurements (Fig. 8A) employing the shorter 8 psec excitation pulses in the visible, was interpreted by Shuvalov et al. (1978) to indicate that excitation of BPh at his wavelength, followed by rapid energy transfer to P, masked the absorbance changes resulting from an intermediate step involving BChl. These conclusions are supported to some extent by additional picosecond studies (Akhamanov et al., 1980) and a nanosecond investigation (Shuvalov and Parson, 1981) of *Rps. sphaeroides* reaction centers having Q reduced.

3.2.2. Multiphoton absorption

Several groups of investigators have found that additional short-lived absorbance changes near 800 nm (and near 830 nm in *Rps. viridis*) are produced when reaction centers are excited with an excessive number of photons. Studies with 8 psec flashes at 530 or 600 nm by Rockley et al. (1975) and Holten et al. (1980a) have revealed a component to the bleaching at 800 nm in *Rps. sphaeroides* that recovers with a 30 psec time constant. A similar observation was made at 830 nm in *Rps. viridis* (Holten et al., 1978a). Studies of the initial amplitude of the absorbance decrease at 800 nm in *Rps. sphaeroides* as a function of either 530 or 600 nm excitation intensity indicate that the state responsible for the 30 psec step does not have the same quantum yield as the formation of P^+BPh^- and that the state also forms when P is oxidized prior to excitation (Holten et al., 1980a). These observations suggest that the transient does not lie along the primary electron transfer sequence. Akhamanov et al. (1978) also concluded that at least part of the transient bleaching observed near 800 nm following excitation of *Rds. rubrum* reaction centers with intense 50 psec 870 nm flashes was due to a secondary reaction. Akhamanov et al. (1980) studied two photon effects in *Rps. sphaeroides* reaction centers in which electron transfer from BPh to Q was blocked. They proposed that the intermediary state $[P^+(B\ BPh)^-]$ would absorb a second 880 nm photon, shifting most of the electron density onto BPh, resulting in additional bleaching in the BPh band at 760 nm.

It is clear that low excitation intensities must be used if meaningful conclusions are to be drawn from the near-infrared absorption changes. This region of the spectrum is important in determining the participation and rates of reaction of B, BPh and also possibly the fourth BChl and second BPh in the initial electron transfer steps.

3.2.3. Picosecond studies with pre-reduced electron acceptors

As discussed above, the electron acceptor complex (B BPh) can be trapped in the reduced state if reaction centers having Q pre-reduced are illuminated continuously in the presence of reduced cytochrome (cf. Tiede et al., 1976; Shuvalov and Klimov, 1976; Okamura et al., 1979). Under these conditions most of the added electron density probably resides on the BPh (Davis et al., 1979; Okamura et al., 1979). Thus, prior to flash excitation, the reaction centers are in the state (cyt$^+$ P B BPh$^-$Q$^-$). One might envisage that excitation of P in such reaction centers could cause electron transfer from P* to B. Evidence for the reduction of B was not found in picosecond studies of reaction centers of *Rps. viridis* (containing bound cytochrome) (Holten et al., 1978a) or *Rps. sphaeroides* (with added cytochrome *c*) (Schenck et al., 1981a) having both BPh and Q reduced.

In *Rps. viridis* neither the extra bleaching at 830 nm that would be expected if B were reduced to B$^-$ (Holten et al., 1978a), nor the absorbance increase at 1310 nm expected if P were oxidized to P$^+$ (Netzel et al., 1977), was observed. Another short-lived state was found that decayed with a 20 psec time constant (Holten et al., 1978a). On the basis of the transient spectrum, it was suggested that the state might be the excited singlet state P* and that interactions between P* and BPh$^-$ were responsible for the unexpectedly short decay time.

Similar studies on *Rps. sphaeroides* (Schenck et al., 1981a) revealed a new state that did not have a sharp bleaching at 800 nm, which one might expect if P$^+$B$^-$ were generated. The transient was found to decay with a time constant of 340 psec. It was concluded that the transient was not P*, since the fluorescence yield in the reaction centers containing reduced BPh and Q did not increase significantly over the yield obtained when BPh and Q were in their normal state prior to excitation. The new transient state was formed with high quantum yield, suggesting that it might be an intermediate in the primary electron transfer sequence. One possibility suggested was that it was a charge transfer state in which an electron has passed from one of the BChls of P to the other. However, several alternatives, including a triplet state, could not be excluded. These studies suggest that B is not an intermediary electron acceptor. However, some degree of caution must be taken on this interpretation in view of the strong interactions between P and B and between B and BPh. Prior reduction of BPh could disrupt these interactions and prevent electron transfer from P* to B.

3.2.4. *Subpicosecond investigations*

Holten et al. (1980a) studied the absorbance changes following excitation of *Rps. sphaeroides* reaction centers with excitation flashes lasting 0.7 psec and at 610 nm. The initial ($\sim$ 1 psec rise time) absorbance changes included a broad absorbance increase across the entire visible spectrum (480–720 nm), broken by a small dip (but still an absorbance increase) centered at 545 nm (Fig. 9). Spectra in the 750–850 nm region were also measured. With a time constant of approximately 4 psec, the dip at 545 nm was replaced by a strong bleaching, and the broad absorbance increase in the 650–680 nm region became fully developed (Fig. 10). The absorbance changes were attributed to the reduction of BPh to BPh$^-$. Since the initial broad absorbance increases appeared to decay in parallel with the reduction of BPh, it was suggested that they might be due to P* and that electron transfer could occur directly from P* to BPh.

However, that the initial absorbance increases were due, at least in part, to P^+B^- could not be excluded, as long as electron transfer from P^* to B is complete in less than 4 psec. Still another possibility is that excited states of B are involved, since the 0.7 psec 610 nm excitation flashes are probably absorbed by B as well as by P. Following excitation of B, B^* could transfer an electron to BPh, followed by electron transfer to B^+. The resultant final state would still be P^+ B BPh^-, but absorbance changes that otherwise would have occurred if B^- had been formed would not have been observed.

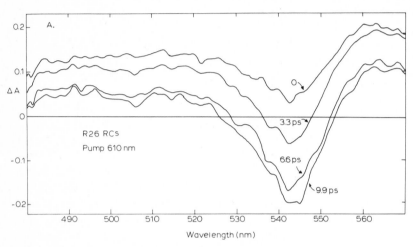

Fig. 9. Spectra of transient absorption changes observed at various delay times following excitation of *Rps. sphaeroides* reaction centers with 0.7 psec 610 nm pulses. Spectra taken at delays of 13.2 and 16.5 psec (not shown) were the same within error as shown for a 9.9 psec delay. The bleaching near 545 nm is due to the reduction of BPh (Holten et al., 1980a).

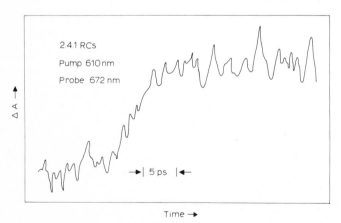

Fig. 10. Kinetics of the absorbance increase at 672 nm produced by excitation of *Rps. sphaeroides* reaction centers with 0.7 psec pulses at 610 nm. The absorbance increase was attributed to BPh^- (Holten et al., 1980a).

3.2.5. Temperature-dependent electron transfer

The observation by DeVault and Chance (1966) that the rate of electron transfer from the low potential cytochrome to P^+ in *Chromatium vinosum* chromatophores is temperature dependent has led to a number of theoretical approaches to the problem of electron transfer in biological systems. Blankenship and Parson (1978) have reviewed most of the pertinent early literature. More recent articles are referenced in the following discussion.

Sarai (1979, 1980) has treated the most general situation of non-adiabatic multiphoton theory (Jortner, 1976), with an arbitrary number of both molecular (quantum) modes and low-frequency (soft) solvent modes which can be activated during the electron transfer processes (see also Jortner, 1980; Kakitani and Kakitani, 1981). This description introduces the possibility that low-frequency solvent (protein and lipid) vibrations become progressively more important as the temperature

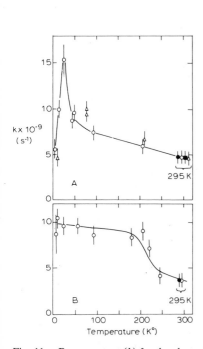
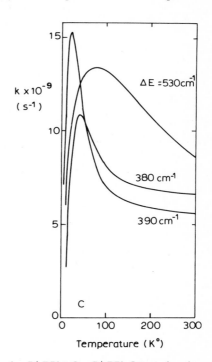

Fig. 11. Rate constant (k) for the electron transfer reaction $P^+ BPh^- Q \rightarrow P^+ BPh Q^-$ as a function of temperature. k was calculated from measurements of absorbance changes at 650 nm. Open symbols are for reaction centers in polyvinyl alcohol, and filled symbols are for reaction centers in TL buffer. Panel A: triangles represent ^1H 2.4.1 reaction centers, circles represent ^1H R-26 reaction centers, and the square represents ^1H R-26 reaction centers in ^2H_2O TL buffer. Panel B: ^2H R-26 reaction centers. The length of the error bars are 2 times the derived sample standard deviation of k. The curves are drawn to connect the points shown with open symbols, and do not represent theoretical fits. Panel C: Theoretical plots of k as a function of temperature, calculated from Eqn. 1 of Sarai (1980) $J = 10$ cm^{-1}; $H\omega_1 = 400$ cm^{-1}; $h\omega_2 = 50$ cm^{-1}; $h\omega_3 = 10$ cm^{-1}; $S_1 = S_2 = S_3 = 1$; ΔE varied as indicated. A positive ΔE means that the reaction is exergonic (Schenck et al., 1981b).

is taken sufficiently low, as the higher-frequency vibrational modes are "frozen-out". An interesting prediction of Sarai's theory is that at very low temperatures, the rate of electron transfer might reverse its temperature dependence. Recently, such a turning-point was experimentally observed in a picosecond study of the temperature dependence of the BPh^- to Q reaction in polyvinyl alcohol films of *Rps. sphaeroides* reaction centers (Schenck et al., 1981b). The experimental data for protonated and for partially deuterated reaction centers are presented in Fig. 11A and B. These results are in disagreement with a previous, less extensive, series of measurements on protonated reaction centers by Peters et al. (1978), who concluded that this reaction was independent of temperature.

Fig. 11C shows several curves computed to fit the kinetics in protonated reaction centers. These were calculated from the non-adiabatic electron transfer theory of Sarai (1979, 1980) with the inclusion of one quantum vibration of 400 cm^{-1} and two soft (protein or solvent) modes of 10 and 50 cm^{-1}. ΔE is the change in electronic potential energy that accompanies the reaction, a value that is not well known for the BPh^- to Q reaction. Although the fit is very good for $\Delta E = 400$ cm^{-1}, it is not a unique solution. Different combinations of high and low-frequency modes, coupling strengths, and ΔEs yield theoretical curves that reasonably reproduce the experimental data for protonated reaction centers shown in Fig. 11A. Similar considerations apply to fitting the data for deuterated reaction centers. More extensive experimental investigations are needed to determine the frequencies, overlap factors and the number of molecular and protein vibrations involved more precisely, and to determine which theoretical approaches are most appropriate. Such studies may provide more information about the molecular and environmental factors that influence the rate and yields of the initial electron transfer reactions.

3.3. Green-plant photosynthesis

Each of the two photosystems in green plants absorb photons and cooperate in the photosynthetic process via the "Z scheme" (Fig. 12). The strong reductant produced by photosystem I is used to reduce NADP, and ultimately to reduce carbon dioxide. The strong oxidant generated by photosystem II drives the oxidation of water to molecular oxygen, a process that appears to involve a manganese enzyme. The weak reductant of PSII is connected to the weak oxidant of PSI via an electron transport chain involving plastocyanin and cytochrome f.

The light-harvesting pigments of PSI are mainly Chl a and the bulk of the carotenoids, while those of PSII are mainly Chl a, Chl b, and a few carotenoids. In addition, both photosystems contain various accessory pigments which, together with the other antenna pigments, harvest the incident photons and transfer the energy to the reaction centers, where the initial charge separation processes take place. The reaction centers of both PSI and PSII appear to contain only Chl a.

Only recently have picosecond absorption techniques been applied to the primary electron transfer reactions in green-plant photosynthesis. Undoubtedly this has resulted from the difficulty in obtaining purified reaction centers of PSI and

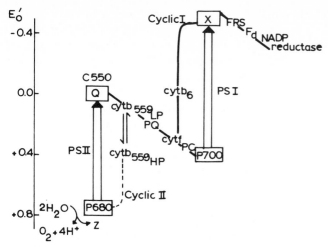

Fig. 12. The Z scheme for flow in green-plant photosynthesis. Studies on intermediates in both PSI and PSII charge separation reactions are discussed in the text. (Williams, 1977)

PSII. The ultrafast spectroscopic studies on PSI particles have been carried out on triton-fractionated PSI subchloroplast fragments (TSF-I) containing 30–40 antenna chlorophylls per reaction center.

3.3.1. Photosystem I

The primary electron donor of PSI is called P700, because of the position of its long-wavelength absorption band. Photochemical oxidation of P700 to $P700^+$ results in bleaching at 700, 680 and 430 nm, broad absorbances through the visible and in the region of 815 nm (cf. Shuvalov et al., 1979a). The band at 815 nm is characteristic of a Chl a^+ radical. The 700 and 680 nm bands exhibit similar kinetic and redox properties and are believed to represent two excitonic components of P700 (cf. Katz and Norris, 1973). Circular dichroism measurements and comparison of ESR and ENDOR data with those of Chl a in vitro are consistent with a dimeric structure for P700 comprised of two interacting molecules of Chl a (Phillipson et al., 1972; Norris et al., 1974; Parson and Ke, 1981).

The identity of the "primary" electron aceptor in PSI has been the subject of much debate, but there is evidence from ESR studies that it is a bound iron-sulfur protein, a ferridoxin (Malkan and Bearden, 1971). In fact, it appears that there are two stable primary electron acceptors in PSI and that both are iron-sulfur centers (ISCs), termed ISC-A and ISC-B. A difference spectrum for reduction of one acceptor in PSI includes bleaching at 430 nm detectable at room temperature and is, therefore, called P430. This absorbance change and the redox properties of the species giving rise to it are also consistent with a bound ferridoxin. Whether or not the optical and ESR observations belong to the same acceptor has been difficult to determine, since the optical measurements are typically carried out at room temperature, while the ESR studies are made at cryogenic temperatures, where back reactions are hindered (Parson and Ke, 1981). Although the measurements are in agreement with the assignment of P430 as one of the final stable electron acceptors in PSI reaction center photochemistry, the results of recent kinetic studies using quick-freeze techniques coupled with low-temperature ESR suggest that P430 may in fact lie before the primary acceptor(s) in the electron transfer sequence (Hiyama and Frock, 1980).

Picosecond studies have been undertaken to examine the initial photooxidation and any intermediary electron carriers between P700 and the final ISC. When PSI subchloroplast fragments were excited with an 8 psec flash from a mode-locked Nd:glass laser, an absorbance decrease at 700 nm appeared within 10 psec (Fenton et al., 1979). This was attributed to the photooxidation of P700. The initial absorbance changes also contained an absorbance increase over the range 740–840 nm. At 800 nm the signal partially relaxed with a time constant of about 40 psec. This fast step was believed to be due to the antenna pigments, since the process appeared sensitive to whether or not P700 was oxidized with background illumination prior to flash excitation. A component with a 40 psec decay at 730 nm was assigned to reduction of an acceptor during the oxidation of P700.

Shuvalov et al. (1979b) studied TSF-I particles with 50 psec, 694 nm pump and probe flashes from a mode-locked ruby laser. The initial bleaching observed at 694 nm was twice as large as that attributable to P700 oxidation alone. This was accounted for by the reduction in <60 psec of an acceptor $A_{I,1}$ that had an extinction coefficient equal to P700 at 694 nm. The 694 nm bleaching decayed to the level for the steady-state concentration of $P700^+$ with a time constant of about 200 psec at room temperature, taken to reflect transfer of the electron from $A_{I,1}^-$ to the next acceptor, $A_{I,2}$, believed to be an iron-sulfur protein (see below). The observations were taken to arise from PSI reaction center photochemistry, since the initial absorbance change was drastically reduced when P700 was maintained in the oxidized state with weak illumination prior to and during the 50 psec flash.

When the bound iron-sulfur proteins were chemically reduced by poising the TSF-I particles at very low potential, decay of the 694 nm bleaching at room temperature became biphasic, the components having time constants of 10 nsec and 3 μsec. The 10 nsec lifetime was ascribed to the sum of two processes, decay of $[P700^+ A_{I,1}^-]$ via charge recombination within the radical pair, as well as interconversion to a state with some triplet character (possibly a triplet radical pair) that decayed in 3 μs at room temperature and 1.3 msec at 5°K (Shuvalov et al., 1979a; Mathis et al., 1978). Comparison of the spectrum of the 1.3 msec component at 5°K with those of model systems (Fujita et al., 1978) has suggested that the intermediary electron carrier $A_{I,1}$ is a Chl a dimer (Shuvalov et al., 1979a, b). ESR studies also indicate that $A_{I,1}$ is some form of Chl a (Freisner et al., 1979).

This view of $A_{I,1}$ has received support from recent picosecond investigations of TFS-I particles employing 689 and 708 nm excitation flashes of 30 psec duration (Shuvalov et al., 1979c). Pulses at these wavelengths are produced by stimulated Raman scattering in C_2D_5OD and liquid nitrogen, respectively, from the 532 nm second harmonic of a Nd:YAG laser. (The 689 nm radiation is the second harmonic of the Raman shift.)

Bleaching at 694 nm appeared within 30 psec of excitation with a 708 nm pulse and showed a biphasic decay (Fig. 13A (a)). The two components disappeared with time constants of 45 and 210 psec. Kinetics at 484 nm (Fig. 13B) showed only a 210 psec component, as did those at 800 nm. When P700 was preoxidized by continuous illumination, the 210 psec component was reduced in amplitude, but the 45 psec component was unaltered (Fig. 13A (b)). Comparison of these observations with the previous picosecond investigations, suggests that the 45 psec step was due to the antenna Chls (Fenton et al., 1979) while the 210 psec decay accompanied transfer of the electron from $A_{I,1}^-$ to $A_{I,2}$ (Shuvalov et al., 1979b). The quantum yield for the formation of $P700^+$ was determined to be about 0.9 for low-intensity flashes.

Difference spectra measured 150 and 800 psec after excitation with 30 psec flashes at 688 and 708 nm are shown in Fig. 14A, B (Shuvalov et al., 1979c). The dashed spectrum is that for P700 oxidation by continuous illumination, and is seen to coincide reasonably well with the spectrum measured at 800 psec. Fig. 14C shows the difference between the 150 and 800 psec difference spectra, which was taken to give the difference spectrum for the reduction of $A_{I,1}$. This reasoning is based on assumptions that the 150 and 800 psec spectra contain equal contributions from the photooxidation of P700 $[P700^+ - P700]$, and that both spectra have equal (or negligible) contributions for the reduction of the second electron acceptor, $A_{I,2}$, over the regions covered by the picosecond measurements. The dashed curve in Fig. 14C is the difference spectrum for the formation of Chl a^- in vitro, shifted to the red by

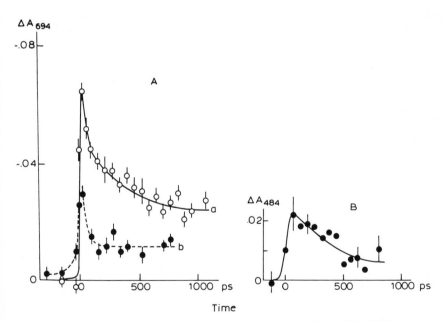

Fig. 13. Kinetics of the absorbance changes at 694 nm (A) and 484 nm (B) in TSF-I particles following excitation with 30 psec 708 nm flashes. The sample for A(b) had its P700 pre-oxidized by continuous illumination (Shuvalov et al., 1979c).

314

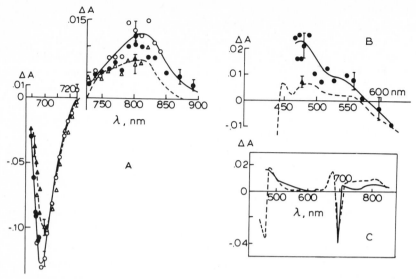

Fig. 14. Spectra of the absorbance changes in the red (A) and shorter-wavelength (B) regions measured 150 psec (O and ●) and 800 psec (△ and ▲) after excitation with 30 psec 708 nm (O and △) or 689 nm (● and ▲) pulses in TSF-I particles. The dashed curve shows the difference spectrum of P700 measured in the same sample by continuous illumination. Panel C shows the difference between the 150 and 800 psec spectra; the dashed curve is the difference spectrum for the formation of Chl a anion radical, shifted toward the red by $\sim$ 30 nm to coincide with the measured difference spectrum (Shuvalov et al., 1979c).

about 30 nm so that the bleaching at 700 nm would coincide with the picosecond results. As in earlier studies (Shuvalov et al., 1979a, b) the necessity of red shifting the in vitro Chl a^- spectrum (Fujita et al., 1978) to align with the in vivo spectrum at 700 nm was taken to indicate that $A_{I,1}$ is a Chl a dimer. However, recent ESR and ENDOR work suggest that $A_{I,1-}$ is more likely monomeric Chl a^- (Fajer et al., 1980). If this is correct, then the red shift of the $A_{I,1}$ bands in vivo is probably due to interactions with the protein and other components in the reaction center.

Optical and ESR studies have revealed a second electron carier ($A_{I,2}$) between P700 and the stable primary acceptor. This species has a lifetime of 250 μsec at 20°C and 130 msec at 5°K (Mathis et al., 1978; Shuvalov et al., 1979a; Parson and Ke, 1981). Spectra ascribed to the reduction of $A_{I,2}$ in TSF-I particles at low potential show bleaching near 450 nm and the development of bands near 670 and 690 nm that cannot be accounted for by P700 photooxidation alone. These optical data correlated with the EPR signal of a species called X (McIntosh et al., 1974), indicating that $A_{I,2}$ and X are the same component, probably an iron-sulfur center (Shuvalov et al., 1979a). Recent EPR studies suggest that $A_{I,2}$ and P430 may be the same species (Hiyama and Frock, 1980). The time of electron transfer from $A_{I,2}$ to the next iron-sulfur center is thought to be < 100 nsec (Parson and Ke, 1981).

3.3.2. Photosystem II

The primary electron donor in PSII is called P680, and is less well characterized than P700 of PSI. Excitation of PSII or chemical oxidation with ferricyanide causes bleaching at 680 and 435 nm, and absorbance increases at 400 and 825 nm. The latter probably reflects formation of a Chl a cation. Comparison with the difference spectrum for the oxidation of the BChl a dimer has led to the speculation that P680 is a Chl a dimer also (Van Gorkom et al., 1974, 1975; Borg et al., 1970). ESR results also suggest that P680$^+$ may be a Chl a^+ dimer (Malkan and Bearden, 1971).

The primary electron acceptor of PSII reaction centers is now thought to be plastocyanin (PQ). There is evidence that two reaction center quinones are alternatively reduced and shuttle electrons to the external plastocyanin pool, in analogy to bacteria (Van Gorkom et al., 1974). Optical and ESR studies on PSII fragments having PQ chemically reduced, indicate that pheophytin (Ph) is an electron carrier between P680 and PQ (Klimov et al., 1977, 1980; Shuvalov et al., 1980). All of these observations are similar to those made in bacteria.

Excitation of TSF-II particles, having PQ chemically reduced, with 5 nsec flashes at 694 nm from a Q-switched ruby laser produce absorbance changes (Fig. 15) ascribed to the formation of the radical pair [P680$^+$ Ph$^-$], in analogy to state P^F in bacteria (Shuvalov et al., 1980). The radical pair, and the delayed fluorescence, decay with a time constant of about 4 nsec at room temperature as the electron returns from Ph$^-$ to P680$^+$. When Q was in its normal oxidized state prior to flash excitation, the transient absorbance changes (and the delayed fluorescence) were reduced at least 10-fold. These observations suggested that electron transfer from Ph$^-$ to PQ occurred in less than 400 psec. Picosecond-resolved measurements on the formation and decay of [P680$^+$ Ph$^-$] in TSF-II particles have not been made at the time of writing of this review. However, it appears that the photochemical be-

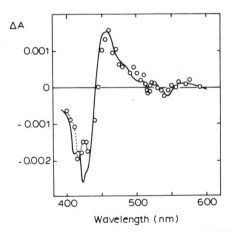

Fig. 15. Spectrum of absorbance changes in TSF-II particles with a lifetime of approximately 4 nsec. Reaction centers had the primary acceptor pre-reduced and were excited with 5 nsec flashes at 694 nm from a Q-switched ruby laser. Solid curve shows the sum of spectra for P680$^+$ and Ph$^-$ formation, normalized at their red maximal bleachings (Shuvalov et al., 1980).

316

havior in reaction centers of PSII is very similar to that observed in bacterial reaction centers.

3.4. Emission studies

The majority of picosecond fluorescence measurements on preparations from photosynthetic organisms have focused on the rates and mechanism of energy transfer in the antenna systems. A continued re-evaluation of previous experimental observations has taken place as new studies are performed, and as a better understanding of multiphoton effects on fluorescence lifetimes and quantum yields emerges. Much of the early literature in this field has been critically reviewed by Campillo and Shapiro (1977), Holten and Windsor (1978), and Pellegrino et al. (1978).

The first emission lifetimes from antenna systems were made with phase fluorometers, nanosecond flash techniques, or continuous excitation with trains of pulses from mode-locked ion lasers. Fluorescence lifetimes ranging from several hundred picoseconds to several nanoseconds have been obtained in a number of different preparations from bacteria (Merkelo et al., 1969; Govindjee et al., 1972, 1975), algae (Mar et al., 1972), and green plants (Müller et al., 1969; Briantanis et al., 1972; Moya et al., 1977). Higher light intensities tend to reduce the fluorescence lifetimes and quantum yields. Background illumination, chemical poisoning, or low temperatures increase the lifetimes by hindering and/or blocking reaction center photochemistry, thus preventing efficient trapping of the excitation as it migrates through the antenna systems. In the low-intensity limit, the fluorescence lifetimes obtained for green plant intact cell or chloroplast preparations tend to center about the value of 600 psec (Pellegrino et al., 1978).

Borisov and Godik (1972a, b), and Borisov and Ilina (1973) observed components with emission lifetimes of about 50 psec in a number of bacterial and green plant PSI preparations with a phase fluorometer. Such short lifetimes were interpreted as a rising from exciton diffusion in the light-harvesting pigments and trapping in the reaction center.

Short-lived components having emission lifetimes of 100 psec and less have also been obtained in picosecond studies employing Kerr gates or streak cameras (section 2.5.2). Exciting with trains of 30 psec pulses from a mode-locked Nd:YAG laser and detecting fluorescence with a Kerr gate, Seibert et al. (1973) and Seibert and Alfano (1974) observed two component decays at room temperature with lifetimes of 10 psec and 200–300 psec in spinach or escarole chloroplasts. The two components were attributed to PSI and PSII respectively. These results were supported by similar investigations by Yu et al. (1975) on enriched PSI and PSII preparations from spinach chloroplasts. Yu et al. (1977) made similar observations and, in addition, found a dependence on detection wavelength and on temperature. More details are reported by Pellegrino et al. (1978).

Early investigations with streak cameras also employed pulse trains for excitation. A pulse near the center of the train was used to trigger the streak camera, and the fluorescence excited by one of the pulses was observed. Kollman et al. (1975) and Shapiro et al. (1975a) found a concentration dependence of the fluorescence lifetimes of Chl a and Chl b in vitro, decreasing to about 10 psec at 1.1 M concentration. In addition, they observed fluorescence decays of 75 and 41 psec from cells of the blue-green alga *Anacystis nidulans* and the green alga *Chlorella pyrenoidosa*, respectively. These results were interpreted as arising from concentration quenching via a Forster energy transfer process both in vivo and in vitro. Beddard et al. (1975) found lifetimes on the order of 100 psec in several algae preparations and in spinach chloroplasts. They proposed a specific form of concentration quenching (see Beddard and Porter, 1977; Altmann et al., 1978). Paschenko et al. (1975) observed a three-component decay in pea chloroplasts, attributed to PSI (80 psec), PSII (300 psec), and free Chl (4.5 nsec).

Consideration of multiphoton effects was prompted by the work of Mauzerall (1976a, b) who observed a decrease in fluorescence quantum yield from Chlorella following excitation with 7 nsec flashes. Mauzerall interpreted the results in terms of excited state interactions and multitrap effects. Campillo et al. (1976a, b) made similar observations with single 20 psec excitation flashes. They argued that with high-intensity single pulses a singlet–singlet annihilation process resulted in the short ($<$ 100 psec) lifetimes. Using a Stern–Volmer relationship to describe this process, along with quantum yield and streak camera data, they obtained a value of 650 ± 150 psec for the Chl lifetime in vivo, in good agreement with the phase fluorometer (low-intensity) measurements discussed above.

Subsequently, excited state annihilation effects on lifetimes and quantum yields were investigated extensively. Breton and Geacintov (1976), Geacintov and Breton (1977), and Geacintov et al. (1977) studied emission from spinach chloroplasts with picosecond pulse trains of various lengths and intensities, single picosecond pulses, and microsecond flashes. Some of these studies employed a passively mode-locked flashlamp-pumped dye laser. They concluded that singlet–triplet fusion or singlet–ion interactions reduced emission lifetimes in multiple-pulse (and long-duration single-pulse) experiments while singlet–singlet annihilation predominated in studies with high-intensity, short-duration single pulses. Preferential quenching of PSII fluorescence was observed in multiple-pulse experiments. An exciton annihilation theory (Swenberg et al., 1976b) was used to explain some of the results. Swenberg et al. (1976a) studied the effect of pigment heterogeneity on emission lifetimes and yields for PSI and PSII. The light-harvesting system of PSI is mainly Chl a, whereas that for PSII contains Chl a and Chl b. The PSII fluorescence lifetimes were longer than those found for PSI. A Chl b "antitrap" argument was proposed.

Studies on chloroplasts of Chlorella by Harris et al. (1976), Searle et al. (1977) and Tredwell et al. (1977) and on enriched PSI and PSII preparations from spinach chloroplasts by Porter et al. (1977) gave further evidence for the effect of excited state interactions at high pulse intensities. The possibility of singlet–singlet, singlet–ion, and singlet–triplet interactions were examined with several different modes of excitation. Studies were made at room temperature and at 77°K. It was shown that at single-pulse excitation intensities of 3×10^{13} photons/cm^2 or less, the observed

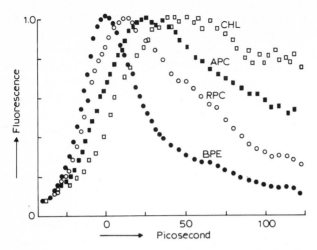

Fig. 16. Energy transfer in the antenna system of *Porphyridium cruentum*, measured with a streak camera. BPE (phycoerythrin) was excited with a 6 psec 530 nm pulse. It fluoresces at 578 nm and decays by transfer in 70 psec to RCP (phycocyanin) which fluoresces at 640 nm. This transfers in 90 psec to ACP (allophycocyanin) which fluoresces at 660 nm. Finally, this transfers in 120 psec to Chl *a* which fluoresces at 685 nm and decays in 175 psec by energy transfer to the reaction center (trap) (Porter et al., 1978).

lifetimes were comparable to those obtained by low-intensity methods. Most of the fluorescence in these studies could be accounted for by an expression of the form $I(t) = I_0 \exp\left[-At + Bt^{1/2}\right]$. The authors argue that this decay law is expected for random diffusion via Forster energy transfer of excitations in a single-harvesting array, with trapping at random sites.

Recently, Geacintov et al. (1979) demonstrated that the reduction in quantum yields with increasing excitation intensity observed in the in vivo studies was not due to excited state absorption and stimulated emission, as had been suggested to explain the observation of stimulated emission by Chl and BChl in solution (Hindman et al., 1977, 1978).

The rate of energy transfer between the accessory harvesting light pigments, and that of transfer to the PSII reaction center in the red alga, *Porphyridium cruetum*, has been studied by Porter et al. (1978) and Searle et al. (1978). Fig. 16 shows the emission risetimes and decay kinetics at room temperature. The pigment B-phycoerythrin was excited directly with a single 6 psec, 530 nm pulse from a (frequency-doubled) mode-locked Nd:glass laser. The emission lifetimes for the sequence were found to be B-phycoerythrin (70 psec), R-phycocyanin (90 psec), allophycocyanin (118 psec), and Chl *a* (175 psec). The emission risetimes for the last three pigments were also resolved. Only B-phycoerythrin showed a non-exponential decay. When the PSII reaction centers were closed by chemical treatment and preillumination, the Chl *a* decay became exponential with a lifetime of about 840 psec. Brody et al. (1981) made similar observations and extended them to *Anacystis nidulans*. The energy transfer rates were found to be temperature dependent.

Barber et al. (1978) studied the effect of $MgCl_2$ on the Chl fluorescence yield and lifetime in broken chloroplasts in which PSII reaction centers were closed. Upon addition of 10 mM $MgCl_2$, the fluorescence yield doubled and the decay rate decreased. The fluorescence decay changed from the $t^{1/2}$ behavior in the absence of cation, to a single exponential with a time constant of 1.6 nsec. The results were taken to support the hypothesis that cations alter the rate of energy transfer from PSII to PSI (spillover), rather than to changes in the original partitioning of the excitation energy between the two photosystems.

Bacterial systems have also been investigated with mode-locked lasers and streak cameras. Campillo et al. (1977) studied the fluorescence quantum efficiency versus excitation intensity in chromatophores from several strains of photosynthetic bacteria, including one strain that lacked reaction centers. Analysis of the results with a Stern–Volmer approach to the exciton annihilation process supported the "lake" model of photosynthesis, in which the antenna system of the chromatophores is pictured as an extended array that supplies excitons to many different reaction centers. In the alternate "puddle" model each independent photosynthetic unit has its own reaction center and associated antenna system. Previous investigations with other techniques have also supported the lake model.

Bacterial reaction center fluorescence lifetimes have been reported by Paschenko et al. (1977) and Rubin (1978). Isolated reaction centers from *Rps. sphaeroides* excited with a 694 nm pulse train from a mode-locked ruby laser exhibited two components with lifetimes of 15 ± 8 and 250 psec. In oxidized reaction centers the fast component was absent and the longer-lived one increased to about 700 psec. The 15 psec decay was attributed to P*, and the slower one to BPh (and/or B). Chromatophores and whole cells were also studied. A 200 psec component was attributed to the antenna system.

3.5. Model systems

Studies of electron transfer from photoexcited chlorophylls and related porphyrins to electron acceptors in solution have been carried out to probe various aspects of photosynthetic electron transfer and to devise model systems in which unwanted reverse electron transfer processes are inhibited (Tollin, 1976; Holten and Windsor, 1978; Seeley, 1978; Blankenship and Parson, 1979). In solution, diffusion of the reactants and products are important factors in determining the rates and yields of forward and reverse reactions. Diffusion of the excited donor and acceptor to within the critical distance required for reaction is not necessary in model systems where the reactants are covalently linked. Neither is it required in reaction centers, since the chromophores are bound to proteins. In the reaction center an additional avenue open to the radical pair is, of course, subsequent electron transfer to the next acceptor. However, in both covalently linked complexes and in the reaction center some reorientation of the pigments may be required before electron transfer can take place.

The immediate products of electron transfer from either excited singlet or triplet state donors to suitable electron acceptors are singlet or triplet radical pairs involving the oxidized donor and the reduced acceptor. It has been suggested that reverse electron transfer within singlet radical pairs can be extremely fast due to good vibrational overlap with the ground state, if there is a shape change in one or both of the participants (Gouterman and Holten, 1977). This is proposed to be the case in model systems with small electron acceptors such as quinones, which are known to have relatively large changes in bond lengths and angles upon being reduced (Efrima and Bixon, 1974). Using picosecond spectroscopy, Holten et al. (1976) observed that although p-benzoquinone efficiently quenched the excited singlet state of BPh, charge-separated ions could be detected only from the triplet state reaction. Radical ions were not detected following quenching of BPh* by m-dinitrobenzene, an acceptor whose redox potential is not low enough to permit reaction with BPh^T (Holten et al., 1978b). The dication methyl viologen, has a one-electron reduction potential similar to that of m-dinitrobenzene. In solution studies, a low yield of radical ions could be detected in a reaction with BPh*, presumably because the singlet radical pair $[BPh^+ MV^+]$ is repulsive and the ions fly apart (Holten et al., 1978b). Huppert et al. (1976) were unable to detect ion formation from the excited singlet state of Chl a quenched by 2,6-dimethylbenzoquinone following picosecond excitation. However, it is well known from previous measurements that the excited triplet state electron transfer reactions of Chl a and BPh a lead to high yields of charge-separated ions (cf. Tollin et al., 1979; Castelli et al., 1979; Holten et al., 1976; Seeley et al., 1977). Reverse electron transfer within triplet radical pairs to give the ground singlet state is a spin-forbidden process, giving the ions a greater opportunity to separate (Gouterman and Holten, 1977). These observations are supported by slower time scale flash photolysis and magnetic resonance experiments (Lamola et al., 1975; Andreeva et al., 1978; Seeley, 1977, 1978; Harriman et al., 1979).

Resonance Raman and X-ray crystallographic measurements indicate that porphyrins and chlorophylls show comparatively smaller structural changes upon oxidation or reduction than do acceptors the size of quinone (Spaulding et al., 1974; Cotton and Van Duyne, 1978). The Franck–Condon factors for reverse electron transfer within the initial singlet radical pairs in bacterial reaction centers might be expected to be poor due to the large sizes of the molecules involved [BChl dimer (P), BChl, BPh; Fig. 6]. The excited singlet state of P is believed to be the initial electron donor. It is also possible that for certain orientations of the chromophores the forward reactions might be favored and reverse reactions inhibited because of the different types and symmetries of porphyrin orbitals involved in the two processes (Holten et al., 1978b; Blankenship and Parson, 1979). The environment in the reaction must also be considered, as movements by the charged groups on the protein accompanying electron transfer could inhibit the reverse process.

In support of one or more of the contentions, "stable" singlet radical pairs have been observed in recent studies of cofacial porphyrin dimers (Netzel et al., 1979). The dimers consist of two basic porphyrin subunits held in a face-to-face geometry 4 Å apart by a pair of bridging groups. Only excited singlet and triplet state difference spectra were observed in THF solutions for the free base dimers (H_2–H_2) or those with magnesium in both rings (Mg–Mg). However, for the mixed dimer (Mg–H_2) in CH_2Cl_2 containing tetraethylammonium chloride, which complexes with Mg and lowers its redox potential, absorbance changes attributable to a charge transfer (radical pair) state (Mg^+–H_2^-) were observed. It decayed with 620 psec kinetics. In the absence of the salt the lifetime was 380 psec, taken to reflect increased Coulomb attraction in the radical pair and a smaller energy gap between the radical pair and the triplet (Netzel et al., 1979).

The effect of redox potential (energetics), solvent dielectric constant, and geometry of the complex have also been shown to be important in picosecond studies of pyrochlorophyll (PChl) and pyropheophytin (PPheo) complexes. In an equilibrium mixture of the two molecules in the appropriate solvent, it was reported that electron transfer occurred from PChl to PPheo in $\sim$ 6 psec and that the radical pair lived for 10–20 nsec (Pellin et al., 1979). Netzel et al. (1980) investigated PChl and PPheo covalently linked dimers and trimers. Only in the case of $(PChl)_2$–PPheo complex was evidence found for a radical pair state. The photoproduct was formed in low yield and exhibited formation and decay kinetics of 110 psec and 3 nsec, respectively.

The observation in these various complexes of radical pair states having lifetimes of several hundred picoseconds or greater is significant, since in the studies of model systems in solution involving BPh and various small acceptors, it was estimated that deactivation by reverse electron transfer probably requires 10 psec or less (Holten et al., 1978b). In bacterial and plant reaction centers, the radical pairs consisting of the BChl (Chl) dimer and BPh (Ph and Chl) live for several hundred picoseconds under normal conditions and about 10 nsec when electron transfer to the next acceptor is blocked (sections 3.2 and 3.3).

Pellin et al. (1980) also showed solvent and temperature effects to be important in a study of the excited state absorption and decay kinetics of a pyrochlorophyllide dimer. No charge transfer states were reported, but a model incorporating a singlet manifold of one fluorescent and one non-fluorescent state was necessary to explain the experimental results.

Studies in micelles also appear promising toward understanding the role of the environment in assisting charge separation and hindering back reactions. It is well known that with the proper choice of donor, acceptor, and micelle, recombination of separated ions can be inhibited (cf. Fendler and Fendler, 1975). However, it is unclear from picosecond studies on photosynthetic pigments in micelles whether the formation of charge-separated ions from an excited singlet state reaction can be significantly increased (Kalyanasundaram and Porter, 1978; Holten et al., 1980b; Shah et al., 1980).

4. Visual pigments and the purple membrane

4.1. Rhodopsin

The visual pigment rhodopsin consists of a polyene chromophore called retinal complexed to an apoprotein called opsin. Free retinal absorbs in the region of 350 nm, but complexing with opsin shifts the absorption spectrum into the visible region. The active visual chromophore in vertebrates (cattle) and invertebrates (squid) is the 11-*cis* retinal and the primary step after light excitation is believed to be conversion to the all-*trans* configuration. Opsin can also accommodate the 9-*cis* form of retinal. This complex is called isorhodopsin. Photochemical studies are usually made on rod outer segments suspended in a detergent such as lauryl dimethylamine oxide (LDAO). The primary photochemical event is the generation of bathorhodopsin (formerly called prelumirhodopsin), probably by way of an earlier intermediate called hypsorhodopsin. A series of other red-shifted intermediates are formed in the dark and involve thermal rearrangements that culminate, for vertebrate rhodopsin, in release of retinal from opsin. During this process an ion is released which shuts off the in-flow of sodium ions through the plasma membrane, generating an electrical signal. The series of intermediates, shown in Fig. 17 for

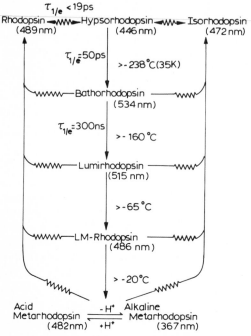

Fig. 17. Photobleaching processes of squid rhodopsin. Photochemical reactions are denoted by wavy lines and thermal (dark) reactions by straight lines. Absorption maxima are shown in parentheses (Kobayashi et al., 1978).

squid rhodopsin, can be trapped at low temperatures, or resolved by picosecond or slower time scale flash photolysis techniques at ambient temperature. For general reviews on rhodopsin photochemistry and models for the mechanism of visual transduction see Lewis (1978), Honig et al. (1979), Mathies (1979), Applebury (1980), or Birge (1981).

Investigators using picosecond spectroscopy have set out to answer a number of questions on the initial photochemical events. These include (1) whether bathorhodopsin is generated directly from the excited singlet state of rhodopsin or whether hypsorhodopsin is a necessary intermediate; (2) what changes in the chromophore accompany these steps (i.e. the structure of bathorhodopsin); (3) the formation kinetics of bathorhodopsin; (4) whether excitation of rhodopsin or isorhodopsin generates the same intermediates; and (5) similarities or differences in the initial photochemistry between squid and cattle rhodopsin.

Busch et al. (1972), Sundstrom et al. (1977) and Peters et al. (1977) reported that at room temperature bathorhodopsin forms within the instrumental resolution (~ 6 psec) of exciting cattle rhodopsin with a 6 psec, 530 nm pulse from a mode-locked Nd:glass laser. Bathorhodopsin was found to decay with a time constant of about 30 nsec at room temperature. At 20°K, Peters et al. (1977) could resolve the rise time of the absorbance changes attributed to bathorhodopsin. No evidence was found for a prior step involving hypsorhodopsin. From the temperature dependence of the formation kinetics and an observed deuterium isotope effect, it was proposed that proton translocation accompanied the formation of bathorhodopsin. Previous work had indicated that a *cis–trans* isomerization occurs, a process which Peters et al. (1977) suggested might be too slow to explain the ~ 6 psec formation of bathorhodopsin at room temperature. From model system studies on protonated 11–*cis* retinylidene Schiff bases, Huppert et al. (1977) concluded that either *cis–trans* isomerization does not accompany bathorhodopsin formation, or compexation with opsin accelerates the rearrangement. Calculations indicate that isomerization on the picosecond time scale is possible (Warshel, 1976; Birge and Hubbard, 1980). This view is supported also by the results of recent picosecond resonance Raman measurements by Hayward et al. (1980). Picosecond transient absorbance studies by Green et al. (1977), and Monger et al. (1979) indicate that (cattle) bathorhodopsin is formed within 3 psec of excitation of either rhodopsin or isorhodopsin with 530 nm flashes, and that the initial photochemical event involves *cis–trans* isomerization. The short formation time is consistent with the instrument-limited (< 12 psec) rise and decay of fluorescence attributed to excited states of rhodopsin and isorhodopsin observed by Doukas et al. (1981).

Picosecond studies on squid rhodopsin have been carried out mainly with 20 psec, 347 nm excitation pulses from a frequency-doubled mode-locked ruby laser (Shichida et al., 1977, 1978; Kobayashi et al., 1978; Suzuki et al., 1980b). Squid hypsorhodopsin was found to form within 19 psec of excitation, followed by conversion to bathorhodopsin in about 50 psec at room temperature. Subsequently, bathorhodopsin was found to convert to lumirhodopsin with a time constant of 300 nsec (Fig. 17). These times are slower than those just described for the cattle rhodopsin intermediates, where hypsorhodopsin was not observed and where bathorhodopsin was found to form in < 3 psec and decay in ~ 30 nsec. Kobayashi et al. (1978) suggested that these differences might be due to the difference in excitation wavelengths (347 versus 530 nm), detergents used, or to inherent differences in the opsin structure. However, Kobayashi (1979) reinvestigated the initial events in cattle rhodopsin with 530 nm, 6 psec flashes and found weak transient absorbance attributable to hypsorhodopsin (formation time < 10 psec) that decayed with the 50 psec kinetics found in squid to reflect conversion to bathorhodopsin. That hypsorhodopsin had not been found previously in cattle preparations was ascribed to the complicated spectral behavior at early times after excitation due to the presence of the excited singlet state of rhodopsin, hypsorhodopsin, and bathorhodopsin. Thus it appears that the scheme of Fig. 17 applies to both cattle and squid rhodopsin photochemistry, but with hyposorhodopsin forming in < 10 psec, and with bathorhodopsin being a distorted all-*trans* configuration, whose formation is accompanied by an isomerization followed by proton translocation.

4.2. *Bacteriorhodopsin*

In response to light and oxygen deprivation, bacteria such as *Halobacterium halobium* develop purple membrane patches that contain bacteriorhodopsin. This pigment is similar to rhodopsin in that it consists of a retinal bound via a Schiff base linkage to an opsin-like protein. Bacteriorhodopsin (bR) exists in two stable forms, a dark-adapted conformer bR_{548} containing 13-*cis* retinal and a light-adapted form bR_{568} in which retinal is in the all-*trans* configuration. (The subscript refers to the absorption maximum of the state.) Upon photon absorption, light-adapted bR_{568} undergoes a photochemical cycle consisting of several intermediates (Fig. 18) that concludes with the regeneration of bR_{568} in about 10 msec at room temperature.

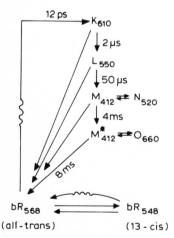

Fig. 18.   The photochemical cycle of light-adapted bacteriorhodopsin (bR_{568}). The exact chronology of the dark reactions is not yet precisely known. Vertical position in the figure is intended to approximately represent the relative free energy of the intermediate (Birge, 1981).

The first step, generating K_{610}, is induced by light while the remainder involve a series of conformational changes in the dark. During the cycle, a proton is released, setting up a potential as it traverses the membrane. Unlike cattle (vertebrate) rhodopsin, but like squid (invertebrate) rhodopsin, the cycle does not lead to the release of retinal from the protein. More details on the photochemical cycle and models for the light-driven proton translocation can be found in a number of articles (Warshel, 1978, 1979; Lewis, 1978; Honig, 1978; Ottolenghi, 1980; Birge, 1981).

Both picosecond transient absorption and emission measurements have been carried out to investigate the photochemical step in bacteriorhodopsin. Kaufmann et al. (1976b) observed absorbance changes attributed to the rapid formation (< 15 psec) of K_{610} following excitation with 6 psec flashes at 530 nm. An additional transient absorbing at 580 nm was tentatively assigned to the excited singlet state of bacteriorhodopsin. More detailed studies by Applebury et al. (1978) supported these conclusions and set the decay time for the excited state and the rise time of K_{610} formation at 11 psec. From the temperature dependence of the rate and a deuterium isotope effect, it was concluded that the formation of K_{610} involved proton transfer, which occurs by tunneling at low temperatures. Similar conclusions had been drawn on the initial step in rhodopsin (Peters et al., 1977; section 4.2). Kryukov et al. (1978) also found evidence for an excited state precursor to K_{610}. The formation time of K_{610} was found to be relatively constant (~ 80 psec) between liquid helium and liquid nitrogen temperatures, but decreased to less than 25 psec at room temperature. Using subpicosecond transient absorption techniques, Ippen et al. (1978) measured the formation time of K_{610} to be 1.0 ± 0.5 psec.

Alfano et al. (1976) were unable to resolve the emission lifetime of the bacteriorhodopsin-excited state at room temperature, but estimated it to be 3 psec using excitation with 530 nm pulse train and Kerr gate detection. This estimate was based on a lifetime of 40 psec measured at 90°K and relative quantum yields determined at the two temperatures. Hirsch et al. (1976) reported an emission lifetime of 15 psec at room temperature, using pulse train excitation and detection with a frequency upconversion gate (sections 2.4 and 2.5.1). Shapiro et al. (1978) made a detailed study of the emission with single 532 nm excitation pulses and streak camera detection. Possible multiphoton effects were investigated. Together with a lifetime of 60 psec at 77°K and a study of the emission quantum yield as a function of temperature, it was concluded that the data were consistent with the 1 psec formation time for K_{610} reported by Ippen et al. (1978).

Thus, the initial photochemical step in rhodopsin and bacteriorhodopsin appears similar in some respects and different in others. The first intermediate in both systems is formed via an excited singlet state of the stable precursor in a few picoseconds or less and involves an isomerization followed by proton transfer. In rhodopsin this process proceeds from a stable 11-*cis* retinal complex, through hypsorhodopsin to a distorted intermediate, bathorhodopsin, on the other hand, photon absorption by the stable light-adapted form of bacteriorhodopsin (bR_{568}) containing all-*trans* retinal, leads to the formation of what appears to be an 11-*cis* intermediate, K_{610}. In both systems it is possible that the isomerization occurs in an excited state of the first intermediate, followed by the proton transfer step in the relaxed form (Birge, 1981).

5. Hemoglobin

Hemoglobin (Hb) reacts with oxygen to form oxyhemoglobin (HbO_2) and with carbon monoxide to form carbonmonoxyhemoglobin (HbCO). The binding site is the central Fe atom of the metalloporphyrin, heme. Hemoglobin is comprised of four subunits, each of which contains the Fe porphyrin active site. Myoglobin (Mb) is a monomeric heme protein that also forms complexes with O_2 and CO. In general, its chemistry is simpler than that found for hemoglobin. The photodissociation rates, yields and recombination rates of these compounds has been an intensive area of study over the past several years (cf. Packhurst, 1979; Moffat, 1979). The quantum yield for dissociation of HbCO under steady illumination is reasonably high ($\sim$ 50%), while that for HbO_2 is only about 3% (cf. Chernoff et al., 1980). The corresponding dissociation yields measured with long-duration flashes for MbCO and MbO_2 are $\sim$ 100% and $\sim$ 3%, respectively (cf. Reynolds et al., 1981). One aim of picosecond studies has been to elucidate the reason for the low photodissociation yields of the oxy complexes. Two possibilities have been considered. The first involves rapid energy relaxation in the initially excited complex, which competes favorably with dissociation. The second is rapid recombination of O_2 and heme before they can diffuse apart (geminate recombination). The nature of the initial excited states has also been of interest.

Shank et al. (1976) studied HbO_2 and HbCO with 615 nm pump and probe pulses (5 nJ, 10^6/sec repetition rate) from a passively mode-locked cw dye laser (section 2.1). It is known that Hb absorbs more strongly at 615 nm than do HbO_2 and HbCO. Thus, HbCO showed an increase in absorbance immediately after excitation with an instrument-limited rise time of 0.5 psec. There was no hint of recombination on the time scale of 20 psec, in agreement with previous observations indicating little recombination over a period of several microseconds. On the other hand, although a prompt increase in absorbance at 615 nm was observed following excitation of HbO_2, the signal recovered completely with a time constant of about 2.5 psec. The same response was seen in a sample void of O_2. These observations suggested that in the case of HbO_2 the low dissociation yield is due to the formation and rapid relaxation of an initial excited state, and not dissociation followed by geminate recombination.

Noe et al. (1978) used single 530 nm excitation flashes, and also 530 nm trains of pulses, to study HbCO. Unligated Hb was monitored by an absorbance increase at 440 nm with part of a picosecond continuum. With single excitation flashes, they reported a longer (11 psec) predissociation lifetime for HbCO than reported by Shank et al. (1976), and a millisecond or slower recombination rate that depended on the excitation conditions. Eisert et al. (1979) studied MbCO and MbO_2 with 530 nm flashes. They observed that the signal for dissociated MbCO at 400 nm recovered by about 15% with a lifetime of 125 ± 50 psec, but no decay for the dissociated MbO_2 (low yield) in the first 450 psec following excitation.

Greene et al. (1978) investigated transient absorption spectra in HbO_2 and HbCO with 8 psec flashes at 353 nm. The dissociation products appeared in both cases within 8 psec and showed broadened unligated heme-like spectra that persisted for at least 680 psec with less than 10% recombination of HbCO and less than 20% recombination of HbO_2 over this period. Similar results were found for HbCO with 530 nm excitation. Chernoff et al. (1980) observed similar spectral effects using 530 nm excitation flashes. The transient product (< 8 psec formation time) observed for HbO_2 showed about 40% recovery through geminate recombination with a lifetime of 200 psec. The photoproduct of HbCO photolysis showed no recovery up to 1200 psec. The broadening in the initial spectra observed by both Greene et al. (1978) and Chernoff et al. (1980) was discussed in terms of conformational and spin states in the dissociated species due to the high level of excitation. In both studies, the low quantum yield for HbO_2 dissociation was attributed to fast relaxation in non-dissociative states rather than to significant geminate recombination. Similar conclusions were drawn by Reynolds et al. (1981), who studied MbO_2 and MbCO with 6 psec 353 and 530 nm excitation pulses. From the rapid bleaching (3.5–5 psec) in the ligated (MbO_2 and MbCO) band at 420 nm and the increase in absorbance (10–112 psec risetime) in the unligated band at 440 nm, it was concluded that there was a 7–10 psec intermediate (excited) state population before dissociation. It was concluded that in the case of MbCO this intermediate was dissociative, reflecting the near-unity photodissociation quantum yield. The predominant fate of the intermediate in MbO_2, however, was judged to be relaxation to the ground state resulting in the low (3%) yield of dissociation.

Picosecond resonance Raman studies by Terner et al. (1981) have examined the structure of the HbCO photoproduct on the picosecond time scale using 30 psec excitation flashes from a cavity-dumped synchronously pumped dye laser system (section 2.1). They suggest that in the HbCO photoproduct the Fe(II) is in a high-spin state, closer to the heme plane than in the deoxyhemoglobin. It is suggested that photodissociation occurs from a quintet ligand field excited state of HbCO. A similar initial photoproduct was found for HbO_2 photolysis (Nagumo et al., 1981). The low yield of photodissociation was attributed to rapid geminate recombination. The resonance Raman spectrum of the intermediate was also measured by Coppey et al. (1980).

6. Additional studies and concluding remarks

Picosecond techniques have been used for investigations in a number of other areas of biological interest. These include energy transfer (Shapiro et al., 1975b) and torsional dynamics (Millar et al., 1980; Robbins et al., 1980) in DNA, photoreactions of nucleic acid bases (Kryukov et al., 1978; Angelov et al., 1980), photoisomerization of bilirubin (Greene et al., 1981), energy transfer in phycocyanins (Kobayashi et al., 1979), transient states in bile pigments (Lippitsch et al., 1980), fluorescence lifetimes of tryptophan and some simple peptides (Beddard et al., 1980b), and proton release from the stentor photoreceptor (Song et al., 1981).

Picosecond spectroscopy has also made a strong impact in a number of other areas of chemistry and physics. Introductions to research in these areas can be found in the review articles cited in the text. Picosecond instrumentation has progressed rapidly over the past decade or so. A variety of laser systems and experimental techniques have been introduced (see section 2) and certainly will continue to be developed as detailed picosecond studies in new areas of chemistry and biology are initiated. Although picosecond spectroscopy has contributed enormously to our understanding of the photochemistry in a number of complex biological systems, notably photosynthesis, visual pigments and bacteriorhodopsin, and hemoglobin. It is clear that in these areas and in others, additional studies are needed and certainly will be forthcoming. Strong interplay between physicists, chemists and biologists applying ultrashort pulses to relevant problems should continue to make this one of the more dynamic areas of research in the years ahead.

Acknowledgements

The author wishes to acknowledge the Research Corporation, the Petroleum Research Fund, administered by the American Chemical Society, and the Camille and Henry Dreyfus Foundation for financial support during the preparation of this review. Discussions with Dr. G.A. Massey on nonlinear optics and with Dr. W.W. Parson on bacterial photosynthesis were very helpful.

References

Adams, M.C., Bradley, D.J., Sibbett, W. and Taylor, J.R. (1980) Synchronously pumped continuous wave dye lasers, in: D.J. Bradley, G. Porter and M.H. Key (Eds.), Ultra-Short Laser Pulses, Royal Society, London, pp. 7–13.

Akhamanov, S.A., Borisov, A.Y., Danielius, R.V., Kozlowski, V.S., Piskarskas, A.S. and Razjivin, A.P. (1978) Primary photosynthesis selectively excited by tunable picosecond parametric oscillator, in: C.V. Shank, E.P. Ippen and S.L. Shapiro (Eds.), Picosecond Phenomena, Springer Series in Chemical Physics, Vol. 4, Springer, Berlin, pp. 134–139.

Akhamanov, S.A., Borisov, A.Y., Danielius, R.V., Gadonas, R.A., Kozlowski, V.S., Piskarskas, A.S., Razjivin, A.P. and Shuvalov, V.A. (1980) One- and two-photon picosecond processes of electron transfer among the porphyrin molecules in bacterial reaction centers, FEBS Lett., 114, 149–152.

Alfano, R.R. and Shapiro, S.L. (1970) Emission in the region 4000 to 7000 Å via four-photon coupling in glass, Phys. Rev. Lett., 24, 584–587.

Alfano, R.R. and Shapiro, S.L. (1971) Picosecond spectroscopy using the inverse Raman effect, Chem. Phys. Lett., 8, 631–633.

Alfano, R.R., Yu, W., Govindjee, R., Becher, B. and Ebery, T.G. (1976) Picosecond kinetics of the fluorescence from the chromophore of the purple membrane protein of *Halobacterium halobium*, Biophys. J., 16, 541–545.

Altmann, J.A., Beddard, G.S. and Porter, G. (1978) A theoretical investigation of the dynamics of energy trapping in a two-dimensional model of the photosynthetic unit, Chem. Phys. Lett., 58, 54–57.

Andreeva, N.E., Zakharova, G.V., Shubin, V.V. and Chibisov, A.K. (1978) The role of the triplet state in chlorophyll photooxidation, Chem. Phys. Lett., 53, 317–320.

Angelov, D.A., Gruzadyan, G.G., Kryukov, G.P., Letokhov, V.S., Nikogosyan, D.N. and Oraevsky, A.A. (1980) Highpower UV ultrashort laser action on DNA and its components, in: R. Hochstrasser, W. Kaiser and C.V. Shank (Eds.), Picosecond Phenomena II, Springer Series in Chemical Physics, Vol. 14, Springer, Berlin, pp. 336–339.

Applebury, M.L. (1980) The primary processes of vision: A view from the experimental side, Photochem. Photobiol., 32, 425–431.

Applebury, M.L., Peters, K.S. and Rentzepis, P.M. (1978) Primary intermediates in the photochemical cycle of bacteriorhodopsin, Biophys. J., 23, 375–382.

Auston, D.H. (1977) Picosecond nonlinear optics, in: S.L. Shapiro (Ed.), Ultrashort Light Pulses, Topics in Applied Physics, Vol. 18, Springer, Berlin, pp. 123–201.

Baldwin, G.C. (1974) An Introduction to Nonlinear Optics, Plenum, New York.

Barber, J., Searle, G.F.W. and Tredwell, C.J. (1978) Picosecond time-resolved study of $MgCl_2$-induced chlorophyll fluorescence yield changes from chloroplasts, Biochim. Biophys. Acta, 501, 174–182.

Beddard, G.S. and Porter, G. (1977) Excited state annihilation in the photosynthetic unit, Biochim. Biophys. Acta, 462, 63–72.

Beddard, G.S., Porter, G., Tredwell, C.J. and Barber, J. (1975) Fluorescence lifetimes in the photosynthetic unit, Nature (London), 258, 166–168.

Beddard, G.S., Doust, T. and Windsor, M.W. (1980a) Picosecond studies of electronic relaxation in triphenylmethane dyes by fluorescence-upconversion, in: R. Hochstrasser, W. Kaiser and C.V. Shank (Eds.), Picosecond Phenomena II, Springer Series in Chemical Physics, Vol. 14, Springer, Berlin, pp. 167–170.

Beddard, G.S., Fleming, G.R., Porter, G. and Robbins, R.J. (1980b) Time-resolved fluorescence from biological systems; tryptophan and simple peptides, in: D.J. Bradley, G. Porter and M.H. Key (Eds.), Ultra-Short Laser Pulses, Royal Society, London, 111–124.

Beesley, M.J. (1971) Lasers and Their Applications, Taylor and Francis, London.

Birge, R.R. (1981) Photophysics of light transduction in rhodopsin and bacteriorhodopsin, Annu. Rev. Biophys. Bioeng., 10, 315–354.

Birge, R.R. and Hubbard, L.M. (1980) Molecular dynamics of cis–trans isomerization in rhodopsin, J. Am. Chem. Soc., 102, 2195–2205.

Blankenship, R.E. and Parson, W.W. (1978) The photochemical electron transfer reactions of photosynthetic bacteria and plants, Annu. Rev. Biochem., 47, 635–653.

Blankenship, R.E. and Parson, W.W. (1979) Kinetics and thermodynamics of electron transfer in bacterial reaction centers, in: J. Barber (Ed.), Photosynthesis in Relation to Model Systems, Elsevier, Amsterdam, pp. 71–114.

Borg, D.C., Fajer, J., Felton, R.H. and Dolphin, D. (1970) The π-cation radical of chlorophyll a, Proc. Natl. Acad. Sci. (U.S.A.), 67, 813–820.

Borisov, A.Y. and Godik, V.I. (1972a) Energy transfer in bacterial photosynthesis, I. Light intensity dependences of fluorescence lifetimes, Bioenergetics, 3, 211–220.

Borisov, A.Y. and Godik, V.I. (1972b) Energy transfer to the reaction centers in bacterial photosynthesis, II. Bacteriochlorophyll fluorescence lifetimes and quantum yields for some purple bacteria, Bioenergetics, 3, 515–523.

Borisov, A.Y. and Ilina, M.D. (1973) The fluorescence lifetime and energy migration mechanism in photosystem I of plants, Biochim. Biophys. Acta, 305, 364–371.

Born, M. and Wolf, E. (1965) Principles of Optics, 3rd edn., Pergamon, New York.

Bradley, D.J. (1977) Methods of generation, in: S.L. Shapiro (Ed.), Ultrashort Light Pulses, Topics in Applied Physics, Vol. 18, Springer, Berlin, pp. 17–81.

Bradley, D.J. and New, G.H.C. (1974) Ultrashort pulse measurements, Proc. IEEE, 62, 313.

Bradley, D.J., Jones, K.W. and Sibbett, W. (1980) Picosecond and femtosecond streak cameras: present and future designs, in: D.J. Bradley, G. Porter and M.H. Key (Eds.), Ultra-Short Laser Pulses, Royal Society, London, pp. 71–75.

Breton, J. and Geacintov, N.E. (1976) Quenching of fluorescence of chlorophyll in vivo by long-lived excited states, FEBS Lett., 86–89.

Breton, J. and Geacintov, N.E. (1980) Picosecond fluorescence kinetics and fast energy transfer processes in photosynthetic membranes, Biochim. Biophys. Acta, 594, 1–32.

Briantais, J.M., Govindjee and Merkelo, H. (1972) Lifetime of the excited state (τ) in vivo, III. Chlorophyll during fluorescence induction in *Chlorella pyrenoidosa*, Photosynthetica, 6, 133–141.

Brody, S.S., Treadwell, C. and Barber, J. (1981) Picosecond energy transfer in *Porphyridium cruentum* and *Anacystis nidulans*, Biophys. J., 34, 439–449.

Busch, G.E., Applebury, M.L., Lamola, A.A. and Rentzepis, P.M. (1972) Formation and decay of pre-lumirhodopsin at room temperatures, Proc. Natl. Acad. Sci. (U.S.A.), 69, 2802–2806.

Campillo, A.J. and Shapiro, S.L. (1977) Picosecond relaxation measurements in biology, in: S.L. Shapiro (Ed.), Ultrashort Light Pulses, Topics in Applied Physics, Vol. 18, Springer, Berlin, pp. 317–376.

Campillo, A.J., Shapiro, S.L., Kollman, V.H., Winn, D.R. and Hyer, R.C. (1976a) Picosecond exciton annihilation in photosynthetic systems, Biophys. J., 16, 93–97.

Campillo, A.J., Kollman, V.H. and Shapiro, S.L. (1976b) Intensity dependence of the fluorescence lifetime of in vivo chlorophyll excited by a picosecond light pulse, Science, 193, 227–229.

Campillo, A.J., Hyer, R.C., Monger, T.C., Parson, W.W. and Shapiro, S.L. (1977) Light collection and harvesting processes in bacterial photosynthesis investigated on a picosecond time scale, Proc. Natl. Acad. Sci. (U.S.A.), 74, 1997–2001.

Campillo, A.J., Hyer, R.C. and Shapiro, S.L. (1979a) Broadly tunable picosecond infrared source, Optics Lett., 4, 325–327.

Campillo, A.J., Hyer, R.C. and Shapiro, S.L. (1979b) Picosecond infrared-continuum generation by three-photon parametric amplification in $LiNbO_3$, Optics Lett., 4, 357–359.

Castelli, F., Cheddar, G., Rizzuto, F. and Tollin, G. (1979) Laser photolysis studies of quinone reduction by pheophytin *a* in alcohol solution, Photochem. Photobiol., 29, 153–163.

Chernoff, D.A., Hochstrasser, R.M. and Steele, A.W. (1980) Geminate recombination of O_2 and hemoglobin, Proc. Natl. Acad. Sci. (U.S.A.), 77, 5606–5610.

Clayton, R.K. and Sistrom, W.R. (Eds.) (1978) The Photosynthetic Bacteria, Plenum, New York.

Colles, M.J. (1969) Efficient stimulated Raman scattering from picosecond pulses, Optics Commun., 1, 169–172.

Coppey, M., Tourbey, H., Valat, P. and Alpert, B. (1980) Nature (London), 284, 568–570.

Cotton, T.M. and Van Duyne, R.P. (1978) Resonance Raman spectroelectrochemistry of bacteriochlorophyll and bacteriochlorophyll cation radical, Biochem. Biophys. Res. Commun., 82, 424–433.

Davis, M.S., Forman, A., Hanson, L.K., Thornber, J.P. and Fajer, J. (1979) Anion and cation radicals of bacteriochlorophyll and bacteriopheophytin *b*. Their role in the primary charge separation of *Rhodopseudomonas viridis*, J. Phys. Chem., 83, 3325–3332.

DeMaria, A.J., Glenn Jr., W.H., Brienza, M.J. and Mack, M.E. (1969) Picosecond laser pulses, Proc. IEEE, 57, 2–24.

DeVault, D. and Chance, B. (1966) Studies of photosynthesis using a pulsed laser, I. Temperature dependence of cytochrome oxidation route in Chromatium, Evidence for tunneling, Biophys. J., 6, 825–847.

Doukas, A.G., Lu, P.Y. and Alfano, R.R. (1981) Fluorescence relaxation kinetics from rhodopsin and isorhodopsin, Biophys. J., 35, 547–550.

Duguay, M.A. and Hansen, J.W. (1969) An ultrafast light gate, Appl. Phys. Lett., 15, 192–194.

Dutton, P.L., Kaufmann, K.J., Chance, B. and Rentzepis, P.M. (1975) Picosecond kinetics of the 1250 nm band of the *Rps. sphaeroides* reaction center, The nature of the primary photochemical intermediary state, FEBS Lett., 60, 275–280.

Efrima, S. and Bixon, M. (1974) On the role of vibrational excitation in electron transfer reactions with large negative free energies, Chem. Phys. Lett., 25, 34–37.

Eisert, W.G., Degenkolb, E.O., Noe, L.J. and Rentzepis, P.M. (1979) Kinetics of carboxymyoglobin and oxymyoglobin studied by picosecond spectroscopy, Biophys. J., 25, 455–464.

Fajer, J., Davis, M.S., Brune, D.C., Spaulding, L.D., Borg, D.C. and Forman, A. (1977) Chlorophyll radicals and primary events: old and new speculations, in: J.M. Olson and G. Hind (Eds.), Chlorophyll-Proteins, Reaction Centers, and Photosynthetic Membranes, Brookhaven Symposium Biology, Vol. 28.

Fajer, J., Davis, M.S., Forman, A., Klimov, V.V., Dolan, E. and Ke, B. (1980) Primary electron acceptors in plant photosynthesis, J. Am. Chem. Soc., 102, 7143–7145.

Feher, G., Hoff, A.J., Isaacson, R.A. and Ackerson, L.C. (1975) Endor experiments on chlorophyll and bacteriochlorophyll in vitro and in the photosynthetic unit, Ann. N.Y. Acad. Sci., 244, 239–259.

Fendler, J.H. and Fendler, E.J. (1975) Catalysis in Micellar and Macromolecular Systems, Academic Press, New York.

Fenton, J.M., Pellin, M.J., Govindjee and Kaufmann, K.J. (1979) Primary photochemistry of the reaction center of photosystem I, FEBS Lett., 100, 1–4.

Freisner, R., Dismukes, G.C. and Sauer, K. (1979) Development of electron spin polarization in photosynthetic electron transfer by the radical pair mechanism, Biophys. J., 25, 277–294.

Fujita, I., Davis, M.S. and Fajer, J. (1978) Anion radicals of pheophytin and chlorophyll *a*: their role in the primary charge separations of plant photosynthesis, J. Am. Chem. Soc., 100, 6280–6282.

Geacintov, N.E. and Breton, J. (1977) Exciton annihilation in the two photosystems in chloroplasts at 100°K, Biophys. J., 17, 1–15.

Geacintov, N.E., Breton, J., Swenberg, C.E. and Paillotin, G. (1977) A single pulse picosecond laser study of excition dynamics in chloroplasts, Photochem. Photobiol., 26, 629–638.

Geacintov, N.E., Husiak, D., Kolubayev, T., Breton, J., Campillo, A.J., Shapiro, S.L., Winn, K.R. and Woodbridge, P.K. (1979) Exciton annihilation versus excited state absorption and stimulated emission effects in laser studies of fluorescence quenching of chlorophyll in vitro and in vivo, Chem. Phys. Lett., 66, 154–158.

Gouterman, M. and Holten, D. (1977) Electron transfer from photoexcited singlet and triplet bacteriopheophytin-II, Theoretical, Photochem. Photobiol., 25, 85–92.

Govindjee (Ed.) (1975) Bioenergetics of Photosynthesis, Academic Press, New York.

Govindjee, Hammond, J.H. and Merkelo, H. (1972) Lifetime of the excited state in vivo, II. Bacteriochlorophyll in photosynthetic bacteria at room temperature, Biophys. J., 12, 809–813.

Govindjee, Hammond, J.H., Smith, W.R., Govindjee, R. and Merkelo, H. (1975) Lifetime of the excited states in vivo, IV. Bacteriochlorophyll and bacteriopheophytin in *Rhodospirillum rubrum*, Photosynthetica, 9, 216–219.

Green, B.H., Monger, T.G., Alfano, R.R., Alton, B. and Callender, R.H. (1977) *Cis–trans* isomerisation in rhodopsin occurs in picoseconds, Nature (London), 269, 179–180.

Greene, B.I., Hochstrasser, R.M., Weisman, R.B. and Eaton, W.A. (1978) Spectroscopic studies of oxy- and carbonmonoxyhemoglobin after pulsed optical excitation, Proc. Natl. Acad. Sci. (U.S.A.), 75, 5255–5259.

Greene, B.I., Hochstrasser, R.M. and Weisman R.E. (1979) Picosecond transient spectroscopy of molecules in solution, J. Chem. Phys., 70, 1247–1259.

Greene, B.I., Lamola, A.A. and Shank, C.V. (1981) Picosecond primary photoprocesses of bilirubin bound to human serum albumin, Proc. Natl. Acad. Sci. (U.S.A.), 78, 2008–2012.

Hallidy, L.A. and Topp, M.R. (1977) Picosecond luminescence detection using type-II phase-matched frequency conversion, Chem. Phys. Lett., 46, 8–14.

Harriman, A., Porter, G. and Searle, N. (1979) Reversible photo-oxidation of zinc tetraphenylporphine by benzo-1,4-quinone, J.C.S. Faraday II, 75, 1515–1521.

Harris, L., Porter, G., Synowiec, J.A., Tredwell, C.J. and Barber, J. (1976) Fluorescence lifetimes of *Chlorella pyrenoidosa*, Biochim. Biophys. Acta, 449, 329–339.

Hayward, G., Carlsen, W., Siegman, A. and Stryer, L. (1980) Picosecond resonance Raman spectroscopy: The initial photolytic event in rhodopsin and isorhodopsin, in: R. Hochstrasser, W. Kaiser and C.V. Shank (Eds.), Picosecond Phenomena II, Springer Series in Chemical Physics, Vol. 14, Springer, Berlin, pp. 377–379.

Heritage, J.P. and Jain, R.K. (1978) Subpicosecond pulses from a tunable cw mode-locked dye laser, Appl. Phys. Lett., 32, 101–103.

Hindman, J.C., Kugel, R., Svirmickas, A. and Katz, J.J. (1977) Chlorophyll lasers: stimulated light emission by chlorophylls and Mg-free chlorophyll derivatives, Proc. Natl. Acad. Sci. (U.S.A.), 74, 5–9.

Hindman, J.C., Kugel, R., Svirmickas, A. and Katz, J.J. (1978) Stimulated fluorescence quenching in chlorophyll a and bacteriochlorophyll a, Chem. Phys. Lett., 53, 197–200.

Hirsch, M.D., Marcus, M.A., Lewis, A., Mahr, H. and Frigo, N. (1976) A method for measuring picosecond phenomena in photolabilie species, The emission lifetime of bacteriorhodopsin, Biophys. J., 16, 1399–1409.

Hiyama, T. and Frock, D.C. (1980) Kinetic identification of component X as P430: a primary electron acceptor of photosystem I, Arch. Biochim. Biophys., 199, 488–496.

Ho, P.P. and Alfano, R.R. (1977) Temperature dependence of the rotational relaxation times of aniso-tropic molecules in neat and mixed binary liquids, Chem. Phys. Lett., 50, 74–80.

Holten, D. and Windsor, M.W. (1978) Picosecond flash photolysis in biology and biophysics, Annu. Rev. Bioeng., 7, 189–277.

Holten, D. and Windsor, M.W. (1980) Automated picosecond spectroscopy with a standard two-dimen-sional optical multichannel analyzer and an inexpensive microprocessor, Photobiochem. Photo-biophys., 1, 243–252.

Holten, D., Gouterman, M., Parson, W.W., Windsor, M.W. and Rockley, M.G. (1976) Electron trans-fer from photoexcited singlet and triplet bacteriopheophytin, Photochem. Photobiol., 23, 415–423.

Holten, D., Windsor, M.W., Parson, W.W. and Thornber, J.P. (1978a) Primary photochemical proces-ses in isolated reaction centers of $Rhodopseudomonas$ $viridis$, Biochim. Biophys. Acta, 501, 112–126.

Holten, D., Windsor, M.W., Parson, W.W. and Gouterman, M. (1978b) Models for bacterial photosyn-thesis: Electron transfer from photoexcited singlet bacteriopheophytin to methyl viologen and m-di-nitrobenzene, Photochem. Photobiol., 28, 951–961.

Holten, D., Hoganson, C., Windsor, M.W., Schenck, C.C., Parson, W.W., Migus, A., Fork, R.L. and Shank, C.V. (1980a) Subpicosecond and picosecond studies of electron transfer intermediates in $Rhodopseudomonas$ $sphaeroides$ reaction centers, Biochim. Biophys. Acta, 592, 461–477.

Holten, D., Shah, S.S. and Windsor, M.W. (1980b) Electron transfer from photoexcited bac-teriopheophytin to p-benzoquinone in cationic micelles, in: R. Hochstrasser, W. Kaiser and C.V. Shank (Eds.), Picosecond Phenomena II, Springer Series in Chemical Physics, Vol. 14, Springer, Berlin, pp. 317–321.

Honig, B. (1978) Light energy transduction in visual pigments and bacteriorhodopsin, Annu. Rev. Phys. Chem., 29, 31–57.

Honig, B., Ebery, T., Callender, R.H., Dinur, U. and Ottolenghi, M. (1979) Photoisomerization, energy, storage, and charge separation: A model for light energy transduction in visual pigments and bacteriorhodopsin, Proc. Natl. Acad. Sci. (U.S.A.), 76, 2503–2507.

Huppert, D., Rentzepis, P.M. and Tollin, G. (1976) Picosecond kinetics of chlorophyll and chlorophyll/quinone solutions in ethanol, Biochim. Biophys. Acta, 440, 356–364.

Huppert, D., Rentzepis, P.M. and Kliger, D.S. (1977) Picosecond and nanosecond isomerization kine-tics of protonated 11-cis retinylidene Schiff bases, Photochem. Photobiol., 25, 193–197.

Huston, A.E. and Helbrough, K. (1980) The synchroscan picosecond streak camera system, in: D.J. Bradley, G. Porter and M.H. Key (Eds.), Ultra-Short Laser Pulses, Royal Society, London, pp. 77–83.

Ippen, E.P. and Shank, C.V. (1975) Dynamic spectroscopy and subpicosecond pulse compression, Appl. Phys. Lett., 27, 488–491.

Ippen, E.P. and Shank, C.V. (1977) Techniques for measurement, in: S.L. Shapoiro (Ed.), Ultrashort Light Pulses, Topics in Applied Physics, Vol. 18, Springer, Berlin, pp. 83–122.

Ippen, E.P. and Shank, C.V. (1979) Picosecond spectroscopy, in: C.L. Tang (Ed.), Quantum Elec-tronics, Methods of Experimental Physics, Vol. 15 (B), Academic Press, New York, pp. 185–209.

Ippen, E.P., Shank, C.V., Lewis, A. and Marcus, M.A. (1978) Subpicosecond spectroscopy of bac-teriorhodopsin, Science, 200, 1279–1281.

Ippen, E.P., Shank, C.V., Wiesenfeld, J.M. and Migus, A. (1980) Subpicosecond pulse techniques, in:

D.J. Bradley, G. Porter and M.H. Key (Eds.), Ultra-Short Laser Pulses, Royal Society, London, pp. 15–22.

Jain, R.K. and Heritage, J.P. (1978) Generation of synchronized cw trains of picosecond pulses at two independently tunable wavelengths, Appl. Phys. Lett., 32, 41–44.

Jortner, J. (1976) Temperature dependent activation energy for electron transfer between biological molecules, J. Chem. Phys., 64, 4860–4867.

Jortner, J. (1980) Dynamics of the primary events in bacterial photosynthesis, J. Am. Chem. Soc., 102, 6676–6686.

Kakitani, K. and Kakitani, H. (1981) A possible new mechanism of temperature dependence of electron transfer in photosynthetic systems, Biochim. Biophys. Acta, 635, 498–514.

Kalyanasundaram, K. and Porter, G. (1978) Proc. Roy Soc. London A, 364, 29–44.

Kano, K., Takuma, K., Ikeda, T., Nakajima, D., Tsusui, Y. and Matsuo, T. (1978) Zinc tetraphenyl-porphyrin-sensitized photoreduction of anthraquinonesulfonate in aqueous micellar solutions, Photochem. Photobiol., 27, 695–701.

Katz, J.J. and Norris Jr., J.R. (1973) Chlorophyll and light energy transduction in photosynthesis, in: D.R. Sanadi and L. Packer (Eds.), Current Topics in Bioenergetics, Vol. 5, Academic Press, New York, pp. 41–75.

Katz, J.J., Norris, J.R., Shipman, L.L., Thurnauer, M.C. and Wasielewski, M.R. (1978) Chlorophyll function in the photosynthetic reaction center, Annu. Rev. Biophys. Bioeng., 7, 393–434.

Kaufmann, K.J. (1979) Picosecond spectroscopy applied to the study of chemical and biological reactions, in: CRC Critical Reviews in Solid State and Material Sciences, Vol. 8, Chemical Rubber Company, Cleveland, OH, pp. 265–316.

Kaufmann, K.J., Dutton, P.L., Netzel, T.L., Leigh, J.S. and Rentzepis, P.M. (1975) Picosecond kinetics of events leading to reaction center bacteriochlorophyll oxidation, Science, 188, 1301–1304.

Kaufmann, K.J., Petty, K.M., Dutton, P.L. and Rentzepis, P.M. (1976a) Picosecond kinetics in reaction centers of Rps. sphaeroides and the effects of ubiquinone extraction and reconstitution, Biochem. Biophys. Res. Commun., 70, 839–845.

Kaufmann, K.J., Rentzepis, P.M. and Stoeckenius, W. (1976b) Primary photochemical processes in bacteriorhodopsin, Biochem. Biophys. Res. Commun., 68, 1109–1115.

Klimov, V.V., Klevanik, A.V., Shuvalov, V.A. and Krasnovsky, A.A. (1977) Reduction of pheophytin in the primary light reaction of photosystem II, FEBS Lett., 82, 183–186.

Klimov, V.V., Dolan, E. and Ke, B. (1980) EPR properties of an intermediary electron acceptor (pheophytin) in photosystem-II reaction centers at cryogenic temperatures, FEBS Lett., 112, 97–100.

Kobayashi, T. (1979) Hypsorhodopsin: the first intermediate of the photochemical process in vision, FEBS Lett., 106, 313–316.

Kobayashi, T., Shichida, Y. Yoshizawa, T. and Nagakura, S. (1978) First observation of the formation process of squid hypsorhodopsin by picosecond spectroscopy, in: A. Zewail (Ed.), Advances in Laser Chemistry, Springer Series in Chemical Physics, Vol. 3, Springer, Berlin, pp. 179–186.

Kobayashi, T., Degenkolb, E.O., Bersohn, R., Rentzepis, P.M., MacColl, R. and Berns, D.S. (1979) Energy transfer among the chromophores in phycocyanins measured by picosecond kinetics, Biochemistry, 18, 5073–5078.

Kollman, V.H., Shapiro, S.L. and Campillo, A.J. (1975) Photosynthetic studies with a 10-psec resolution streak camera, Biochem. Biophys. Res. Commun., 63, 917–923.

Kranitzky, W., Ding, K., Seilmeier, A. and Kaiser, W. (1980) Parametric generation of shortened, narrow-band picosecond pulses using a YAG-pump laser, Optics Commun., 34, 483–487.

Kryukov, P.G., Letokhov, V.S., Matveetz, Y.A., Nikogosian, D.N. and Sharkov, A.V. (1978) Picosecond research of some biomolecules. In: C.V. Shank, E.P. Ippen and S.L. Shapiro (Eds.) Picosecond Phenomena, Springer Series in Chemical Physics, Vol. 4, Springer Verlag, Berlin-Heidelberg-New York, pp. 158–166.

Kuhl, J., Lambrick, R. and von der Linde, D. (1977) Generation of near-infrared picosecond pulses by modelocked synchronous pumping of a jet-stream dye laser, Appl. Phys. Lett., 31, 657–658.

Kung, A.H. (1974) Generation of tunable picosecond VUV radiation, Appl. Phys. Lett., 25, 653–654.

334

Lamola, A.A., Manion, M.L., Roth, H.D. and Tollin, G. (1975) Photooxidation of chlorins by quinones studied by nuclear magnetic resonance techniques (Chlorophyll, CIDNP, Photoredox reaction), Proc. Natl. Acad. Sci. (U.S.A.), 72, 3265–3269.

Laubereau, A., Fendt, A., Seilmeier, A. and Kaiser, W. (1978) High power, nearly bandwidth limited, tunable picosecond pulses in the visible and infrared, in: C.V. Shank, E.P. Ippen and S.L. Shapiro (Eds.), Picosecond Phenomena, Springer Series in Chemical Physics, Vol. 4, Springer, Berlin, pp. 89–95.

Lewis, A. (1978) The molecular mechanism of excitation in visual transduction and bacteriorhodopsin, Proc. Natl. Acad. Sci. (U.S.A.), 75, 549–553.

Lippitsch, M.E., Leitner, A., Riegler, M. and Aussenegg, F.R. (1980) Picosecond studies on bile pigments, in: R. Hochstrasser, W. Kaiser and C.V. Shank (Eds.), Picosecond Phenomena II, Springer Series in Chemical Physics, Vol. 14, Springer, Berlin, pp. 327–330.

Magde, D. and Windsor, M.W. (1974) Picosecond flash photolysis and spectroscopy: 3,3'-diethyloxadicarbocyanine iodide (DODCI), Chem. Phys. Lett., 27, 31–36.

Malkan, R. and Bearden, A.J. (1971) Primary reactions of photosynthesis: Photoreduction of a bound chloroplast ferredoxin at low temperature as detected by Epr spectroscopy, Proc. Natl. Acad. Sci. (U.S.A.), 68, 16–19.

Mar, T., Govindjee, Singhal, G.S. and Merkelo, H. (1972) Lifetime of the excited state in vivo, I. Chlorophyll a in algae, at room and at liquid nitrogen temperatures; rate constants of radiationless deactivation and trapping, Biophys. J., 12, 797–808.

Mathies, R. (1979) Biological applications of resonance Raman spectroscopy in the visible and ultraviolet: Visual pigments, purple membrane, and nucleic acids, in: C.B. Moore (Ed.), Chemical and Biochemical Applications of Lasers, Academic Press, New York, pp. 55–99.

Mathis, P., Sauer, K. and Remy, R. (1978) Rapidly reversible flash-induced electron transfer in a P-700 chlorophyll–protein complex isolated with SDS, FEBS Lett., 88, 275–278.

Mauzerall, D. (1976a) Multiple excitations in photosynthetic systems, Biophys. J., 16, 87–91.

Mauzerall, D. (1976b) Fluorescence and multiple excitation in photosynthetic systems, J. Phys. Chem., 80, 2306–2309.

McIntosh, A.R., Chu, M. and Bolton, J.R. (1974) Flash photolysis electron spin resonance studies of the electron acceptor species at low temperature in photosystem I of spinach subchloroplast particles, Biochim. Biophys. Acta, 376, 308–314.

Merkelo, H., Hartman, S.R., Mar, T., Singhal, G.S. and Govindjee (1969) Mode-locked lasers: measurements of very fast radiative decay in fluorescent systems, Science, 164, 301–302.

Millar, D.P., Robbins, R.J. and Zewail, A.H. (1980) Direct observation of the torsional dynamics of DNA and RNA by picosecond spectroscopy, Proc. Natl. Acad. Sci. (U.S.A.), 77, 5593–5597.

Moffat, (1979) A structural model for the kinetic behavior of hemoglobin, Science, 206, 1035–1042.

Monger, T.G., Alfano, R.R. and Callender, R.H. (1979) Photochemistry of rhodopsin and isorhodopsin investigated on a picosecond time scale, Biophys. J., 27, 105–115.

Moskowitz, E. and Malley, M.M. (1978) Energy transfer and photooxidation kinetics in reaction centers on the picosecond time scale, Photochem. Photobiol., 27, 55–59.

Moya, I., Govindjee, Vernotte, C. and Briantais, M.J. (1977) Antagonistic effect of mono- and divalent-cations on lifetime (τ) and quantum yield of luminescence (ϕ) in isolated chloroplasts, FEBS Lett., 75, 13–18.

Müller, A., Lumry, R. and Walker, M.S. (1969) Light-intensity dependence of the in vivo fluorescence lifetime of chlorophyll, Photochem. Photobiol., 9, 113–126.

Nagumo, M., Nicol, M. and El-Sayed, M. (1981) Polarized resonance Raman spectroscopy of the photointermediate of oxyhemoglobin on the picosecond time scale, J. Phys. Chem., 85, 2435–2438.

Nakashima, N. and Mataga, N. (1975) Picosecond flash photolysis and transient spectral measurements over the entire visible, near ultraviolet and near infrared regions, Chem. Phys. Lett., 35, 487–492.

Netzel, T.L., Rentzepis, P.M. and Leigh, J.S. (1973) Picosecond kinetics of reaction centers containing bacteriochlorophyll, Science, 182, 238–241.

Netzel, T.L., Rentzepis, P.M., Tiede, D.M., Prince, R.C. and Dutton, P.L. (1977) Effect of reduction of the reaction center intermediate upon the picosecond oxidation reaction of the bacteriochlorophyll dimer in *Chromatium vinosum* and *Rhodopseudomonas viridis*, Biochim. Biophys. Acta, 460, 467–479.

Netzel, T.L., Kroger, P., Chang, C.K., Fujita, I. and Fajer, J. (1979) Electron transfer reactions in cofacial diporphyrins, Chem. Phys. Lett., 67, 223–228.

Netzel, T.L., Bucks, R.R., Boxer, S.G. and Fujita, J. (1980) A report on picosecond studies of electron transfer in photosynthetic models, in: R. Hochstrasser, W. Kaiser and C.V. Shank (Eds.), Picosecond Phenomena II, Springer Series in Chemical Physics, Vol. 14, Springer, Berlin, pp. 322–326.

New, G.H.C. (1980) Mode-locked laser systems: theoretical models, in: D.J. Bradley, G. Porter and M.H. Key (Eds.), Ultra-Short Laser Pulses, Royal Society, London, pp. 37–46.

Noe, L.J., Eisert, W.G. and Rentzepis, P.M. (1978) Picosecond photodissociation and subsequent recombination processes in carbon monoxide hemoglobin, Proc. Natl. Acad. Sci. (U.S.A.), 75, 573–577.

Norris, J.R., Uphaus, R.A., Crespi, H.L. and Katz, J.J. (1971) Electron spin resonance of chlorophyll and the origin of signal I in photosynthesis, Proc. Natl. Acad. Sci. (U.S.A.), 68, 625–628.

Norris, J.R., Scheer, H., Druyan, M.E. and Katz, J.J. (1974) An electron-nuclear double resonance (endor) study of the special pair model for photo-reactive chlrophyll in photosynthesis, Proc. Natl. Acad. Sci. (U.S.A.), 71, 4897–4900.

Okamura, M.Y., Issacson, R.A. and Feher, G. (1979) Spectroscopic and kinetic properties of the transient intermediate acceptor in reaction centers of *Rhodopseudomonas sphaeroides*, Biochim. Biophys. Acta, 546, 394–417.

Olson, J.M. and Hind, G. (Eds.) (1977) Chlorophyll-Proteins, Reaction Centers, and Photosynthetic Membranes, Brookhaven Symposium Biology, Vol. 28.

Ottolenghi, M. (1980) The photochemistry of rhodopsins, in: J.N. Pitts Jr., G.S. Hammond, K. Gollnick and D. Grosjean (Eds.), Advances in Photochemistry, Vol. 12, John Wiley, New York, pp. 97–200.

Packhurst, L.J. (1979) Hemoglobin and myoglobin ligand kinetics, Annu. Rev. Phys. Chem., 30, 503–546.

Parson, W.W. and Ke, B. (1981) Primary photochemical reactions, in: Govindjee (Ed.), Integrated Approach to Plant and Bacterial Photosynthesis, Academic Press, New York.

Parson, W.W., Clayton, R.K. and Cogdell, R.J. (1975) Excited states of photosynthetic reaction centers at low redox potentials, Biochim. Biophys. Acta, 387, 265–278.

Paschenko, V.Z., Protasov, S.P., Rubin, A.B., Timofeev, K.N., Zamazova, L.M. and Rubin, L.B. (1975) Probing the kinetics of photosystem I and photosystem II fluorescence in pea chloroplasts on a picosecond pulse fluorometer, Biochim. Biophys. Acta, 408, 143–153.

Paschenko, V.Z., Kononenko, A.A., Protasov, S.P., Rubin, A.B., Rubin, L.B. and Uspenskaya, N.K. (1977) Probing the fluorescence emission kinetics of the photosynthetic apparatus of *Rhodopseudomonas sphaeroides*, strain 1760-1, on a picosecond pulse fluorometer, Biochim. Biophys. Acta, 461, 403–412.

Pellegrino, F., Yu, W. and Alfano, R.R. (1978) Fluorescence kinetics of spinach chloroplasts measured with a picosecond optical Kerr gate, Photochem. Photobiol., 28, 1007–1012.

Pellin, M.J., Wraight, C.A. and Kaufmann, K.J. (1978) Modulation of the primary electron transfer rate in photosynthetic reaction centers by reduction of a secondary acceptor, Biophys. J., 3, 361–369.

Pellin, M.J., Kaufmann, K.J. and Wasielewski, M.R. (1979) In vitro duplication of the primary light-induced charge separation in purple photosynthetic bacteria, Nature (London), 278, 54–55.

Pellin, M.J., Wasielewski, M.R. and Kaufmann, K.J. (1980) Picosecond photophysics of covalently linked pyrochlorophyllide *a* dimer, Unique kinetics within the singlet manifold, J. Am. Chem. Soc., 102, 1868–1873.

Peters, K., Applebury, M.L. and Rentzepis, P.M. (1977) Primary photochemical event in vision: Proton translocation, Proc. Natl. Acad. Sci. (U.S.A.), 74, 3119–3123.

Peters, K.S., Avouris, P. and Rentzepis, P.M. (1978) Picosecond dynamics of primary electron-transfer processes in bacterial photosynthesis, Biophys. J., 23, 207–217.

Phillipson, K.P., Satoh, V.L. and Sauer, K. (1972) Exciton interaction in photosystem I reaction center from spinach chloroplasts, Absorption and circular dichroism difference spectra, Biochemistry, 11, 4591–4594.

Porter, G., Synowiec, J.A. and Tredwell, C.J. (1977) Intensity effects on the fluorescence of in vivo chlorophyll, Biochim. Biophys. Acta, 459, 329–336.

Porter, G., Tredwell, C.J., Searle, G.F.W. and Barber, J. (1978) Picosecond time-resolved energy transfer in *Porphyridium cruentum*, Part I. In the intact alga, Biochim. Biophys. Acta, 501, 232–245.

Rentzepis, P.M., Kaufmann, K.J., Avouris, P., Kobayashi, T. and Degenkolb, E.O. (1978) Picosecond studies of some ultrafast events in bacterial photosynthesis, in: A Zewail (Ed.), Advances in Laser Chemistry, Springer Series in Chemical Physics, Vol. 3, Springer, Berlin, pp. 126–144.

Reynolds, A.H., Rand, S.D. and Rentzepis, P.M. (1981) Mechanisms for excited state relaxation and dissociation of oxymyoglobin and carboxymyoglobin, Proc. Natl. Acad. Sci. (U.S.A.), 78, 2292–2296.

Robbins, R.J., Millar, D.P. and Zewail, A.H. (1980) Picosecond torsional dynamics of DNA, in: R. Hochstrasser, W. Kaiser and C.V. Shank (Eds.), Picosecond Phenomena II, Springer Series in Chemical Physics, Vol. 14, Springer, Berlin, pp. 331–335.

Robinson, G.W., Caughey, T.A. and Auerbach, R.A. (1978) Picosecond emission spectroscopy with an ultraviolet sensitive streak camera, in: A. Zewail (Ed.), Advances in Laser Chemistry, Springer Series in Chemical Physics, Vol. 3, Springer, Berlin, pp. 108–125.

Rockley, M.G., Windsor, M.W., Cogdell, R.J. and Parson, W.W. (1975) Picosecond detection of an intermediate in the photochemical reaction of bacterial photosynthesis, Proc. Natl. Acad. Sci. (U.S.A.), 72, 2251–2255.

Rubin, A.B. (1978) Picosecond fluorescence and electron transfer in primary photosynthetic processes, Photochem. Photobiol., 28, 1021–1028.

Ruddock, I.S. and Bradley, D.J. (1976) Bandwidth-limited subpicosecond generation in mode-locked cw dye lasers, Appl. Phys. Lett., 29, 296–297.

Sarai, A. (1979) Energy-gap and temperature dependence of electron and excitation transfer in biological systems, Chem. Phys. Lett., 63, 360–366.

Sarai, A. (1980) Possible role of protein in photosynthetic electron transfer, Biochim. Biophys. Acta, 589, 71–83.

Schenck, C.C., Parson, W.W., Holten, D. and Windsor, M.W. (1981a) Transient states in reaction centers containing reduced bacteriopheophytin, Biochim. Biophys. Acta, 635, 383–392.

Schenck, C.C., Parson, W.W., Holten, D., Windsor, M.W. and Sarai, A. (1981b) Temperature dependence of electron transfer between bacteriopheophytin and ubiquinone in protonated and deuterated reaction centers of *Rhodopseudomonas sphaeroides*, Biophys. J., 36, 479–489.

Schneider, S. (1980) Flashlamp-pumped mode-locked dye lasers, in: D.J. Bradley, G. Porter and M.H. Key (Eds.), Ultra-Short Laser Pulses, Royal Society, London, pp. 23–35.

Searle, G.F.W., Barber, J., Harris, L., Porter, G. and Tredwell, C.J. (1977) Picosecond laser study of fluorescence lifetimes in spinach chloroplast photosystem I and photosystem II preparations, Biochim. Biophys. Acta, 459, 390–401.

Searle, G.F.W., Barber, J., Porter, G. and Tredwell, C.J. (1978) Picosecond time-resolved energy transfer in *Porphyridium cruentum*, Part II. In the isolated light harvesting complex (phycobilisomes), Biochim. Biophys. Acta, 501, 246–256.

Seeley, G.R. (1977) Chlorophyll in model systems: Clues to the role of chlorophyll in photosynthesis, in: J. Barber (Ed.), Primary Processes of Photosynthesis, Elsevier, Amsterdam, pp. 1–53.

Seeley, G.R. (1978) The energetics of electron-transfer reactions of chlorophyll and other compounds, Photochem. Photobiol., 27, 639–654.

Seibert, M. and Alfano, R.R. (1974) Probing photosynthesis on a picosecond time scale, Evidence for photosystem I and photosystem II fluorescence in chloroplasts, Biophys. J., 14, 269–283.

Seibert, M., Alfano, R.R. and Shapiro, S.L. (1973) Picosecond fluorescent kinetics of in vivo chlorophyll, Biochim. Biophys. Acta, 292, 493–495.

Seilmeier, A., Spanner, K., Laubereau, A. and Kaiser, W. (1978) Narrow-band tunable infrared pulses with sub-picosecond time resolution, Optics Commun., 24, 237–342.

Shah, S.S., Holten, D. and Windsor, M.W. (1980) Models for photosynthetic charge separation: Excited state electron transfer reactions bacteriopheophytin in micelles, Photobiochem. Photo-biophys., 1, 361–373.

Shank, C.V., Ippen, E.P. and Bersohn, R. (1976) Time-resolved spectroscopy of hemoglobin and its complexes with sub-picosecond optical pulses, Science, 193, 50–51.

Shank, C.V., Fork, R.L., Leheny, R.F. and Shah, J. (1979) Dynamics of photoexcited Ga As band-edge absorption with subpicosecond resolution, Phys. Rev. Lett., 42, 112–121.

Shapiro, S.L., Kollman, V.H. and Campillo, A.J. (1975a) Energy transfer in photosynthesis: Pigment concentration effects and fluorescent lifetimes, FEBS Lett., 54, 358–362.

Shapiro, S.L., Campillo, A.J., Kollman, V.H. and Goad, W.B. (1975b) Exciton transfer in DNA, Optics Commun., 15, 308–310.

Shapiro, S.L., Campillo, A.J., Lewis, A., Perreault, G.J., Spoonhower, J.P., Clayton, R.K. and Stoeckenius, W. (1978) Picosecond and steady state, variable intensity and variable temperature emission spectroscopy of bacteriorhodopsin, Biophys. J., 23, 383–393.

Shichida, Y., Yoshizawa, T., Kobayashi, T., Ohtani, H. and Nagakura, S. (1977) Squid hypsorhodopsin and bathorhodopsin by a picosecond laser photolysis, FEBS Lett., 80, 214–216.

Shichida, Y., Kobayashi, T., Ohtani, H., Yoshizawa, T. and Nagakura, S. (1978) Picosecond laser photolysis of squid rhodopsin at room temperatures, Photochem. Photobiol., 27, 335–341.

Shuvalov, V.A. and Klimov, V.V. (1976) The primary photoreactions in the complex cytochrome-P-890·P-760 (bacteriopheophytin$_{760}$) of Chromatium minutissium at low redox potentials, Biochim. Biophys. Acta, 440, 587–599.

Shuvalov, V.A. and Parson, W.W. (1981) Energies and kinetics of radical pairs involving bacteriochlorophyll and bacteriopheophytin in bacterial reaction centers, Proc. Natl. Acad. Sci. (U.S.A.), 78, 957–961.

Shuvalov, V.A., Klevanik, A.V., Sharkov, A.V., Matveetz, J.A. and Krukov, P.G. (1978) Picosecond detection of BChl-800 as an intermediate electron carrier between selectively-excited P$_{870}$ and bacteriopheophytin in Rhodospirillum rubrum reaction centers, FEBS Lett., 81, 135–139.

Shuvalov, V.A., Dolan, E. and Ke, B. (1979a) Spectral and kinetic evidence for two early electron acceptors in photosystem I, Proc. Natl. Acad. Sci. (U.S.A.), 76, 770–773.

Shuvalov, V.A., Ke, B. and Dolan, E. (1979b) Kinetic and spectral properties of the intermediary electron acceptor A$_1$ in photosystem I, FEBS Lett., 100, 5–8.

Shuvalov, V.A., Klevanik, A.V., Sharkov, A.V., Kryukov, P.G. and Ke, B. (1979c) Picosecond spectroscopy of photosystem I reaction centers, FEBS Lett., 107, 313–316.

Shuvalov, V.A., Klimov, V.V., Dolan, E., Parson, W.W. and Ke, B. (1980) Nanosecond fluorescence and absorbance changes in photosystem II at low redox potential, FEBS Lett., 118, 279–282.

Song, P-S., Walker, E.B., Auerbach, A. and Robinson, G.W. (1981) Proton release from Stentor photoreceptors in the excited states, Biophys. J., 35, 551–555.

Spaulding, L.D., Eller, P.G., Bertrand, J.A. and Felton, R.H. (1974) Crystal and molecular structure of the radical perchloratotetraphenylporphinatozinc (II), J. Am. Chem. Soc., 96, 982–987.

Sundstrom, V., Rentzepis, P.M., Peters, K. and Applebury, M.L. (1977) Kinetics or rhodopsin at room temperature measured by picosecond spectroscopy, Nature (London), 267, 645–646.

Suzuki, Y., Tsuchiya, Y., Kinoshita, K. and Sugiyama, M. (1980a) Recent developments in picosecond streak camera systems, in: D.J. Bradley, G. Porter and M.H. Key (Eds.), Ultra-Short Laser Pulses, Royal Society, London, pp. 85–92.

Suzuki, K., Kobayashi, T., Ohtani, H., Yesaka, H., Nagakura, S., Shichida, Y. and Yoshizawa, T. (1980b) Observation of the picosecond time-resolved spectrum of squid bathorhodopsin at room temperature, Photochem. Photobiol., 32, 809–811.

Swenberg, C.E., Geacintov, N.E. and Pope, M. (1976a) Bimolecular quenching of excitons and fluorescence in the photosynthetic unit, Biophys. J., 16, 1447–1452.

Swenberg, C.E., Dominijanni, R. and Geacintov, N.E. (1976b) Effects of pigment heterogeneity on fluorescence in photosynthetic units, Photochem. Photobiol., 24, 601–604.

Tang, C.L. (Ed.) (1979) Quantum Electronics, Methods of Experimental Physics, Vol. 15 (A), Academic Press, New York.

Terner, J., Strong, J.D., Spiro, T.G., Nagumo, M., Nicol, M. and El-Sayed, M.A. (1981) Picosecond

338

resonance Raman spectroscopic evidence for excited-state spin conversion in carbonmonoxy-hemo-globin photolysis, Proc. Natl. Acad. Sci. (U.S.A.), 78, 1313–1317.

Tiede, D.M., Prince, R.C. and Dutton, P.L. (1976) EPR and optical spectroscopic properties of the electron carrier intermediate between the reaction center bacteriochlorophylls and the primary acceptor in *Chromatium vinosum*, Biochim. Biophys. Acta, 449, 447–469.

Tollin, G. (1976) Model systems for photosynthetic energy conversion, J. Phys. Chem., 80, 2274–2277.

Tollin, G., Castelli, F., Cheddar, G. and Rizzuto, F. (1979) Laser photolysis studies of quinone reduction by chlorophyll *a* in alcohol solution, Photochem. Photobiol., 29, 147–152.

Topp, M.R., Rentzepis, P.M. and Jones, R.P. (1971) Time resolved picosecond emission spectroscopy of organic dye lasers, Chem. Phys. Lett., 9, 1–5.

Tredwell, C.J., Porter, G., Synowiec, J., Barber, J., Searle, G.F.W. and Harris, L. (1977) Picosecond laser spectroscopy of the photosynthetic unit, in: M.A. West (Ed.), Lasers in Chemistry, Elsevier, Amsterdam, pp. 304–310.

Uhl, R. and Abrahamson, E.W. (1981) Dynamic processes in visual transduction, Chem. Rev., 81, 291–312.

Van Best, J.A. and Duysens, L.N.M. (1977) A one microsecond component of chlorophyll luminescence suggesting a primary acceptor of system II of photosynthesis different from Q, Biochim. Biophys. Acta, 459, 187–206.

Van Gorkom, H.J., Tamminga, J.J., Haveman, J. and Van Der Linden, I.K. (1974) Primary reactions, plastoquinone and fluorescence yield in subchloroplast fragments prepared with deoxycholate, Biochim. Biophys. Acta, 347, 417–438.

Van Gorkom, H.J., Pulles, M.P.J. and Wessels, J.S.C. (1975) Light-induced changes of absorbance and electron spin resonance in small photosystem II particles, Biochim. Biophys. Acta, 408, 331–339.

Warshel, A. (1976) Bicycle-pedal model for the first step in the vision process, Nature (London), 260, 679–683.

Warshel, A.W. (1978) Charge stabilization mechanism in the visual and purple membrane pigments, Proc. Natl. Acad. Sci. (U.S.A.), 75, 2558–2562.

Warshel, A. (1979) Conversion of light energy to electrostatic energy in the proton pump of *Halobacterium halobium*, Photochem. Photobiol., 30, 285–290.

Weber, M.J. (1979) Solid state lasers, in: C.L. Tang (Ed.), Quantum Electronics, Methods of Experimental Physics, Vol. 15 (A), Academic Press, New York, pp. 167–207.

Wiesenfeld, J.M. and Ippen, E.P. (1979) Tunable probe subpicosecond spectroscopy: spectral relaxation dynamics, Chem. Phys. Lett., 67, 213–217.

Williams, W.P. (1977) The two photosystems and their interactions, in: J. Barber (Ed.), Primary Processes of Photosynthesis, Elsevier, Amsterdam, pp. 99–147.

Yarborough, J.M. and Massey, C.A. (1971) Efficient high-gain parametric generation in ADP continuously tunable across the visible spectrum, Appl. Phys. Lett., 18, 438–440.

Yariv, A. (1971) Introduction to Optical Electronics, Holt, Rinehart and Winston, New York.

Yariv, A. (1975) Quantum Electronics, 2nd edn., John Wiley, New York.

Young, M. (1977) Optics and Lasers, Springer Series in Optical Sciences, Vol. 5, Springer, Berlin.

Yu, W., Ho, P.P., Alfano, R.R. and Seibert, M. (1975) Fluorescent kinetics of chlorophyll in photosystem I and II enriched fractions of spinach, Biochim. Biophys. Acta, 387, 159–164.

Yu, W., Pellegrino, F. and Alfano, R.R. (1977) Time-resolved fluorescence spectroscopy of spinach chloroplast, Biochim. Biophys. Acta, 460, 171–181.

Zernike, F. (1979) Nonlinear optical devices, in: C.L. Tang (Ed.), Quantum Electronics, Methods of Experimental Physics, Vol. 15 (B), Academic Press, New York, pp. 143–183.

Zernike, F. and Midwinter, J.E. (1973) Applied Nonlinear Optics, Wiley, New York.

Fluorescence photobleaching as a probe of translational and rotational motions

D.E. KOPPEL

Department of Biochemistry,
University of Connecticut Health Center, Farmington, CT 06032, U.S.A.

Contents

1. Introduction and perspective 339
2. Theoretical principles 342
    2.1. Translational diffusion on a planar surface 342
    2.2. Translational diffusion on a spherical surface 345
    2.3. Rotational diffusion 349
          2.3.1. Isotropic samples 350
          2.3.2. Alligned planar membranes 352
3. Experimental methods 353
    3.1. Optical system 353
    3.2. Sample data 355
    3.3. Statistical accuracy 358
4. Radiation-induced artifacts 362
    4.1. Heating during photobleaching 362
    4.2. Photochemical damage 362
5. Prospects and conclusions 364
References 365

1. Introduction and perspective

A classical approach to the study of molecular translational or rotational diffusion is to form an initial non-equilibrium distribution of concentration or orientation, and follows the dynamics of the subsequent relaxation processes. Hydrodynamic measurements of this type (free diffusion at a boundary, flow birefringence, electric dichroism) have yielded valuable information on the size and shape of purified biological macromolecules in solution (e.g., see Cantor and Schimmel, 1980).

Recently, the same general approach has been extended to the study of molecu-

R.I. Sha'afi and S.M. Fernandez (Eds.), Fast Methods in Physical Biochemistry and Cell Biology
© *1983 Elsevier Science Publishers*

lar dynamics in situ, down to the level of specific molecules in single living cells. The extraordinary sensitivity and specificity required for such measurements are supplied by the use of extrinsic fluorescent probes. Non-equilibrium distributions of concentration and orientation are produced by fluorescence photobleaching.

In general, "photobleaching" consists of a variety of photo-induced processes. Some of these, such as the intersystem crossing to long-lived triplet states, are reversible. Others, such as the oxidation of excited triplets, are irreversible. For a given system, the bleaching processes that one observes depends upon the time scale of the experiment. The time range of lateral diffusion measurements (generally $> 10^{-2}$ sec) is much longer than triplet state lifetimes in solution, even in anaerobic preparations. For all practical purposes, the bleaching observed in these experiments is irreversible. On the time scale of rotational measurements (the microsecond to millisecond range), however, a significant fraction of the bleaching observed may be reversible. Data analysis must take this into account in an appropriate way.

With a few noteworthy exceptions (e.g., Peters et al., 1974) fluorescence photobleaching experiments have used a laser light source for both monitoring and bleaching. Although several technical refinements (e.g., laser-spot scanning and image intensification) have added to the power of the technique, the basic features of the method can be understood best, in a general way, with examination of the simplest experimental design. We consider the case of monitoring and bleaching illuminations with the same polarizations and relative intensity distributions, differing only in the levels of absolute intensity. The fluorescence intensity excited by the monitoring beam is recorded before photobleaching, and as a function of time after bleaching (see Fig. 1). If the fluorescent probe molecules rotate on the time scale of the measurement, or move laterally over beam intensity variations, then one will observe the characteristic fluorescence "recovery" after photobleaching, illustrated schematically in Fig. 1. This recovery is the direct consequence of one of the funda-

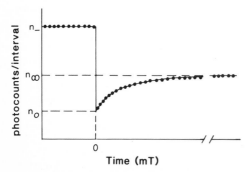

Fig. 1. Schematic representation of fluorescence "recovery" after photobleaching. The sample fluorescence is monitored with an attenuated laser beam. Momentarily increasing the intensity by a few orders of magnitude in a brief interval at time zero bleaches a substantial fraction of the fluorescent molecules. Subsequent redistribution or reorientation of the unbleached molecules leads to an increase in the observed level of fluorescence monitored after the bleaching pulse.

mental aspects of the method: both fluorescence excitation and the initiation of bleaching are mediated through the same absorption transition dipole moments. As a result, those fluorphores which emit the least fluorescence during the initial monitoring stage by virtue of an unfavorable location (in a non-uniform beam) or an unfavorable orientation (relative to the excitation polarization), are also those which are least likely to be bleached. The subsequent redistribution or reorientation of the unbleached molecules after the bleaching pulse will thus increase the emitted fluorescence intensity.

Fluorescence photobleaching methods were first applied to the study of translational motion in cell plasma membranes (Peters et al., 1974). Localized or non-uniform photobleaching was developed as a convenient, general way of providing the kind of non-uniform cell surface labeling achieved in earlier experiments by such means as heterokaryon fusion (Frye and Edidin, 1970) and spot labeling (Edidin and Fambrough, 1973). For a recent review of these and other techniques for the study of membrane protein lateral mobility, see Koppel (1982).

More recently (L.M. Smith et al., 1981; Johnson and Garland, 1981), fluorescence photobleaching has proven to be a promising approach to the study of rotational motion in membranes as well. Measurements of fluorescence intensities excited with polarized monitoring beams, can be used to determine the fluorescence-detected absorption anisotropy as a function of time after bleaching. This approach is directly analogous to earlier studies of transient linear dichroism with naturally occurring membrane chromophores (Cone, 1972), and extrinsic triplets probes (Cherry, 1978, 1979). Fluorescence-detected absorption, however, offers the possibility of large increases in sensitivity (Johnson and Garland, 1981).

Similar rotational information can be obtained from measurements of time-resolved *emission* anisotropy (Rigler and Ehrenberg, 1973) (see Chapter 9). The essential difference, however, is the matter of time scale. Fluorescence emission anisotropy reports on motion limited to times comparable to the lifetime of the excited singlet state ($\sim 10^{-8}$ sec). Phosphorescence emission (Austin et al., 1979; Garland and Moore, 1979) extends the range to the lifetime of excited triplet states (as long as $\sim 10^{-3}$ sec in anaerobic solutions), but has an extremely low quantum yield. Fluorescence photobleaching extends the accessible time range essentially without limit.

It is the object of this article to try to present a fresh perspective on certain selected aspects of the theory and experimental methods of fluorescence photobleaching techniques. No attempt has been made to present a summary of experimental results. This can be found in several recent reviews (Cherry, 1979; Peters, 1981; Edidin, 1981).

2.   Theoretical principles

2.1.   Translational diffusion on a planar surface

Several specialized treatments of fluorescence photobleaching theory for a planar geometry have been published. Some of these (Axelrod et al., 1976; Koppel, 1979) have dealt with spatially localized photobleaching pulses, placing a primary emphasis on the analysis of distributions in real coordinate space. Others (B.A. Smith and McConnell, 1978; B.A. Smith et al., 1979; L.M. Smith et al., 1979; Lanni and Ware, 1982) have introduced extended periodic pattern photobleaching, shifting the emphasis to Fourier transform space. It is the object of this section to present a simple, general analysis of the theory in both coordinate and Fourier space, utilizing the general mathematical formalism of linear shift invariant systems analysis (Cooper and McGillem, 1967; Bracewell, 1978).

Consider a general "system" which maps a set of input functions, $f(\underline{r})$, into a set of output functions, $g(\underline{r})$; a process which can be represented schematically as (see Fig. 2A):

$$f(\underline{r}) \rightarrow g(\underline{r}) \tag{1}$$

As we shall see, a "system", in this context, can take the form of a physical or chemical process, or a measuring instrument monitoring that process. For the purpose of this section, $\underline{r}$ can be thought of as a two-dimensional coordinate on the surface of the plane.

(A)

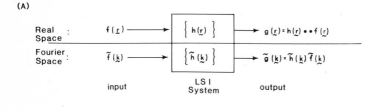

(B)

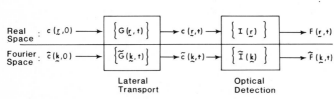

Fig. 2.   The application of linear, shift-invariant systems analysis to the analysis of lateral diffusion on a plane. (A) For a general LSI system, the input function, $f(r)$ in real space, is transformed into the output function, $g(r)$, by convolution with the impulse response function, $h(r)$. In Fourier transform space, $\tilde{g}(k)$ is the product of $\tilde{f}(k)$ with $\tilde{h}(k)$, the transfer function of the system. (B) Lateral diffusion in a plane can be analyzed in terms of two sequential LSI processes. $G(r,t)$ and $I(r)$ are the impulse response functions associated with lateral transport and optical detection respectively.

Valuable, general relationships exist between $f(\underline{r})$ and $g(\underline{r})$ if we place certain conditions on the properties of the transformation. We require that the system be linear and shift invariant (LSI). A system is linear if

$$A_1 f_1(\underline{r}) + A_2 f_2(\underline{r}) \to A_1 g_1(\underline{r}) + A_2 g_2(\underline{r}) \tag{2}$$

whenever

$$f_1(\underline{r}) \to g_1(\underline{r})$$

$$f_2(\underline{r}) \to g_2(\underline{r})$$

The system is shift invariant if the condition of Eqn.1 implies

$$f(\underline{r} - \underline{r}') \to g(\underline{r} - \underline{r}') \tag{3}$$

There are two LSI transformations associated with the photobleaching measurements on the plane (see Fig. 2B). The first arises from the lateral transport process itself; cf.

$$c(\underline{r},0) \to c(\underline{r},t), \tag{4}$$

where $c(\underline{r},0)$ is the initial concentration distribution of label, conveniently (but not necessarily) defined as that immediately after photobleaching, and $c(\underline{r},t)$ is the distribution a time t later. The second process which can be described in terms of an LSI system is that of the optical detection of $c(\underline{r},t)$. In this case,

$$c(\underline{r},t) \to F(\underline{r},t), \tag{5}$$

where $F(\underline{r},t)$ is the distribution of detected fluorescence intensity at time t.

LSI systems, in general, are completely characterized by a function, $h(\underline{r})$, called the impulse response function. It is straightforward to show that given the properties of Eqns. 2 and 3, $g(\underline{r})$ and $f(\underline{r})$ are related through the two-dimensional convolution operation (see Chapter 1):

$$g(\underline{r}) = \int h(\underline{r} - \underline{r}') f(\underline{r}') d^2 \underline{r}' \tag{6}$$

often written in shorthand notation as

$$g(\underline{r}) = h(\underline{r}) ** f(\underline{r})$$

It is clear from Eqn. 6 how $h(\underline{r})$ is an 'impulse response'. When the input, $f(\underline{r})$, is a Dirac delta function, the output, $g(\underline{r})$, is simply equal to $h(\underline{r})$.

344

In Fourier space, the relationship between input and output is even more straightforward. Consider $\tilde{g}(\underline{k})$, the two-dimensional Fourier transform of $g(\underline{r})$:

$$\tilde{g}(\underline{k}) \equiv \frac{1}{2\pi} \int g(\underline{r}) \exp(-i\underline{k} \cdot \underline{r}) d^2\underline{r} \tag{7}$$

Plugging Eqn. 6 into Eqn. 7, applying the well known relationship for the Fourier transform of a convolution, we have

$$\tilde{g}(\underline{k}) = \tilde{h}(\underline{k})\tilde{f}(\underline{k}) \tag{8}$$

$\tilde{h}(\underline{k})$, the Fourier transform of the impulse response function, is known as the transfer function of the system.

Applying these general results to the specific problem at hand, we can, first of all, identify a function $G(\underline{r},t)$ as the impulse response function of the lateral transport process, i.e. (see Fig. 2B):

$$c(\underline{r},t) = G(\underline{r},t)**c(\underline{r},0) \tag{9}$$

$G(\underline{r},t)$ is the probability per unit area that a labeled molecule at an arbitrary origin at time 0 will be at point $\underline{r}$ at time t. For simple, isotropic, two-dimensional diffusion with coefficient D,

$$G(\underline{r},t) = (4\pi Dt)^{-1} \exp(-|\underline{r}|^2/4Dt) \tag{10}$$

Similarly, we can designate the impulse response function of the optical detection process as $I(\underline{r})$, so that

$$F(\underline{r},t) = I(\underline{r})**c(\underline{r},t) \tag{11}$$

This relationship is a general one that is applicable to the different types of detection methods that have been employed. If one applies a uniform monitoring illumination to the sample after photobleaching and records the fluorescence image (e.g., B.A. Smith and McConnell, 1978; B.A. Smith et al., 1979), then $I(\underline{r})$ is the combined 'point spread function' of the imaging and recording systems, and $F(\underline{r},t)$ is the resulting image. Alternatively, if one moves a non-uniform monitoring beam across the sample in a series of scans after photobleaching (e.g., Koppel, 1979; Lanni and Ware, 1982), then $I(-\underline{r})$ corresponds to the monitoring beam intensity distribution, and $F(\underline{r},t)$ is the total fluorescence intensity detected when the monitoring pattern is centered at position $\underline{r}$.

Substituting Eqn. 9 into Eqn. 11 one obtains

$$F(\underline{r},t) = I(\underline{r})**[G(\underline{r},t)**c(\underline{r},0)] \tag{12}$$

Thus, theoretical expressions for $F(\underline{r},t)$ can be derived for specific assumed functional forms of $I(\underline{r})$ and $c(\underline{r},0)$ (as in Axelrod et al., 1976). Rearranging the order of convolution in Eqn. 12 one can write

$$F(\underline{r},t) = G(\underline{r},t)**F(\underline{r},0) \tag{13}$$

where

$$F(\underline{r},0) = I(\underline{r})**c(\underline{r},0)$$

Thus, expected values of $F(\underline{r},t)$ can also be derived from assumed or measured values of $F(\underline{r},0)$ (as in Koppel, 1979).

The problem is simplified significantly if the data is analyzed in Fourier transform space (as in B.A. Smith and McConnell, 1978; B.A. Smith et al., 1979; Lanni and Ware, 1982). Transforming both sides of Eqn. 13 gives

$$\widetilde{F}(k,t) = \widetilde{F}(\underline{k},0)\widetilde{G}(\underline{k},t) \tag{14}$$

where for the $G(\underline{r},t)$ of Eqn. 10

$$\widetilde{G}(\underline{k},t) = \exp(-D|\underline{k}|^2 t) \tag{15}$$

Thus, each Fourier component decays as a simple exponential. One need not know or specify the exact forms of $I(\underline{r})$ and $c(\underline{r},0)$ or $F(\underline{r},0)$. These affect only the amplitude of the observed signal.

2.2. Translational diffusion on a spherical surface

Two approaches have been devised specifically for the study of lateral diffusion on a spherical surface. The first of these (Koppel et al., 1980) employs high resolution fluorescence scans across the membrane surface, and is applicable to vesicles and cells much larger than the resolution limit of the microscope. The second (L.M. Smith and McConnell, 1981) is applicable in principle to vesicles whose radii are less than the wavelength of light, but requires that the absorption transition dipole moment of the fluorescent label have a distinctly non-random orientation relative to the membrane surface.

The general solution in spherical coordinates for isotropic diffusion on the surface of a sphere of radius r can be written as (Huang, 1973; L.M. Smith and McConnell, 1981)

$$C(\theta,\varphi,t) = \sum_{l=0}^{\infty} \sum_{m=-l}^{l} A_{lm} Y_{lm}(\theta,\varphi) \exp[-l(l+1)Dt/r^2] \tag{16}$$

The spherical harmonics, $Y_{lm}(\theta,\varphi)$, form a complete orthogonal set on the surface of the unit sphere (e.g., see Jackson, 1962), normalized such that

$$\int Y^*_{l'm'}(\theta,\varphi)Y_{lm}(\theta,\varphi) \sin \theta d\theta d\varphi = \delta_{l'l}\delta_{m'm} \tag{17}$$

where δ_{ij} is the Kronecker delta function defined as

$$\delta_{ij} = \begin{cases} 0 & i \neq j \\ 1 & i = j \end{cases}$$

For a uniform concentration distribution, all coefficients A_{lm} are zero except for the constant term $Y_{00} = (4\pi)^{-\frac{1}{2}}$. Non-uniform photobleaching reduces the value of A_{00}, and introduces non-zero higher order terms. The values of A_{lm} after bleaching are determined by the initial post-bleach concentration distribution

$$A_{lm} = \int c(\theta,\varphi,0)Y^*_{lm}(\theta,\varphi) \sin \theta d\theta d\varphi \tag{18}$$

The problem simplifies considerably if we assume that $c(\theta,\varphi,0)$ is azimuthally symmetric, i.e. independent of angle φ. In this case all terms in the series with $m \neq 0$ vanish, and we are left with a single sum expansion in terms of

$$Y_{l0}(\theta,\varphi) = [(2l + 1)/4\pi]^{\frac{1}{2}}P_l(\cos \theta), \tag{19}$$

where $P_l(\cos \theta)$ is the ordinary Legendre polynominal of order l. For the first few values of l:

$$P_0(z) = 1$$
$$P_1(z) = z$$
$$P_2(z) = (\tfrac{1}{2})(3z^2 - 1)$$
$$P_3(z) = (\tfrac{1}{2})(5z^3 - 3z)$$

For a large vesicle or spherical cell, one can apply a localized photobleaching pulse and subsequently map out the redistribution of unbleached label as a function of time with a series of linear fluorescence scans along the axis of azimuthal symmetry (see section 3.2 below). For experimental estimates of $c(z,t)$ (where $z \equiv \cos \theta$ is proportional to the distance along the scan axis) one can compute the "moments" or normal-mode amplitudes of the distribution at time t:

$$M_l(t) \equiv \int_{-1}^{1} P_l(z)c(z,t)dz \tag{20}$$

Combining Eqns. 16, 17, 19 and 20, we have

$$M_l(t)/M_l(0) = \exp[-l(l + 1)Dt/r^2], \tag{21}$$

independent of the particular functional form of $c(z,0)$. The distribution of bleaching is usually designed to maximize either A_{10} (and therefore $M_1(0)$), A_{20} (and therefore $M_2(0)$) or both.

Recently, L.M. Smith and McConnell (1981) have described an alternative method based upon the polarization photoselection process. Bleaching the sample with a short burst of uniform linearly polarized light produces an anisotropic orientation distribution of chromophore transition dipoles (see section 2.2 below). Orientation anisotropy implies a non-uniform surface distribution, provided the absorption transition dipole moments have a non-random orientation relative to the membrane surface. Translational diffusion subsequently leads to the relaxation of the non-uniform surface distribution (see Eqn. 16), and hence reduces the orientation anisotropy.

A quantitative analysis of the process can be performed through a consideration of the absorption anisotropy function (see section 2.3.1)

$$r(t) = \overline{\langle P_2(z_{A3})\rangle}_t \tag{22}$$

Here z_{A3} is the cosine of the angle (θ_A in Fig. 3) between the absorption transition dipole ($\hat{\mu}_A$ in Fig. 3) and the laboratory fixed axis of bleaching polarization ($\hat{x}_3$ in Fig. 3), the bar is a time average over all reorientation processes fast compared to the bleaching rate, and $\langle\rangle_t$ signifies an ensemble average over all the bleached molecules at time t. $r(t)$ can be determined experimentally from measurements of polarized fluorescence depletion (see section 2.3.1).

To couple $r(t)$ to translational diffusion, we can rewrite Eqn. 22 in terms of z_{n3}, the cosine of the angle between $\hat{n}$, the time-averaged (during the bleach) direction

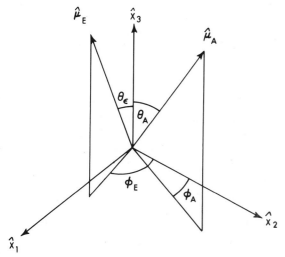

Fig. 3. Spherical coordinates of fluorescent probe absorption transition dipoles ($\hat{\mu}_A$) and emission transition dipoles ($\hat{\mu}_E$) with respect to a laboratory-fixed Cartesian coordinate system.

of $\hat{\mu}_A$, and polar axis $\hat{x}_3$. For probe molecules free to rotate in the membrane rapidly compared to the bleaching rate, $\hat{n}$ is the normal to the membrane surface at the position of the probe, and z_{n3} is equivalent to the cosine of angle θ in Eqn. 16. Moreover, $\hat{\mu}_A$ is azimuthally symmetric about $\hat{n}$, so that (see Petersen and Chan, 1977):

$$\overline{P_2(z_{A3})} = \overline{P_2(z_{An})}P_2(z_{n3}) \tag{23}$$

where z_{An} is the cosine of the equatorial angle between $\hat{\mu}_A$ and $\hat{n}$. Hence

$$r(t) = S\langle P_2(z_{n3})\rangle_t \tag{24}$$

where we have identified

$$S \equiv \overline{P_2(z_{An})} \tag{25}$$

as the order parameter for $\hat{\mu}_A$ (see, for example, Seelig, 1977). Thus by Eqns. 16, 17, 19 and 25, we have

$$r(t)/r(0) = \exp(-6Dt/r^2) \tag{26}$$

analogous to Eqn. 21.

The magnitude of $r(0)$ is of interest in itself. The ensemble average in Eqn. 24 can be expressed as an integral over $W(z_{n3},t)$, the distribution of z_{n3} for the bleached component:

$$r(t) = S\int W(z_{n3},t)P_2(z_{n2})dz_{n3}/ \ W(z_{n3},t)dz_{n3} \tag{27}$$

We assume that bleaching proceeds as a first-order process with a rate constant proportional to

$$\overline{|\hat{\mu}_A \cdot \hat{x}_3|^2} = \overline{z_{A3}^2} = (1/3)[1 + SP_2(z_{n3})] \tag{28}$$

with the latter equality derived from a combination of Eqns. 23 and 25. After a bleaching pulse of duration T, we then have:

$$W(z_{n3},0) = (1/2)[1 - \exp\{-\Gamma T[1 + SP_2(z_{n3})]\}] \tag{29}$$

where Γ is the photobleaching rate constant for an isotropic distribution. Expanding in powers of ΓT, substituting back into Eqn. 27, gives finally

$$r(0) = (2/5)S^2\{1 - [(1/2) + (2/7)S - (2/5)S^2]\Gamma T + \dots +\} \tag{30}$$

Thus one can calculate the order parameter S from measured values of $r(0)$, obtaining potentially important structural information. Note that in a totally 'ordered' system, in the absence of motion of any kind on the time scale of bleaching, $z_{An} = 1$, and $S = 1$. For a 'disordered' system, in which μ_A is free to move isotropically about $\hat{n}$, during bleaching $\overline{z^2_{An}} = 1/3$, and $S = 0$. In general, $-1/2 \leq S \leq 1$.

2.3. Rotational diffusion

We seek to characterize the decay of the orientation anisotropy induced by a polarized photobleaching pulse with time-resolved measurement of fluorescence-detected absorption. A suitably defined anisotropy function should fulfill certain minimum requirements. (i) It should be independent of the kinetics of probe chemical recovery in cases of reversible bleaching. (ii) It should be readily measurable within the constraints of practical experimental geometries. (iii) It should be readily interpreted with simple models of probe dynamics.

The first of these requirements is met by considering suitably normalized functions of the anisotropic fluorescence *depletion* (Johnson and Garland, 1981). In this case, analogous to defined *emission* anisotropies (e.g. Rigler and Ehrenberg, 1973), one follows the orientation distribution of the component photoselected (in this case, bleached) by the excitation flash. For a homogeneous probe population, this will depend solely on probe rotation, independent of chemical recovery. In contrast, the chemical recovery of bleached molecules will reduce the anisotropy of the unbleached component, even in the complete absence of probe rotation.

The last two requirements listed above lead to definitions of anisotropy dependent upon the sample geometry and optical configuration. Two limiting cases are considered below.

In general, all information that one can obtain on the orientation distribution of fluorescent molecules in an ordered structure is contained in a set of nine intensity values. These are the functions $F_{ij}(t)$, defined as the fluorescence intensity with polarization along the jth axis of a laboratory-fixed Cartesian coordinate system ($j = 1$–3), excited at time t by an incident beam with polarization along the ith axis, normalized by the incident intensity (Badley et al., 1973). Referring to Fig. 3, the depletion of fluorescence intensity $F_{ij}(t)$,

$$\Delta F_{ij}(t) \equiv F_{ij}(-) - F_{ij}(t) \tag{31}$$

where $F_{ij}(-)$ is the intensity before bleaching, can be expressed as:

$$\Delta F_{ij}(t) = \langle |\hat{\mu}_A \cdot \hat{x}_i|^2 |\hat{\mu}_E \cdot \hat{x}_j|^2 \rangle_t N(t) \tag{32}$$

Here $\hat{\mu}_A$ and $\hat{\mu}_E$ are unit vectors along the fluorophore absorption and emission dipoles, $\hat{x}_i$ and $\hat{x}_j$ are unit vectors of the Cartesian coordinate system, $<>_t$ signifies an ensemble average over the distribution of bleached molecules at time t after bleaching, and $N(t)$ is the number of molecules still bleached at time t.

Whenever possible, it is desirable to eliminate the complicating dependence on the orientation distribution of $\hat{\mu}_E$ (which in general is *not* parallel to $\hat{\mu}_A$). This can be accomplished through determinations of $\Delta F_i(t)$, the total fluorescence emission depletion (for all emitted polarizations) observed with an excitation polarization along $\hat{x}_i$. From the definition of $\Delta F_{ij}(t)$,

$$\Delta F_i(t) = \sum_{j=1}^{3} \Delta F_{ij}(t) = \langle |\mu_A \cdot \hat{x}_i|^2 \rangle_t N(t) \tag{33}$$

$\Delta F_i(t)$ is simply proportional to the polarized absorption depletion determined in the experimental approach of Cherry and coworkers (Cherry, 1978, 1979).

To proceed further, we must specify particular sample geometries. We consider two useful limiting cases: (1) samples isotropic before the bleach, e.g., spherical cells, vesicles or membrane suspensions; and (2) oriented planar membranes. In either case, we identify three functions $\Delta F_\parallel(t)$, $\Delta F_\perp(t)$ and $\Delta F_T(t)$, the suitably defined fluorescence intensity changes for excitation polarization parallel to the polarization of the bleaching flash ($\parallel$), perpendicular to the polarization of the bleaching flash ($\perp$), and the total change (T), independent of excitation polarization. The sample anisotropy is then characterized by the normalized function:

$$r(t) = \frac{\Delta F_\parallel(t) - \Delta F_\perp(t)}{\Delta F_T(t)} \tag{34}$$

2.3.1. Isotropic samples

The polarization of the bleaching beam defines a unique laboratory-fixed reference axis (call it the $\hat{x}_3$ axis). The monitoring beam is incident along an axis orthogonal to $\hat{x}_3$, with polarization either parallel or perpendicular to $\hat{x}_3$.

We can then equate (referring to Eqn. 33 and Fig. 3):

$$\Delta F_\parallel(t) \equiv \Delta F_3(t)$$
$$= \langle \cos^2 \theta_A \rangle_t N(t) \tag{35}$$

$$\Delta F_{\perp(t)} \equiv \Delta F_1(t) = \Delta F_2(t)$$
$$= \tfrac{1}{2} \langle \sin^2 \theta_A \rangle_t N(t) \tag{36}$$

$$\Delta F_T(t) \equiv \sum_{i=1}^{3} \Delta F_i(t)$$
$$= \Delta F_\parallel(t) + 2\Delta F_\perp(t)$$
$$= N(t) \tag{37}$$

Combining Eqns. 34–37,

$$r(t) = \tfrac{1}{2}(3 \langle \cos^2 \theta_A \rangle_t - 1)$$
$$= \langle P_2 (\cos \theta_A) \rangle_t \tag{38}$$

where $P_2(x)$ is the second-order Legendre polynomial.

In a practical experimental configuration, one cannot measure directly all the components of $\Delta F_{ij}(t)$ needed to compute $\Delta F_\parallel(t)$ and $\Delta F_\perp(t)$. From considerations of symmetry, however, all needed components can be deduced from measurable ones. Assume, for example, that fluorescence is collected back along the axis of the monitoring beam (call it the $\hat{x}$ axis), the geometry appropriate for a fluorescence microscope equipped with vertical illumination (see section 3.1 below). In this case one can measure $F_{32}(t)$, $F_{33}(t)$, $F_{22}(t)$ and $F_{23}(t)$, but it is impossible to measure $F_{31}(t)$ and $F_{21}(t)$. By symmetry, however,

$$F_{31}(t) = F_{32}(t) \tag{39}$$

and since

$$\frac{F_{21}(t)}{F_{22}(t)} = \frac{F_{21}(-)}{F_{22}(-)} = \frac{F_{23}(-)}{F_{22}(-)} \tag{40}$$

we have

$$F_{21}(t) = F_{23}(-)F_{22}(t)/F_{22}(-) \tag{41}$$

As a simple example, first consider isotropic rotational diffusion in solution, with rotational diffusion coefficient D_R. Mathematically, this is equivalent to the isotropic lateral diffusion of the "tip" of $\hat{\mu}_A$ on the surface of the unit sphere. The problem has been solved in section 2.2 above with the result (compare with Eqn. 26)

$$r(t)/r(0) = \exp(-6D_R t) \tag{42}$$

The absorption depletion anisotropy has also been evaluated for models of anisotropic rotation in membranes (Cherry and Godfrey, 1981; Kawato and Kinosita, 1981). In the simplest case, if it is assumed that rotation occurs only about the local axis normal to the membrane (with rotational diffusion coefficient $D_\parallel$), then, for linear or circularly symmetric chromophores (Kawato and Kinosita, 1981),

$$r(t)/r(0) = 3 \sin^2 \theta \cos^2 \theta \exp(-D_\parallel t) + (3/4) \sin^4 \theta \exp(-4D_\parallel t) + [P_2 (\cos \theta)]^2 \tag{43}$$

where θ is the angle between the mebrane normal and the absorption dipole moment (or the perpendicular to the plane of the chromophore for the circularly symmetric case). This model has been shown to provide an excellent description of the anisotropic rotation of bacteriorhodopsin in lipid membranes (Cherry and Godfrey, 1981).

352

2.3.2. Alligned planar membranes

Consider a planar membrane or stack of membranes in the $\hat{x}_1,\hat{x}_2$ plane, perpendicular to the optical axis of illumination and fluorescence collection, with the bleaching polarization alligned along the $\hat{x}_2$ axis. In this case, referring again to Fig. 3, θ_A and θ_E now are angles relative to the membrane normal, and φ_A and φ_E are azimuthal rotational angles about the normal.

Experimentally, we have access only to $F_{11}(t)$, $F_{12}(t)$, $F_{21}(t)$ and $F_{22}(t)$. We can then define:

$$\Delta F_{\|}(t) \equiv \Delta F_{21}(t) + \Delta F_{22}(t)$$
$$= \langle \cos^2 \varphi_A \sin^2 \theta_A \sin^2 \theta_E \rangle_t \, N(t) \tag{44}$$

$$\Delta F_{\perp}(t) \equiv \Delta F_{11}(t) + \Delta F_{12}(t)$$
$$= \langle \sin^2 \varphi_A \sin^2 \theta_A \sin^2 \theta_E \rangle_t \, N(t) \tag{45}$$

and

$$\Delta F_T(t) \equiv \Delta F_{\|}(t) + \Delta F_{\perp}(t)$$
$$= \langle \sin^2 \theta_A \sin^2 \theta_E \rangle_t N(t) \tag{46}$$

Consider again the model of probe rotation about the membrane normal, assuming that the distributions of θ_A and θ_E do not change on the time scale of the experiment. Combining Eqns. 34, and 44–46, we then have

$$r(t) = 2 \langle \cos^2 \varphi_A \rangle_t - 1 \tag{47}$$

This can be readily evaluated in terms of rotational diffusion coefficient $D_{\|}$ (Cone, 1972), giving finally,

$$r(t)/r(0) = \exp(-4D_{\|}t) \tag{48}$$

Comparison of Eqns. 43 and 48 demonstrates the advantage of the alligned planar membrane configuration over membrane suspensions. Even under optimal conditions, one needs excellent data to extract meaningful values of $D_{\|}$ from data of the form of Eqn. 43. Analysis is extremely difficult for samples with both a range of $D_{\|}$ values (e.g., in an aggregating system), and a range of transition moment orientation angles (θ) relative to the membrane normal (e.g. for an extrinsic probe labeling several sites on the protein). In contrast, the anisotropy function of Eqn. 48 is independent of probe orientation relative to the membrane normal, and can be readily evaluated for a polydisperse distribution of diffusion rates (Koppel, 1972).

3.   Experimental methods

3.1.   Optical system

Fig. 4 shows a schematic diagram of the optical apparatus that is currently in use in our laboratory with minor modifications (Koppel, 1979, 1982). Designed primarily for measurements of lateral diffusion, it can readily be adapted for measurements of rotational diffusion, and serves as a useful starting point for a general description.

As shown, the apparatus is centered about a modified research microscope, equipped with a fluorescence vertical illuminator. Rotational measurements, of course, can also be performed on bulk suspensions, in a modified spectrofluorometer geometry.

The light source for fluorescence excitation is provided by a CW water-cooled argon ion laser. These lasers provide ample power in several discrete lines ranging from 457.9 nm in the blue to 528.7 nm in the green, with the strongest lines at 488.0 and 514.5 nm suitable for the commonly used fluorescein and rhodamine dyes. UV lines are available at 351.1 and 363.8 nm with special optics.

The laser beam is attenuated to an appropriate degree (by neutral density filter

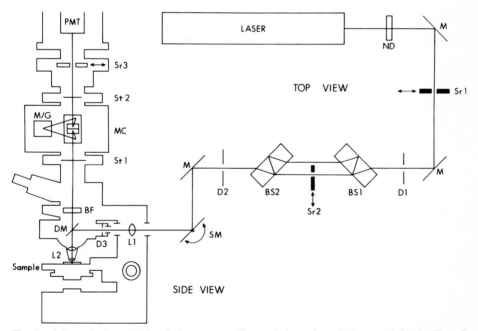

Fig. 4.   Schematic diagram of optical apparatus. The symbols used are: ND, neutral density filter; Sr, shutter; M, mirror; D, diaphragm; BS, beam splitter; SM, scanner mirror; L, lens; DM, dichroic mirror; BF, barrier filter; St, slit; MC, monochromater housing; M/G, mirror or grating; PMT, photomultiplier tube. (Reprinted with permission from Koppel, 1979.)

ND, Fig. 4), and directed by mirrors (M) into the back of the fluorescence illuminator of the microscope. Lens L1 is an auxiliary microscope objective that adjusts the final focus of the laser beam at the sample. The precise orientation of the incident laser beam, and hence the location on the sample of the beam along a scan axis, is modulated by a servo-activated galvanometric optical scanning mirror (SM).

A real image of the fluorescent sample is formed in the plane of the entrance slit (St1) of a monochrometer housing (MC). The resulting spatial discrimination acts to block the major part of background fluorescence excited above or below the in-focus object plane (Koppel et al., 1976). In normal operation, component M/G is a mirror (M) that reflects all the fluorescence through the exit slit St_2. When desired, however, the mirror can be replaced with the regular monochrometer grating (G) for the measurement of fluorescence emission spectra. If spectra are not needed, this whole set-up can be replaced by a simple slit or diaphragm. A final lens (not shown) re-images the back focal plane of the microscope objective onto the photomultiplier tube (PMT), located in a thermoelectrically cooled housing supported above the microscope.

For relaxation processes with recovery times $\gg 1$ msec (i.e. lateral diffusion, or extraordinarily slow rotational diffusion as in L.M. Smith et al., 1981), bleaching pulses can be timed with electro-mechanical shutters. The combination of diaphragms D1 and D2, uncoated beam splitters BS1 and BS2, and shutter Sr2 in Fig. 4 work together to switch the laser intensity between bleaching levels and monitoring levels. With Sr2 open, a nearly unattenuated beam, transmitted through BS1 and BS2 without reflection, passes through diaphragm D2. This provides the bleaching beam. With Sr2 closed, however, only that beam reflected 4 times, once at each glass-to-air interface, is in a position to pass through D2. This attenuates the beam by a factor of $\sim 10^4$ ($\sim 10\%$ is reflected each time) producing the monitoring beam.

Physically separating the bleaching and monitoring beams allows the possibility of selectively modifying one or the other. To bleach a pole of a spherical cell, for example, a cylindrical lens can be placed in the bleaching beam, producing a focused line on the sample perpendicular to the polar axis. In this configuration, bleaching approximates the desired azimuthal symmetry (see section 2.2 above). Monitoring of the subsequent redistribution, on the other hand, can be performed with the attenuated beam focused to a small circularly symmetric spot. This produces scans with sharp edges, giving clear indication of the cell boundaries, and demonstrating the maintenance of sharp focus.

For faster relaxation rates, the mechanical shutter can be replaced by an electro-optic or acousto-optic modulator (e.g., Johnson and Garland, 1981) to yield bleaching pulses in the microsecond range. The extent of bleaching induced by these short pulses will be severly limited, however, so that extensive signal averaging may be needed. To bleach a large fraction of the molecules in a short bleaching pulse, one can use a separate pulsed laser, (e.g. neodymium YAG or flashlamp-pumped dye lasers) with peak intensities far greater than those available from a

CW ion laser. Under these conditions, however, one should beware of possible sample heating during bleaching.

The PMT is protected from the high level of fluorescence emitted during the bleaching pulse by an electronic shutter Sr3. Alternatively, one can prevent damagingly high anode currents by pulsing down the PMT high voltage during bleaching, or applying a reverse bias at the first dynode or group of dynodes (Farinelli and Malvano, 1958; DeMarco and Penco, 1969). Specialized experimental phototubes have a gating grid incorporated between the cathode and the first dynode, analogous to a conventional planar vacuum triode (e.g., Barisas and Leuther, 1980). A control grid bias voltage change of only a few volts can reduce the tube output current by more than a factor of 10^4, in a time of less than 5 nsec.

Additional polarization optics are needed for fluorescence-detected anisotropy experiments. For some measurements, the incident polarization of the monitoring beam must be rotated 90° with respect to the bleaching polarization. This can be accomplished with a half-wavelength plate, a broad-band polarization rotator, or a Pockels cell used without polarizers. In the configuration of Fig. 4, the polarization rotator can be placed between the beam splitters so as to intersect only the monitoring beam, again taking advantage of the physical separation between the bleaching and monitoring beams.

3.2. Sample data

The data presented in this section were selected to demonstrate the features and capabilities of the kind of optical apparatus described above. Specific examples are drawn from measurements of lateral diffusion. Examples of rotational measurements are found in the recent literature (L.M. Smith et al., 1981; Johnson and Garland, 1981).

Fig. 5 presents an example of isotropic lateral diffusion in a plane, analyzed under optimal conditions in a laser spot scanning experiment. The measurement was performed on a reconstituted planar multibilayer membrane composed of a 1 : 1.3 : 1 mixture by weight of phospholipids, matrix protein, and lipopolysaccharide, isolated from the outer membrane of gram negative bacteria *Escherichia coli* and *Salmonella typhimurium*. Before reconstitution, the fluorescent probe *N*-4-nitrobenzo-2-oxa-1,3-diazole phosphatidylethanolamine (NBD-PE) was added to the phospholipid mixture at a molar ratio of 1 : 1000. The preparation and characterization of such membranes are described in detail elsewhere (Schindler et al., 1980b).

The figure shows a set of recovery curves of NBD-PE fluorescence measured with a circularly symmetric laser spot sequentially positioned at 12 equally spaced locations separated by 1.0 μm. The bleaching pulse was centered at the 7th location. The top graph, for simplicity, shows fluorescence data from only 3 locations: coincident with the bleaching pulse (▲), and 5.0 μm on either side of center (+, ×). Note that the recovery curves on either side of center are identical, consistent with an isotropic diffusive transport mechanism.

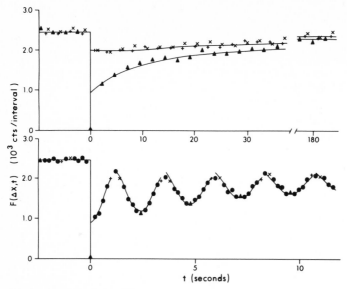

Fig. 5. Fluorescence redistribution after photobleaching data of NBD-PE in a reconstituted multibilayer membrane (1 : 1.3 : 1 by weight phospholipid to matrix protein to lipopolysaccharide), monitored at 12 equally spaced spots (1.0 μM separation) about a single point bleach. The solid curves are a fit to theory with $w = 4.49$ μm, $D = 4.71 \times 10^{-9}$ cm^2/sec. (Top): Recoveries measured at 3 of the 12 locations; coincident with the bleach pulse (▲), and 5.0 μm on either side of center (+, ×). (Bottom): Sequential scans including all 12 locations on expanded time scale. (Reprinted with permission from Koppel, 1979.)

The bottom graph in Fig. 5 shows fluorescence data recorded at all 12 locations plotted on a 4 times expanded time axis. The solid theoretical curves now trace out the individual scans through the bleached region. Fitting the data to the theory (Koppel, 1979) yields values of $\tau_D = w^2/4D$, the characteristic diffusion time, w, the characteristic diffusion distance, and hence D, the diffusion coefficient.

Fig. 6 demonstrates one version of the periodic pattern bleaching method of characterizing diffusion in a plane (L.M. Smith et al., 1979; Lanni and Ware, 1982), in this case a 3T3 fibroblast in culture labeled with fluorescein conjugated succinyl-concanavalin A. A laser illuminated Ronchi ruling is imaged down onto the cell for monitoring and bleaching purposes, forming a periodic pattern of stripes. Bleaching in this configuration produces a periodic pattern of fluorescent molecules. The modulation amplitude of this concentration patern is monitored as a function of time after bleaching with a series of fluorescence scans of the monitoring illumination pattern perpendicular to the direction of the stripes.

Each data point in Fig. 6 is the number of fluorescence photons detected in a 0.2 sec counting interval. The lower right-hand side of Fig. 6 shows the entire linear sequence of 1008 data points (42 scans, 24 points/scan, 4.8 sec/scan, with 4.8 sec pauses between scans), displayed as the series of lines connecting the points. The

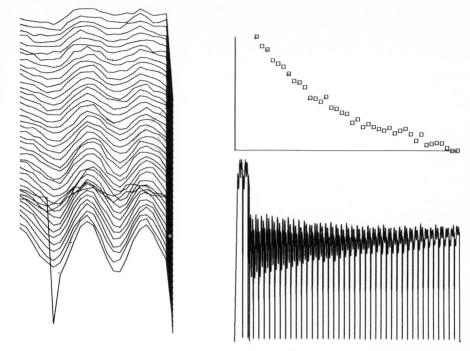

Fig. 6. Fourier transform analysis of periodic pattern photobleaching for a 3T3 fibroblast labeled with fluorescein-conjugated succinyl-concanavalin A. Lower right: Linear sequence of all the data during series of fluorescence scans with a pattern of stripes. Left: Stack of individual 24 point scans recorded at 9.6 sec intervals before and after bleaching. Upper right: Normalized modulation amplitude, above pre-bleach modulation, for each post-bleach scan.

left side of the figure displays the data again rearranged this time into a stack of individual 24 points scans. Bleaching occurs during the closed shutter point of the third scan. Subsequent scans reveal a clear periodicity as the monitoring pattern is move over a total range of 4.3 μm, slightly more than twice the 1.8 μm periodic repeat distance.

Each scan is computer analyzed to give the amplitude and phase of the intensity modulation. The calculated amplitudes, normalized by the average intensities (with the small normalized pre-bleach amplitude subtracted) are plotted in the upper righthand side of the figure. A least-squares fit of these values to a single exponential (see Eqn. 15) yields a diffusion coefficient of 6.4×10^{-12} cm^2/sec.

We conclude this section with an example of the analysis of diffusion on the surface of a sphere, an osmotically swollen human erythrocyte. The erythrocytes were labeled as intact cells with dichlorotriazinylaminofluorescein, a derivative of fluorescein which binds covalently to amino groups, with roughly two-thirds of the label attaching to band 3 protein (Schindler et al., 1980a).

The distribution of fluorescence on the labeled cell was followed with a series of

24 point fluorescence scans (0.24 sec/point with 5.76 sec pauses between scans), each covering a total distance of 14.0 μm. As in Fig. 6, the total linear sequence of data, and the stack of individual scans are displayed in the lower right-hand side and the left side of Fig. 7, respectively. Before photobleaching, the fluorescence distribution is symmetric, with sharp peaks due to the geometrical edge effect. Photobleaching at the leading edge of the cell, during the third scan produces a marked asymmetry, which decays as the labeled protein molecules redistribute over the surface of the sphere.

Each scan is analyzed to give the amplitude of the first-order Legendre polynomial component. Geometrical edge effects for each post-bleach scan are corrected with point-by-point normalizations with the average of the pre-bleach scans. The calculated amplitudes, normalized by the average intensities, are plotted in the upper right-hand side of the figure. A least-squares fit of these values to a single exponential (see Eqn. 21) yields a diffusion coefficients of 9.7×10^{-11} cm^2/sec.

3.3. Statistical accuracy

It is a useful exercise to investigate the limiting accuracy with which mobility measurements can be made, considering signal variance arising solely from photon-

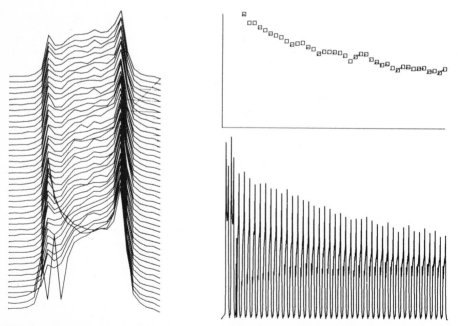

Fig. 7. Normal-mode analysis of diffusion on a spherical surface for an osmotically swollen human erythrocyte covalently labeled with dichlorotriazinylaminofluorescein. Lower right: Linear sequence of all the data during a series of fluorescence scans with a focused laser beam. Left: Stack of individual 24 point scans recorded at 11.52 sec intervals before and after bleaching. Upper right: Normalized amplitude of first-order Legendre polynomial component for each post-bleach scan.

counting statistics. The results of such a calculation can be used to help optimize experimental conditions to balance statistical errors against the possible introduction of systematic effects.

We consider two of the types of fluorescence photobleaching experiments considered above: periodic pattern bleaching on a plane with fundamental period $2\pi/k$, and azimuthally symmetric bleaching on the surface of a sphere of radius r. In either case, the data consist of a series of numbers, $n(x_j, t_m)$, the number of photocounts detected at the jth position of the mth scan, where t_m is the mean time after photobleaching of the mth scan, and x_j represents a Cartesian coordinate in the plane, or the cosine of the equatorial angle on the sphere.

In general*, we can express $n(x_j, t_m)$ as an appropriate expansion in $\psi_l(x)$, the eigenfunctions of the diffusion equation, i.e.,

$$\langle n(x_j, t_m) \rangle = n_0 + n_1(t_m)\Psi_1(x_j) + n_2(t_m)\Psi_2(x_j) + \ldots + \tag{49}$$

where the angular bracket signify an ensemble average, and

$$\Psi_l(x) = \begin{cases} \cos(lkx) & \text{(for the plane)} \\ P_l(x) & \text{(for the sphere)} \end{cases} \tag{50}$$

From the orthogonality property of these eigenfunctions, $n'_1(t_m)$, the experimental estimate of $n_1(t_m)$, can be computed as**

$$n\}(t_m) = \frac{a_1}{M} \sum_{j=1}^{M} \Psi_1(x_j)n(x_j, t_m) \tag{51}$$

where a_1 is a normalization factor approximated as

$$a_1 = M / \sum_{j=1}^{M} [\Psi_1(x_j)]^2 \tag{52}$$

so that

$$a_1 = \begin{cases} 2 & \text{(for the plane)} \\ 2l+1 & \text{(for the sphere)} \end{cases} \tag{53}$$

---

* This is a slight simplification for data of the form of Fig. 7. In this case $n(x_j, t_m)$ must be corrected for geometrical edge effects before Eqn. 49 applies.
** In practice, n'_1 for the plane is actually computed with a complex transform; so that one also derives the relative phase of the Fourier component.

The 'signal' for such data can be appropriately defined as

$$S'_1(t_m) = n'_1(t_m) - n'_1(\infty) \tag{54}$$

where $n'_1(\infty)$ is derived from an average of many sufficiently long-time fluorescence scans. If we assume that $n'_1(\infty)$ can be determined much more accurately than any single $n'_1(t_m)$, then the signal variance is simply

$$\text{var}[S'_1(t_m)] \simeq \text{var}[n'_1(t_m)] \tag{55}$$

$$= \sum_{j=1}^{M} \left[\frac{\partial n_1(t_m)}{\partial n(x_j, t_m)} \right]^2 \text{var}[n(x_j, t_m)]$$

$$= \left(\frac{a_1}{M} \right)^2 \sum_{j=1}^{M} [\Psi_1(x_j)]^2 \langle n(x_j, t_m) \rangle$$

with the last equality resulting from the Poisson photon-counting statistics. For moderate bleaching levels, this expression for the variance will be dominated by the constant term, n_0, in Eqn. 49, so that to a good approximation, combining Eqns. 55 and 52,

$$\text{var}[S'_1(t_m)] \simeq a_1 n_0 / M \tag{56}$$

To procede further we must specify a particular functional form for the signal. We consider a system in which a fraction, β, of the fluorescent label diffuses with diffusion coefficient D, with the remainder immobile on the time scale of the experiment. In this case,

$$\langle S'_1(t_m) \rangle = \beta n_1(0) e^{-\Gamma_1 t_m} \tag{57}$$

where

$$\Gamma_1 = \begin{cases} D(lk)^2 & \text{(for the plane)} \\ Dl(l+1)/r^2 & \text{(for the sphere)} \end{cases} \tag{58}$$

If we now define D' as the experimental estimate of D, determined with an appropriate least-squares data analysis, we can compute the variance in D' arising from the variances of the individual estimations of the signal for each scan.
For scans equally spaced in time (i.e., $t_m = mT$), adapting Eqns. 47, 49 and 51 of Koppel (1974),

$$D/(\text{var } D')^{1/2} = \{n_1(0)/[8a_1n(-)n_0]^{1/2}\}\beta[Mn(-)/\Gamma_1 T]^{1/2} \tag{59}$$

where $n(-)$ is the average number of photocounts per counting interval before bleaching.

The first factor on the right-hand side of Eqn. 59 is a function of the distribution and extent of photobleaching. To optimize the signal-to-noise ratio of the measurement, photobleaching should maximize the quantity $n_1(0)/n_0^{1/2}$.

The ratio $Mn(-)/\Gamma_1 T$ is equal to the total number of photocounts per characteristic diffusion time detected during monitoring. It is proportional to the number of fluorophores illuminated, the monitoring illumination intensity, the fluorescence quantum efficiency, and the efficiency of fluorescence collection and detection. If necessary, the number of fluorophores can be increased manyfold by linking the ligand of interest to a heavily labeled carrier macromolecule. Hirschfeld (1973) bound between 80 and 100 molecules of fluorescein isothiocyanate to molecules of polyethyleneimine, which were then bound to antibody molecules. Schechter et al. (1978) produced active peptide hormone molecules attached to heavily rhodaminated lysozyme molecules.

The monitoring intensity can be increased, but only up to a certain limit, beyond which unacceptable bleaching occurs during monitoring. Ultimately, one is limited by the ratio of the quantum efficiency of bleaching, to the quantum efficiency of fluorescence. The lower the ratio the better it is, as long as one still has the laser power to bleach the label, and can do so without unacceptable sample heating (see section 4.1).

We can put this on a more quantitative level by considering the general relation between the bleaching rate during monitoring (R_B), and the average photon-counting monitoring rate per fluorophore ($Mn(-)/NT$, where N is the total number of fluorophores). We have

$$R_B/[Mn(-)/NT] = \varphi_B/E\varphi_F \tag{60}$$

ϕ_B and ϕ_F are the quantum efficiencies of bleaching and fluorescence, respectively, and E is the efficiency of fluorescence collection and detection. We require that the monitoring intensity be limited so that the fraction of molecules bleached per characteristic diffusion time is only some small fraction (α) of $[n(-)-n_0]/n(-)$, the fraction bleached during the bleaching pulse, i.e.,

$$R_B/\Gamma_1 = \alpha[n(-) - n_0]/n(-) \tag{61}$$

Combining Eqns. 60 and 61, this limits the total number of photocounts per diffusion time to

$$Mn(-)/\Gamma_1 T = \alpha EN\{[n(-) - n_0]/n(-)\}(\varphi_F/\varphi_B) \tag{62}$$

Substitution into Eqn. 59 yields, finally,

$$D/(\text{var } D')^{1/2} = (8a_1)^{-1/2}[n_1(0)/n(-)]\{[n(-)/n_0] - 1\}^{1/2}\beta(\alpha EN\varphi_F/\varphi_B)^{1/2} \tag{63}$$

362

The arguments leading to this last result are quite general. It can be shown (Koppel, manuscript in preparation; see also Johnson and Garland, 1981), that the statistical accuracy of rotational diffusion coefficients determined in fluorescence photobleaching experiments is also proportional to $(\alpha E N \phi_F/\phi_B)^{1/2}$.

4. Radiation-induced artifacts

No description of a technique would be complete without some discussion of the potential technical problems. The sections below consider two effects associated with the high-level irradiation levels used during photobleaching: local heating and photochemical damage. Theoretical calculations indicate that with proper experimental design heating is not ordinarily a serious problem. The extent of dye-mediated photodamage is more difficult to access. Empirical control experiments, in specific cases, have thus far reported no significant effect.

4.1. Heating during photobleaching

Intense illumination leads to a rate of local heat production equal to that part of the incident power that is absorbed without reradiation. Axelrod (1977) calculated the temperature increase produced by the exposure of a planar array of chromophores to a focused Gaussian profile laser beam, as a function of time after the start of illumination. It was shown that, as the heat is conducted away from the illuminated area into the surrounding aqueous medium, the temperature rise approaches a steady state with a characteristic time constant

$$t_c = w^2/2\varkappa \tag{64}$$

where w is the $1/e^2$ radius of the Gaussian laser profile in the plane of the membrane, and $\varkappa$ is the thermal diffusivity of the medium (1.4×10^{-3} cm^2/sec for water). t_c is ordinarily much shorter than the bleaching times used in measuring translational diffusion (since $\varkappa$ is much, much greater than translational diffusion coefficients); but it could be comparable or even longer than the bleaching times needed for measurements of rotational diffusion. The steady-state temperature increase has the maximal value

$$T_S = (\pi/2)^{1/2}qw/4K \tag{65}$$

in the center of the beam, where q is the rate of heat production per unit area in the center of the beam, and K is the thermal conductivity of the medium (6 mW/cm/°K for water). For conditions "typical" for measurements of membrane protein diffusion, as described by Axelrod (1977), T_S is negligible, no more than 0.03°C. T_S could become significant, however, if a large area is bleached at a rapid rate. Caution should thus be exercised during pattern photobleaching and rotational diffusion measurements.

The extent of local sample heating can also be analyzed for uniform whole cell illumination. Consider as a model system an irradiated sphere of radius a, with chromophores distributed uniformly throughout the volume and on the surface. At the thermal steady state, in general,

$$\nabla^2 T(r) = -q(r)/K \tag{66}$$

where $q(r)$ is the rate of heat production per unit volume. Mathematically, this is equivalent to Poisson's equation in electrostatics, with $T(r)$ analogous to the electric potential, and $q(r)/4\pi K$ analogous to the electric charge density (Jackson, 1962). Given the symmetry of our problem, $T_S(r)$, the steady-state temperature rise as a function of the distance from the center of the sphere, can be readily solved by applying the equivalent of Gauss' law, with the result

$$T_S(r) = (a/K)(\sigma + a\varrho/2) - (r^2/6K)\varrho \qquad \text{(inside)} \tag{67}$$

$$T_S(r) = (a^2/Kr)(\sigma + a\varrho/3) \qquad \text{(outside)} \tag{68}$$

where ϱ and σ are the rates of heat production per unit volume within the sphere and per unit area on the sphere surface, respectively. At the surface of the sphere,

$$T_S(a) = Q/4\pi Ka \tag{69}$$

where

$$Q \equiv (4/3)\pi a^3 \varrho + 4\pi a^2 \sigma \tag{70}$$

is the total rate of heat production for the whole sphere. The approach to steady state, in this case, proceeds with a characteristic time constant of $a^2/2\varkappa$.

4.2. Photochemical damage

Despite the widespread application of the photobleaching technique, questions remain as to whether or not results are affected by dye-sensitized photodamage. This possibility was brought home by gel chromatographic analyses demonstrating the extensive cross-linking of membrane proteins induced under certain conditions by the irradiation of fluorescently labeled membranes in bulk suspension (Sheetz and Koppel, 1979; Nigg et al., 1979; Dubbelman et al., 1978). The extent of cross-linking observed was markedly reduced under anaerobic conditions, and with the addition of various reducing agents such as reduced glutathione. Evidence thus points to the operation of a type II photosensitized oxidation (Foote, 1968, 1976), a process characterized by the initial interaction between oxygen and the electronically

364

excited dye triplet state (reached by intersystem crossing from the excited singlet state, the initial product of light absorption). The most probable of the type II interactions leads to the de-excitation of the dye molecule back to the groud state, with the transfer of the excitation producing an electronically excited singlet state of oxygen. A single dye molecule could go through many such cycles, catalytically producing one molecule of singlet oxygen in each cycle. Reactions initiated by the singlet oxygen, in this scenario, would lead to the subsequent photodamage.

Several approaches have been developed to check for the possible effects of photodamage under the actual conditions of a photobleaching experiment. Using the photobleaching technique as its own control, investigators have looked for changes induced by additional bleaching (Jacobson et al., 1978; Wolf et al., 1980), and reported no evidence of the effects of membrane damage. These results are subject to the argument, however, that significant systematic error is introduced, but that it is saturated even at the minimum bleaching levels employed. One would really like to compare the diffusion characteristics of a membrane system determined with and without fluorescence photobleaching.

Two reports of just this type of comparison have recently appeared. Wey et al. (1981) demonstrated that the lateral diffusion characteristics of rhodopsin in rod outer segments determined in fluorescence photobleaching experiments are essentially indistinguishable from the original measurements of diffusion in that system based on rhodopsin absorption bleaching (Poo and Cone, 1974).

Koppel and Sheetz (1981) used cell-fusion techniques to study erythrocyte membranes. One set of experiments was performed on polyethylene glycol-induced fused erythrocyte pairs in which membrane proteins (primarily band 3) of only one member of the couplet were intially fluorescently labeled with dichloro-triazinylaminofluorescein. The intermixing of surface proteins was followed as a function of time after fusion with a series of fluorescence scans. In parallel, another set of experiments was performed on fused pairs of labeled cells, using an intense laser beam to bleach the label on one of the cells. The results indicate that fluorescence photobleaching does not alter the lateral mobility of erythrocyte glycoproteins. The cumulative evidence of all of these studies supports the general validity of the photobleaching technique.

5. Prospects and conclusions

Fluorescence photobleaching techniques allow the accurate quantitation of lateral and rotational motions of a small number of specific molecules in a complex system. Thus far, they have been applied almost exclusively to the study of cell plasma membranes and model membrane systems. In the future look for further applications to dynamics in solution (as in Barisas and Leuther, 1979; Lanni et al., 1981), the cell cytoplasm (Wojcieszyn et al., 1981), and cell organelles (Sowers et al., 1982; Hochman et al., 1982).

Fluorescence photobleaching measurements of lateral diffusion are well established and widespread. In comparison, the measurement of rotation by this technique is in its infancy. Rotational measurements are technically more demanding, requiring improved time resolution and rigidly bound fluorescent probes (see Eqn. 30). Nevertheless, look for a large increase in activity in this area. Rotational and translational measurements, in combination, supply a much more definitive description of a system than either could supply alone.

References

Austin, R.H., Chan, S.S. and Jovin, T.M. (1979) Rotational diffusion of cell surface components by time-resolved phosphorescence anisotropy, Proc. Natl. Acad. Sci. (U.S.A.), 76, 5650–5654.

Axelrod, D. (1977) Cell surface heating during fluorescence photobleaching recovery experiments, Biophys. J., 18, 129–131.

Axelrod, D., Koppel, D.E., Schlessinger, J., Elson, E.L. and Webb, W.W. (1976) Mobility measured by fluorescence photobleaching recovery kinetics, Biophys. J., 16, 1055–1069.

Badley, R.A., Martin, W.G. and Schneider, H. (1973) Dynamic behavior of fluorescent probes in lipid bilayer model membranes, Biochemistry, 12, 268–275.

Barisas, B.G. and Leuther, M.D. (1979) Fluorescence photobleaching recovery measurements of protein absolute diffusion constants, Biophys. Chem., 10, 221–229.

Barisas, B.G. and Leuther, M.D. (1980) Grid-gated photomultiplier photometer with subnanosecond time response, Rev. Sci. Instrum., 51, 74–78.

Bracewell, R.N. (1978) The Fourier Transform and its Application, McGraw-Hill, New York.

Cantor, C.R. and Schimmel, P.R. (1980) Biophysical Chemistry, Part II. Techniques for the Study of Biological Structure and Function, W.H. Freeman and Co., San Francisco, CA.

Cherry, R.J. (1978) Measurements of protein rotational diffusion in membranes by flash photolysis, Methods Enzymol., 54, 47–61.

Cherry, R.J. (1979) Rotational and lateral diffusion of membrane proteins, Biochim. Biophys. Acta, 559, 289–327.

Cherry, R.J. and Godfrey, R.E. (1981) Anisotropic rotation of bacteriorhodopsin in lipid membranes, Biophys. J., 36, 257–276.

Cone, R.A. (1972) Rotational diffusion on rhodopsin in the visual receptor membrane, Nature, New Biol., 236, 39–43.

Cooper, G.R. and McGillem. C.D. (1967) Methods of Signal and System Analysis, Holt, Rinehart and Winston, New York.

DeMarco, F. and Penco, E. (1969) Pulsed photomultipliers, Rev. Sci. Instrum., 40, 1158–1160.

Dubbleman, T.M.A.R., De Goeij, A.F.P.M. and Van Steveninck, J. (1978) Photodynamic effects of protoporphyrin on human erythrocytes, Nature of the cross-linking of membrane proteins, Biochim. Biophys. Acta, 511, 141–151.

Edidin, M. (1981) Molecular motions and membrane organization and function, in: J.B. Finean and R.H. Michell (Eds.), A New Comprehensive Biochemistry, Vol. 1, Membrane Structure, Elsevier/North-Holland, Amsterdam, pp. 37–82.

Edidin, M. and Fambrough, D. (1973) Fluidity of the surface of cultured muscle fibers, Rapid lateral diffusion of marked surface antigens, J. Cell. Biol., 57, 27–37.

Farinelli, U. and Malvano, R. (1958) Pulsing of photomultipliers, Rev. Sci. Instrum., 29, 699–701.

Foote, C.S. (1968) Mechanisms of photosensitized oxidation, Science, 162, 963–970.

Foote, C.S. (1976) Photosensitized oxidation and singlet oxygen: Consequences in biological systems, in: W.A. Pryor (Ed.), Free Radicals in Biology, Academic Press, New York, pp. 85–127.

Frye, L.D. and Edidin, M. (1970) The rapid intermixing of cell suface antigens after formation of mouse–human heterokaryons, J. Cell Sci., 7, 319–335.

Garland, P.B. and Moore, C.H. (1979) Phosphorescence of protein-bound eosin and erythrosin, Biochem. J., 183, 561–572.

Hirschfeld, T. (1973) Staining antibodies, Application for Letters Patent.

Hochman, J., Schindler, M., Lee, J., Matia, J. and Ferguson-Miller, S. (1982) Mobility of cytochrome c on mitochondrial membranes measured by fluorescence recovery after photobleaching, Biophys. J., 37, 401a.

Huang, H.W. (1973) Mobility and diffusion in the plane of cell membranes, J. Theoret. Biol., 40, 11–17.

Jackson, J.D. (1962) Classical Electrodynamics, John Wiley, New York.

Jacobson, K., Hou, Y. and Wojcieszyn, J. (1978) Evidence for lack of damage during photobleaching measurements of the lateral mobility of cell surface components, Exp. Cell Res., 116, 179–189.

Johnson, P. and Garland, P.B. (1981) Depolarization of fluorescence depletion, A microscopic method for measuring rotational diffusion of membrane proteins on the surface of a single cell, FEBS Lett., 132, 252–256.

Kawato, S. and Kinosita, K. (1981) Time-dependent absorption anisotropy and rotational diffusion of proteins in membranes, Biophys. J., 36, 277–296.

Koppel, D.E. (1972) Analysis of macromolecular polydispersity in intensity correlation spectroscopy: The method of cumulants, J. Chem. Phys., 57, 4814–4820.

Koppel, D.E. (1974) Statistical accuracy in fluorescence correlation spectroscopy, Phys. Rev. A, 10, 1938–1944.

Koppel, D.E. (1979) Fluorescence redistribution after photobleaching: A new multipoint analysis of membrane translational dynamics, Biophys. J., 28, 281–291.

Koppel, D.E. (1982) Measurement of membrane protein lateral mobility, in: J. Metcalfe and T.R. Hasketh (Eds.), Techniques in the Life Sciences, Vol. B4, Techniques in Lipid and Membrane Biochemistry, Elsevier/North-Holland, Amsterdam.

Koppel, D.E. and Sheetz, M.P. (1981) Fluorescence photobleaching does not alter the lateral mobility of erythrocyte glycoproteins, Nature (London), 293, 159–161.

Koppel, D.E., Axelrod, D., Schlessinger, J., Elson, E.L. and Webb, W.W. (1976) Dynamics of fluorescence marker concentration as a probe of mobility, Biophys. J., 16, 1315–1329.

Koppel, D.E., Sheetz, M.P. and Schindler, M. (1980) Lateral diffusion in biological membranes: A normal-mode analysis of diffusion on a spherical surface, Biophys. J., 30, 187–192.

Lanni, F. and Ware, B.R. (1982) Polymerization and cross-linking of actin filaments measured by fluorescence photobleaching recovery, Biophys. J., 37, 55a.

Lanni, F., Taylor, D.L. and Ware, B.R. (1981) Fluorescence photobleaching recovery in solutions of labeled actin, Biophys. J., 35, 351–364.

Nigg, E.A., Kessler, M. and Cherry, R.J. (1979) Labeling of human erythrocyte membranes with eosin probes used for protein diffusion measurement, Inhibition of anion transport and photo-oxidative inactivation of acetylcholinesterase, Biochim. Biophys. Acta, 550, 328–340.

Peters, R. (1981) Translational diffusion in the plasma membrane of single cells as studied by fluorescence microphotolysis, Cell Biol. Int. Rep., 5, 733–760.

Peters, R., Peters, J., Tew, K.H. and Bähr, W. (1974) A microfluorimetric study of translational diffusion in erythrocyte membranes, Biochim. Biophys. Acta, 367, 282–294.

Petersen, N.O. and Chan, S.I. (1977) More on the motional state of lipid bilayer membranes: Interpretation of order parameters obtained from nuclear magnetic resonance experiments, Biochemistry, 16, 2657–2667.

Poo, M.M. and Cone, R.A. (1974) Lateral diffusion of rhodospin in the photoreceptor membrane, Nature (London), 247, 438–441.

Rigler, R. and Ehrenberg, M. (1973) Molecular interactions and structure as analysed by fluorescence relaxation spectroscopy, Quart. Rev. Biophys., 6, 139–199.

Schechter, Y., Schlessinger, J., Jacobs, S., Chang, K.J. and Cuatracasas, P. (1978) Fluorescent labeling of hormone receptors in viable cells: Preparation and properties of highly fluorescent derivatives of epidermal growth factor and insulin, Proc. Natl. Acad. Sci. (U.S.A.), 75, 2135–2139.

Schindler, M., Koppel, D.E. and Sheetz, M.P. (1980a) Modulation of membrane protein lateral mobility by polyphosphates and polyamines, Proc. Natl. Acad. Sci. (U.S.A.), 77, 1457–1461.

Schindler, M., Osborn, M.J. and Koppel, D.E. (1980b) Lateral mobility in reconstituted membranes: Comparisons with diffusion in polymers, Nature (London), 283, 346–350.

Seelig, J. (1977) Deuterium magnetic resonance: Theory and application to lipid membranes, Quart. Rev. Biophys., 10, 353–418.

Sheetz, M.P. and Koppel, D.E. (1979) Membrane damage caused by irradiation of fluorescent concanavalin A, Proc. Natl. Acad. Sci. (U.S.A.), 76, 3314–3317.

Smith, B.A. and McConnell, H.M. (1978) Determination of molecular motion in membranes using periodic pattern photobleaching, Proc. Natl. Acad. Sci. (U.S.A.), 75, 2759–2763.

Smith, B.A., Clark, W.R. and McConnell, H.M. (1979) Anisotropic molecular motion on cell surface, Proc. Natl. Acad. Sci. (U.S.A.), 76, 5641–5644.

Smith, L.M. and McConnell, H.M. (1981) Pattern photobleaching of fluorescent lipid vesicles using polarized laser light, Biophys. J., 33, 139–146.

Smith, L.M., Parce, J.W., Smith, B.A. and McConnell, H.A. (1979) Antibodies bound to lipid haptens in model membranes diffuse as rapidly as the lipids themselves, Proc. Natl. Acad. Sci. (U.S.A.), 76, 4177–4179.

Smith, L.M., Weis, R.M. and McConnell, H.M. (1981) Measurement of rotational motion in membranes using fluorescence recovery after photobleaching, Biophys. J., 36, 73–91.

Sowers, A.E., Hoechli, M., Hoechli, L., Derzko, Z., Jacobson, K. and Hackenbrock, C.R. (1982) Lateral diffusion of a lipid analog and the b–c_1 complex in the mitochondrial inner membrane: A fluorescence recovery after photobleaching study, Biophys. J., 37, 278a.

Wey, C.L., Cone, R.A. and Edidin, M.A. (1981) Lateral diffusion of rhodopsin in photoreceptor cells measured by fluorescence photobleaching and recovery, Biophys. J., 33, 225–232.

Wojcieszyn, J.W., Schlegel, R.A., Wu, E.-S. and Jacobson, K.A. (1981) Diffusion of injected macromolecules within the cytoplasm of living cells, Proc. Natl. Acad. Sci. (U.S.A.), 78, 4407–4410.

Wolf, D.E., Edidin, M. and Dragston, P.R. (1980) Effect of bleaching light on measurements of lateral diffusion in cell membrane by fluorescence photobleaching recovery method, Proc. Natl. Acad. Sci. (U.S.A.), 77, 2043–2045.

Subject Index

Absorption anisotropy, 341, 347, 351
— dipole, 349
Acetylcholine channel lifetime, 134
Activation energy, 29
Acute lymphocytic leukemia, 209
Aliasing, 3, 175
D-Amino acid oxidase, 97
Ammonium dihydrogen phosphate, 295
8-Anilino-1-naphthalene sulfonate, 273
Anisotropy, absortion, see Absorption aniso-
    tropy
—, time resolved emission, 260–267, 341
Antenna pigments, 302
Antigen–antibody reaction, 206
Anti-Stokes radiation, 291, 298
Autocorrelation, 5–6, 184, 185, 187, 189, 246

Bacteriorhodopsin, 324–326, 351
Bathorhodopsin, 322, 323
Beer's Law, 17
Benzoate, 98, 99
Bile pigments, transient states in, 328
Bilirubin, photoisomerization of, 327
Birefringence, 293
Blue semiquinone, 96, 97, 99
Bovine serum albumin, 201
Boxcar integrator, 224
Bragg's Law, 141–142, 181
Bragg spacing, 141, 150
Brownian diffusion, 177, 178, 183
t-Butyl alcohol, 91

Cage effect, 121
Carbocyanine, 286
Carotenoids, 310
Catalase, 122
Cell surface topography, 274
Cerenkov radiation, 90, 91
Ceruloplasmin, 103
Cesium hydrogen arsenate, 295
Chromaffin granules, 204
Chronic lymphocytic leukemia, 214
Coherence area, 198
Compound I, 94, 95
Compound II, 94, 95

Compound III, 94
Concanavalin A, fluorescent conjugates, 208,
    211–213, 274, 356
Conformational fluctuations, 45
Constant fraction discriminator, 229, 230
Continuous flow, 16, 40, 64, 65, 67
Convolution, 6–7, 223, 224–245, 248, 343
Copper oxidase, 103–104
Correlation functions, 3–4, 165, 166
Cysteine, 98
Cytochrome c, 93, 102
Cytochrome f, 310
Cytochrome oxidase, 93, 101–102, 125, 126, 127
—, rotational relaxation of, 127
Cytochrome P-450, 96

Dead time, in steady-state kinetics, 14
—, in temperature jump, 26
—, in time-correlated single photon counting,
    231–233
—, in time-resolved X-ray scattering, 147
Debye's formula, 150, 152, 166
Debye–Huckel approximation, 203
Debye–Huckel parameter, 175–176, 177
Debye–Sears modulator, 249
Deconvolution, 6–9, 153, 224, 240–248, 300
Deoxyhemoglobin, 100
Dichlotriazinylamino fluorescein, 357, 364
Dichroic ratio, 127
Dichroism, transient linear, 341
Dielectric constant, 29, 30–32, 34
—, effect of temperature on, 30–32
Diffusion, 183–184
—, coefficient, 32, 33, 185, 188, 190, 203, 344,
    356, 360
—, equation, 120
—, lateral, 340–342, 349, 355
—, layer, 121, 127
—, rotational, 260–262, 266, 269, 273, 341,
    349–351
Diffusion-controlled reactions, see under
    Reaction kinetics
Dioxane, 30, 34, 35
1,6-Diphenyl-1,3,5-hexatriene, 264–265
Dipolar relaxation, 255

R.I. Sha'afi and S.M. Fernandez (Eds.), Fast Methods in Physical Biochemistry and Cell Biology
© 1983 Elsevier Science Publishers

Dirac delta function, 186, 343
Dispersion, 292
Disproportionation reaction, 105
DNA, 202–203
—, torsional dynamics of, 327
Doppler shift, 174, 177–178, 180–181, 183, 190, 198
Drude model, 178

Eigen equation, 33
Electric displacement, 292
Electro-osmosis, 196–197
Electron transfer, reverse, 320–321
Electron-transfer reactions, 93
Electrophoresis, 175–177
Electrophoretic light scattering, 173–215
—, scattering angle, 181, 183
—, sample chamber for, 193–194
Electrophoretic mobility, 176, 177, 188, 190, 197, 203
Emission dipole, 349
Endocytic vesicles, 212–213
Energy transfer, see under Fluorescence
Enthalpy of activation, 29
Entropy of activation, 29
Enzymes immobilized, 118, 121
Eosinophils, 211
Erythrocyte, see Red blood cell
Excitation function, 224
Exciton annihilation theory, 317
Exciton migration, 302
Exocytosis, 204, 213
Exponential series, method of, 245–246

Fabry–Perot interferometer, 283, 288
FAD, 96, 97
Ferridoxin, 311
Ferrihemoproteins, 114, 115, 116, 117
Ferryl myoglobin, 95
Fibrin polymerization, 43–44
Fibroblasts, 214
Filament elongation, 151
Flash lamp, 234
Flash photolysis, 27, 28, 113–134
—, dichroic ratio in, 127
—, low temperature, 125
Flavodoxin, 97–98
Flavoproteins, 96, 97
Flocculation, 206
Fluorescence, see also under Time resolved fluorimetry
—, anisotropy, 253
—, decay curve, 223, 240–241, 274

Fluorescence, (continued)
—, depolarization, 126
—, energy transfer, 253, 269–271
—, labeling living cells, 272
—, lifetime, 7, 222–226, 253–256, 270, 290, 300, 301, 316
—, lifetime of bacterial reaction center, 319
—, — of bacteriorhodopsin, 325
—, microscope, 353–354
—, multiexponential decay, 240, 248, 254, 255, 256, 271
— of flavoproteins, 132
—, parametric, 295
—, photobleaching, see Photobleaching
—, polarization, 254
—, quenching, 253
—, rotational correlation time, 253, 262
FMN, 96
Fourier series, 2
— transform, 174, 175, 184, 188, 189, 198, 245, 248, 283, 342, 343, 357
Frequency doubling, see Wavelength shifting
— mixing, 293
— spectrum, 2
Fresnel equation, 293
Frictional drag coefficient, 175

Gain medium, 282, 285
Gauss' Law, 363
Glucose oxidase, 96
Glutamate dehydrogenase, polymerization, see Relaxation techniques, biological applications
Granulocyte, 213

Hairy cell leukemia, 214
Half-life, see under Kinetics
Harmonics, 2
Heme A, 101, 102
Heme oxygenase, 96
Hemoglobin, 93, 99, 101, 125, 202, 326–327
—, R state, 100–101
—, T state, 100–101
Hemoproteins, 113, 114, 126
Henry's Law, 176
Heterodyne detection, 180, 189, 190–191, 226, 249, 252
Heterokaryon fusion, 341
Histidine, 98
Homodyne, 190
Horseradish peroxidase, 93, 94, 104, 105, 115, 117, 118, 124
Hydrogen peroxide, 95

IgG immune complexes, 211
Impulse response, 343, 344
Indoleamine deoxygenase, 104
Inhomogeneous broadening, 286
Inner filter effect, 41
Interference, 178–181
Internal reflection spectroscopy, 272–273
Intersystem crossing, 340, 346
Ionic strength, 177
Iron–sulfur protein, 311, 312
Isoalloxazine ring, 98
Isorhodopsin, 322
Isotropic rotator, 267–268

Joule heating, 195–196

K_m, see Michaelis–Menten constant
Kerr gate, 300, 301
Kinetics
—, basic theory for, 12
—, first-order reactions, 12, 18, 19
—, half-life, 18
—, maximum velocity, 14
—, Michaelis–Menten, 12, 14
—, pseudo first-order 19, 20, 21
—, rate constant, 12, 13
—, second-order, 19
—, steady-state, 12, 14, 15
—, steady-state approximation, 14
—, steady-state, isotopic labeling, 15
—, stopped-flow, 15, 21
—, transient state, 23
—, zero-order, 13, 19
Kronecker delta function, 346

Laccase, 93, 103
Laplace transformation, 243–244, 248
Laser, 282–285
—, argon ion, 236, 253, 283, 288, 289
—, cavity dumping, 234–238, 287, 290
—, Doppler velocimetry, 174
—, dye, 238, 282, 286, 287, 299
—, gain, 282, 283
—, krypton ion, 236, 285, 288, 289
—, mode locking, 234, 236–238, 275–289
—, neodymium:glass, 282, 286
—, neodymium:YAG, 282, 286, 288, 290, 297
—, nitrogen, 283
—, pulsed, 234–239
—, Q-switching, 234
—, ruby, 286
—, solid state, 282

Laser, (continued)
—, synchronous pumping, 237–238, 289–290, 299
Leading edge timing, 230
Lectin receptors, 274
Legendre polynomials, 346, 351, 358
Ligand binding, 40–43, 119
—, to acetylcholine receptors, 42
—, colchicine–tubulin interaction, 41
—, to hormone receptors, 42
—, nucleotide binding to actin, 41–42
Ligand exchange, 125
Light scattering, see Electrophoretic light scattering
Lithium iodate, 295
Lithium niobate, 295
Local oscillator, 178, 180, 183, 186, 189, 193
Lock-in amplifier, 226, 251–252, 299
Longitudinal modes, 283
Lymphocytes, 208, 209, 213
Lymphokines, 213

Macrophage, 211, 214
Mast cells, 213
Mast cell granules, 205
Maximum velocity, 14
Membrane fluidity, 273
Membranes, structural dynamics of, 264–266
Metarhodopsin equilibrium, see Relaxation techniques, biological applications
Michaelis–Menten constant, 12, 14
Microsomal vesicles, 205
Microtubule
—, assembly, 156–168
—, associated proteins, 157, 167
—, effect of temperature on, 161–162
—, rings, 157, 158, 162, 167
—, α- and β-tubulin, 156
Microwave linear accelerator, 88
Mitochondria, 126, 127
—, electron transfer in, 129
Modulating functions, method of, 245, 248
Moments, 346–347
—, method of, 241–244, 248
Multichannel analyzer, 229–233
Muscle contraction and ATP hydrolysis, 42–43
Myoglobin, 115, 116, 117, 121, 122, 123, 124, 125, 127, 128, 129, 130, 131–133
—, in cardiac cells, 127, 128, 129–133
—, rotational diffusion of, 124
Myosin assembly, see Relaxation techniques, biological applications

Neuraminidase, 209, 214
N-4-Nitrobenzo-2-oxa-1,3-diazole phos-
    phatidylethanolamine (NBD-PE), 355
Noise, 5–6
Non-adiabatic electron transfer theory, 310
Non-adiabatic multiphoton theory, 309
Nucleic acid bases, photoreactions of, 327
Nyquist sampling rate, 3, 174–175

Ohm's law, 195–196
Opsin, 322
Optical amplifier, 288, 291
Optical beating, 178–181, 186
Optical delay, 226
Optical modulators
—, accousto-optic, 249, 354
—, electro-optic, 250–251, 354
Optical multichannel analyzer, 298
Optical parametric amplifier, 291, 293, 294,
    295, 296
Optics, non-linear, 292–296
Order parameter, 263, 348
Oxygen toxicity, 93, 104
Oxyheme, 96
Oxyperoxidase, 94, 95

Parametric down conversion, 293
Parametric generator, 298, 299
Parametric up conversion, 293
Periodic function, 2
Periodic signals, 1–3, 5
Peroxisome, 122
Perrin equation, 253, 268
Phagolysosome, 213
Phase sensitive detector, 226, 248, 251–252
Phase shift, 179
Phase shift fluorimetry, 225–227, 248–252, 255,
    300
—, demodulation in, 225, 226
—, polarized, 267–269
—, spectral measurements with, 259–260
Phenylbutazone, 41
Phosphorescence, 341
Photobleaching, 339–365
—, local heating in, 362
—, periodic pattern, 342, 356, 359
—, photochemical damage in, 363, 364
Photomultiplier tube, 239, 240, 252
—, gated grid, 355
Photon correlation spectroscopy, 202
Photon counting, 358–360
Photon detection, time correlated, 224, 225,
    228, 233, 290, 300

Photon, (continued)
—, dead time, 231–233
—, pulse pile up, 230
Photoselection, 126, 127, 347
Photosynthesis, 302–321
—, bacterial, 303–316
—, electron transfer in, 302, 307, 309, 310, 319,
    321
—, Forster energy transfer in, 318
—, reaction centers in, 302, 306, 310
Phycocyanins, energy transfer in, 327
Phytohemagglutinin, 208
Picosecond continuum, 291, 296, 298–299
Picosecond spectroscopy, 281–328
Picosecond transient absorption spectroscopy,
    296–300, 302
Plastocyanin, 310, 315
Pockels cell, 250, 290–291, 292
Poisson equation, 363
Polarizability, 292
Polydispersity, 194–195
Polyelectrolytes, 202, 203
Polymerization, degree of, 151–152, 162–164
Polymorphonuclear leukocytes, 214
Polystyrene latex spheres, 206–207
Population inversion, 282
Position-sensitive detector, 145–146
Potassium dihydrogen phosphate, 295, 297
Power spectrum, 2, 184, 190
Poynting's theorem, 184, 186
Pressure jump, 51–55
—, light-scattering detection in, 55
Protein–protein interactions, 43
Protein structural dynamics, 255, 266
Pulse-chase, 69
Pulse radiolysis, 88, 107
—, G value, 91
—, spectrographic recording in, 88–90
—, spectrophotoelectric recording in, 88–90
Pump-probe technique, 296, 299
Pyridine nucleotide fluorescence, 132–133

Quasi-elastic light scattering, 174
Quenching (see also Rapid quenching), 68, 69
—, of fluorescence, 253

Radiation damage, 155
Raman scattering, 90–91, 291
Rapid mixing, 146
—, in pulse radiolysis, 91
—, turbulence and Reynolds number in, 67
Rapid-quench, 63–107
—, calibration, 80–85
—, chemical, 68–69, 77, 81

Rapid-quench, (*continued*)
—, electron paramagnetic resonance, 77
—, flow system, 72–77
—, flow velocity, 67, 68
—, isotope trapping, 83–84
—, mixing, 66, 67
—, rapid freezing, 69, 77, 78, 80–81, 84–85
Rayleigh–Benard criterion, 194
Reaction kinetics
—, diffusion controlled, 33, 34
—, effect of dielectric constant on, 29–32
—, — of pH on, 29
—, — of pressure on, 49–50
—, — of solvent parameters on, 29
—, — of temperature on, 29
—, — of viscosity on, 32
—, ping-pong mechanism, 15
—, reaction mechanism, 13–15
—, theory of, 12, 13
Red blood cells, 46–47, 122, 127, 199, 205, 208,
    211, 357, 364
Red semiquinone, 96–99
Relaxation techniques, 23, 24, 25
—, biological applications, 56–58
—, pressure jump, 51–55
—, temperature jump, 25–27
Relaxation time, 24
Resonance energy transfer, 253, 269–271, 274
Retinal, 322
Rhenium dihydrogen phosphate, 295
Rhodopsin, 322–324
Riboflavin, 97
Ribosome assembly, *see* Relaxation techniques,
    biological applications
Ronchi ruling, 356
Rotational correlation time, *see under* Fluores-
    cence
Rotational diffusion, 351–352
Rotational relaxation, 127

Sampling techniques, 3, 224
Saturable absorber, 285, 286, 287, 288
Scattering in solution, degree of polymerization,
    151–152
—, theory of, 150–153
Scattering vector, 182–183, 186, 189
Second harmonic generation, 293, 295
Secretory vesicles, 204
Shannon's theorem, 153
Shot noise, 90, 185, 186
Sialic acid, 209
Siegert's relation, 185
Singlet–singlet annihilation, 317
Singlet–triplet fusion, 317

Smoluchowski equation, 33
Solvation, 258–259
Spectral overlap, 269
Spherical harmonics, 346
Spot labeling, 341
Steady state, 12
Stern–Volmer equation, 253, 317, 319
Stokes–Einstein equation, 32, 33
Stopped flow, 15, 16, 17, 18, 23, 40, 64, 65, 115,
    117, 122, 129
—, apparatus manufacturers for, 22
—, application to studies of membrane
    permeability, 45–49
—, circular dichroism detection in, 44
—, detection techniques in, 21, 22
—, dielectric relaxation detection in, 51
—, fluorescence polarization detection in, 43
—, light-scattering detection in, 43
—, NMR and EPR detection in, 51
—, optical scanning in, 50
—, rapid mixing in, 15, 16
—, X-ray-scattering detection in, 44
Streak camera, 224, 255, 300, 301, 316
Superoxide, 104–107
—, anion, 93
—, dismutase, 93, 104–107
—, radical, 92
Surface charge density, 176–177, 195, 214
Susceptibility tensor, 292, 294
Synaptic vesicles, 205
Synchrotron radiation, 138–140, 237
Systems analysis, 342–345

Temperature jump, 146, 147
Thermal conductivity, 362
Time-correlated single photon detection, *see
    under* Photon detection
Time digitization, 145, 147
Time-resolved absorption spectra, 90
Time-resolved fluorimetry, *see also under*
    Fluorescence, 7–8, 222, 274, 316–319, 379
—, anisotropy measurements, 260–267, 270
—, cross-correlation methods, 227–228
—, deconvolution of decay curves, *see under*
    Deconvolution
—, effect of light scattering on measurements,
    254, 261
—, phase-shift method, *see under* Phase shift
    fluorimetry
—, pulsed method, 223–225
—, pulsed sampling technique, 224–225
—, pulsed light sources, *see also under* Lasers,
    233–239

Time-resolved fluorimetry, (*continued*)
—, spectral measurements, 256–259, 271
—, through the microscope, 273–274
Time-resolved X-ray scattering, 138–169
—, contrast and resolution in, 153–155
—, dead time in, 147
Time-to-amplitude conversion, 145, 229–233
Transfer function, 7
Transient digitizer, 224–225
Triplet probes, 341
Tryptophan fluorescence, 254–256, 328
Tuberculosis, 208
Tubulin
—, α and β, 116
—, binding to colchicine, 41
—, radius of gyration of, 166–167
—, rings, 157–158, 162, 167
Turnover number, 14

Uniaxial crystal, 250, 293

V_{max}, *see under* Kinetics
Valency-hybrid hemoglobin, 100–101
Van de Graaff accelerator, 88
Vidicon detector, 298
Viruses, 203–204
Viscosity, 32, 33, 34
—, shear, 175

Warfarin, 41
Wavelength shifting, 291–296
—, angle and temperature tuning of crystals,
    294–296
Weiner–Khinchine theorem, 184, 189

X-ray scattering, *see* Time-resolved X-ray
    scattering
X-ray optics, 141–144